VOLUME ONE HUNDRED AND NINETEEN

PROGRESS IN MOLECULAR BIOLOGY AND TRANSLATIONAL SCIENCE

Chronobiology: Biological Timing in Health and Disease

VOLUME ONE HUNDRED AND NINETEEN

PROGRESS IN MOLECULAR BIOLOGY AND TRANSLATIONAL SCIENCE

Chronobiology: Biological Timing in Health and Disease

Edited by

MARTHA U. GILLETTE

Department of Cell & Developmental Biology
University of Illinois
URBANA, IL 61801
mgillett@illinois.edu

ELSEVIER

AMSTERDAM • BOSTON • HEIDELBERG • LONDON
NEW YORK • OXFORD • PARIS • SAN DIEGO
SAN FRANCISCO • SINGAPORE • SYDNEY • TOKYO
Academic Press is an imprint of Elsevier

Academic Press is an imprint of Elsevier
The Boulevard, Langford Lane, Kidlington, Oxford, OX5 1GB, UK
32 Jamestown Road, London NW1 7BY, UK
Radarweg 29, PO Box 211, 1000 AE Amsterdam, The Netherlands
225 Wyman Street, Waltham, MA 02451, USA
525 B Street, Suite 1800, San Diego, CA 92101-4495, USA

First edition 2013

Library of Congress Cataloging-in-Publication Data
A catalog record for this book is available from the Library of Congress

British Library Cataloguing in Publication Data
A catalogue record for this book is available from the British Library

ISBN: 978-0-12-396971-2
ISSN: 1877-1173

Printed and bound by CPI Group (UK) Ltd, Croydon, CR0 4YY
Transferred to digital print 2012

CONTENTS

CONTRIBUTORS

Mathias Basner
Division of Sleep and Chronobiology, Department of Psychiatry, Perelman School of Medicine, University of Pennsylvania, Philadelphia, Pennsylvania, USA

Gordon F. Buchanan
Department of Neurology, Yale University School of Medicine, New Haven, and Veteran's Affairs Medical Center, West Haven, Connecticut, USA

Ruud Buijs
Departamento de Biología Celular y Fisiología, Instituto de Investigaciones Biomédicas, Universidad Nacional Autónoma de México, Distrito Federal, Mexico

Etienne Challet
Neurobiology of Rhythms, Institute of Cellular and Integrative Neurosciences, CNRS UPR3212 Associated with University of Strasbourg, Strasbourg, France

Alec J. Davidson
Department of Neurobiology, Morehouse School of Medicine, Atlanta, Georgia, USA

David F. Dinges
Division of Sleep and Chronobiology, Department of Psychiatry, Perelman School of Medicine, University of Pennsylvania, Philadelphia, Pennsylvania, USA

Kristin Eckel-Mahan
Center for Epigenetics and Metabolism, University of California, Irvine, California, USA

Carolina Escobar
Departamento de Anatomía, Facultad de Medicina, Universidad Nacional Autónoma de México, Distrito Federal, Mexico

Jennifer A. Evans
Department of Neurobiology, Morehouse School of Medicine, Atlanta, Georgia, USA

Russell G. Foster
Nuffield Department of Clinical Neurosciences, Nuffield Laboratory of Ophthalmology, Oxford, United Kingdom

Loning Fu
Department of Pediatrics/U.S. Department of Agriculture/Agricultural Research Service/ Children's Nutrition Research Center; Department of Molecular and Cellular Biology, and Dan L. Duncan Cancer Center, Baylor College of Medicine, Houston, Texas, USA

Ying-Hui Fu
Department of Neurology, University of California, San Francisco, California, USA

Namni Goel
Division of Sleep and Chronobiology, Department of Psychiatry, Perelman School of Medicine, University of Pennsylvania, Philadelphia, Pennsylvania, USA

Nicole M. Kettner
Department of Pediatrics/U.S. Department of Agriculture/Agricultural Research Service/Children's Nutrition Research Center, and Department of Molecular and Cellular Biology, Baylor College of Medicine, Houston, Texas, USA

Robert Y. Moore
Department of Neurology, University of Pittsburgh, Pittsburgh, Pennsylvania, USA

Stuart N. Peirson
Nuffield Department of Clinical Neurosciences, Nuffield Laboratory of Ophthalmology, Oxford, United Kingdom

Louis J. Ptáček
Department of Neurology, and Howard Hughes Medical Institute, University of California, San Francisco, California, USA

Hengyi Rao
Division of Sleep and Chronobiology, Department of Psychiatry, and Department of Neurology, Center for Functional Neuroimaging, Perelman School of Medicine, University of Pennsylvania, Philadelphia, Pennsylvania, USA

Till Roenneberg
Institute for Medical Psychology, University of Munich, Munich, Germany

Elizabeth Sabath
Departamento de Biología Celular y Fisiología, Instituto de Investigaciones Biomédicas, Universidad Nacional Autónoma de México, Distrito Federal, Mexico

Roberto Salgado
Departamento de Biología Celular, Facultad de Ciencias, Universidad Autónoma de San Luis Potosí, San Luis Potosí, Mexico

Paolo Sassone-Corsi
Center for Epigenetics and Metabolism, University of California, Irvine, California, USA

Céline Vetter
Institute for Medical Psychology, University of Munich, Munich, Germany

Eva Winnebeck
Institute for Medical Psychology, University of Munich, Munich, Germany

Katharina Wulff
Nuffield Department of Clinical Neurosciences, Nuffield Laboratory of Ophthalmology, Oxford, United Kingdom

Jamie M. Zeitzer
Department of Psychiatry and Behavioral Sciences, Stanford University, and Mental Illness Research Education and Clinical Center, VA Palo Alto Health Care System, Palo Alto, California, USA

Luoying Zhang
Department of Neurology, University of California, San Francisco, California, USA

INTRODUCTION TO BIOLOGICAL TIMING IN HEALTH AND DISEASE

Martha U. Gillette
Center for Advanced Study Professor, Alumni Professor of Cell & Developmental Biology, Departments of Cell & Developmental Biology, Molecular & Integrative Physiology, and Bioengineering, Neuroscience Program, Colleges of Medicine and Liberal Arts & Science, University of Illinois, Urbana, Illinois, USA

Chronobiology, *the biology of time*, seeks to understand a fundamental property of life: intrinsic, near 24-h (*circadian*) rhythms of biological processes. The discovery that the diversity of cells in the brain and body possesses biological timekeeping properties led to questions about the role(s) of this multitude of biological clocks. Why are cellular processes and the emergent behaviors of tissues, organs, and body systems organized around a ~24-h time base? What roles do circadian clocks play in good health? How do they contribute to diseases, either inherited or acquired? The chapters of this volume address current understanding of these issues and emphasize the importance of these fundamental insights for good health and longevity, as well as for ameliorating a range of diseases.

Reshaping the physiological concept of homeostasis from a baseline steady-state to a daily circadian oscillation has been underway since the 1970s but is not yet firmly part of medical understanding. The evidence is now, incontrovertible. Initial predictions of an intrinsic timekeeping system were based on the persistence of patterned functions in constant environments: predictable rhythmic changes in the amplitudes of behavior, physiology, and metabolism continue in the absence of detectable external cues. These rhythms are encoded in the genes and repeat with periods close to, but not exactly, 24 h. Periodic activities anticipate, rather than follow, entraining environment signals. From these observations and the experiments they spawned, hypotheses developed that the integrated circadian timekeeping system enhances health, wellness, and longevity.[1]

A specific brain site, the *suprachiasmatic nucleus* (SCN), is an essential key component of the mammalian circadian system. The SCN generates intrinsic circadian rhythms in neuronal activity[2–4] and peptide release,[5] signals that transmit synchronizing cues to other brain and body regions. The SCN is necessary for daily patterning of behaviors into circadian rhythms. SCN lesion abolishes circadian rhythms of behavior, physiology, and metabolism. Transplantation of fetal SCN into an arrhythmic, SCN-lesioned host restores that organism's rhythms, except for pituitary hormone release, with the period of the transplanted clock tissue.[6,7] Thus, the SCN is the central or master circadian clock.

Approximately, 10,000 cells of each bilaterally paired SCN reside directly above the optic chiasm (hence the name, *suprachiasmatic*), at the base of the third ventricle and hypothalamus. This places the SCN in close proximity to hypothalamic nuclei that control many autonomic and non-cognitive behaviors. These include feeding, drinking, sleep/arousal, body temperature, and sexual/maternal/affiliative behaviors. The location of the SCN enables it to: 1) receive information about the environment via sensory inputs, and 2) coordinate the timing of biological rhythms of the circadian clocks throughout the body via outputs to other brain regions.

Discovering how a 24-h timekeeping system could emerge from orchestrated interactions of molecules and cells was a truly remarkable scientific feat. *Clock genes* are genes whose transcription–translation–posttranslational modification cycles form interacting positive- and negative-feedback loops necessary to generate the 24-h timekeeping system. They were discovered first in *Drosophila* and then in mouse, rat, and man.[8,9] These genes are highly conserved across species, so much so that human clock genes can substitute for fly homologs to generate behavioral rhythms. The clockwork machinery has expanded to include not only the necessary "clock" genes and their products but also the posttranslational elements, such as specific kinases and phosphatases that modify clock proteins, and small molecules, such as Ca^{2+} and cAMP, that regulate posttranslational effectors.[10] Recent evidence indicates that cellular metabolic state at the level of redox molecules oscillates and contributes importantly to circadian timekeeping.[11]

How could a molecular rhythm generator within cells contribute globally to health and disease? Clues that clock genes act outside of the SCN came from studies of mice in which clock genes had been mutagenized, deleted, or genetically engineered. These animals not only exhibit profound rhythms deficits but they have other abnormalities as well. These may include weight gain or -loss, diminished fecundity, arthritis, cardiovascular disease, tumors, sleep–wake cycle disorders, and shortened life span. When Joseph Takahashi and collaborators generated a knock-in mouse with a luminescent protein encoded in tandem with the clock protein PERIOD 2, PERIOD 2::LUCIFERASE bioluminescence oscillated with a circadian beat within all the cells of all tissues examined.[12] Further, coordinated circadian rhythms pulse in the various tissues, but this coordinated beating diminishes over the days the tissue is deprived of interaction with the SCN. Currently, the SCN is the acknowledged conductor of the "clock orchestra," the myriad cellular clocks of the brain and body.

This book addresses the current status of the roles of circadian clocks in key modes of health and disease. It begins with a chapter by Robert Y. Moore, the eminent neuroanatomist who with his collaborators established in 1972 that a direct projection from the eye innervates the SCN and the SCN is required for circadian corticosterone rhythms.[13,14] Moore evaluates current understanding of the SCN within the circadian timing system, which he defines in mammals as the network of brain structures that regulate timing of physiology and behavior. He analyzes the functional organization of the SCN in this context.

The next two chapters focus on how alterations in clock genes result in negative health consequences. Kristin Eckel-Mahan and Paolo Sassone-Corsi evaluate accumulating evidence that epigenetic changes, including dynamic changes in mRNA expression, chromatin state, protein stability, and cellular metabolite levels, regulate gene expression. Also, cellular metabolism interacts with the molecular clockwork, which may offer new strategies for treating metabolic disorders. Luoying Zhang, Louis Ptáček, and Ying-Hui Fu review their investigations of the diversity of human polymorphisms in clock genes. They discuss evidence that these sequence alterations in elements of the basic clock machinery are associated with a range of phenotypes in sleep patterns, metabolism, addiction, mood disorders, and reproductive physiology.

Circadian clocks in peripheral organs and their interactions with both the central SCN clock and organismic metabolism are critical in health and disease. Ruud Buijs, Roberto Salgado, Elizabeth Sabath, and Carolina Escobar discuss how the SCN communicates time of day to clocks in peripheral organs, focusing on the liver because of its essential role in metabolism. They present new evidence that uncoupling of the brain–body networks leads to uncoupling of cellular systems, which compromises health. Etienne Challet evaluates bidirectional interactions between food cues, which may carry reward salience, and the SCN. Disturbances in the coupling of central and peripheral clocks, such as those caused by shift work, may be associated with metabolic dysfunctions and obesity. She looks to chronotherapeutics to minimize risk for metabolic disease.

The alternation between sleep and wakefulness is the most significant of daily rhythmic changes in physiology and behavior. Jamie Zeitzer reviews current understanding of the neurobiology of sleep–wake regulation and how various pathologies emerge when these regulatory mechanisms are disturbed due to alterations and disruptions in neural substrates. He discusses how understanding the pathophysiology of these neural substrates is enabling targeted therapies. Interactions of circadian rhythms, sleep loss, and human performance are evaluated by Namni Goel, Mathias Basner, Hengyi Rao, and David Dinges. Their work has revealed that both sleep homeostasis and circadian systems affect brain metabolism and neural activation. They propose that responses to sleep loss involve genetic components, which may enable identification of biomarkers that predict human performance under sleep loss and circadian perturbation. Gordon Buchanan discusses the bidirectional interactions between breathing, circadian phase, and sleep/vigilance. He highlights the important implications of the interplay between these systems for a range of human disorders, from sleep apnea and asthma to epilepsy and sudden infant death syndrome.

Cancer biology is an emerging area with significant relevance to circadian biology. Loning Fu and Nicole Kettner discuss evidence that normal functioning of the circadian system promotes tumor suppression, whereas loss of circadian homeostasis in energy balance and immune function due to circadian disruption or aging may favor the development of cancers of many types. They consider evidence from humans and animal models for interaction of circadian regulation of these processes at the cellular and organismic levels. They propose that understanding the deregulated processes of cell proliferation and metabolism may provide new options for anticancer therapies.

The broad health consequences of circadian disruption in shift workers are discussed by Jennifer Evans and Alec Davidson. Desynchronization of the hierarchy of the body's circadian clocks is associated with increased risk for a range of health consequences, including cardiovascular dysfunction, immune dysregulation, and premature death. They propose that circadian disruption synergizes with states compromised by stress or disease with detrimental consequences, including accelerated disease progression or severity.

Mental health disorders are often associated with the complex interactions of exogenous influences of societal timing and the brain systems that regulate circadian timing and homeostatic sleep. Russell Foster, Stuart Peirson, Katharina Wulff, Eva Winnebeck, Céline Vetter, and Till Roenneberg discuss the vulnerability of behaviors to sleep and circadian rhythm disruption and the links with mood, cognition, and mental illnesses.

They note that sleep, circadian rhythms, and affective disorders share overlapping neural substrates. They propose that emerging understanding will provide new ways to treat and possibly prevent some mental illnesses.

This volume encompasses key contemporary issues that emphasize the critical role that biological timing plays in health and disease. I hope it provides you with new and valuable insights. Consider how these issues may affect your research, design of therapies, and your own good health.

Support during the preparation of this volume to M. U. G. from the National Heart, Lung, and Blood Institute of the National Institutes of Health (HL 086870, HL 092571Z) and the National Science Foundation (ISO 08-18555) is gratefully acknowledged. Sincere thanks to Maureen Holtz for her exceptional talents in organizing, editing, and guiding the project to completion. Helene Kabes, the volume editor at Elsevier, enhanced the production process; her contributions are sincerely appreciated. Rhanor Gillette generously provided comments on the manuscript. Finally, I thank the members of my lab, especially Jennifer Mitchell, and the students in my class on Biological Rhythms in Health and Disease for many thoughtful, productive discussions.

REFERENCES

1. Davidson AJ, Sellix MT, Daniel J, Yamazaki S, Menaker M, Block GD. Chronic jet-lag increases mortality in aged mice. *Curr Biol.* 2006;16(21):R914–R916.
2. Gillette MU. The suprachiasmatic nuclei: circadian phase-shifts induced at the time of hypothalamic slice preparation are preserved in vitro. *Brain Res.* 1986;379(1):176–181.
3. Inouye ST, Kawamura H. Persistence of circadian rhythmicity in a mammalian hypothalamic "island" containing the suprachiasmatic nucleus. *Proc Natl Acad Sci USA.* 1979;76(11):5962–5966.
4. Prosser RA, Gillette MU. The mammalian circadian clock in the suprachiasmatic nuclei is reset in vitro by cAMP. *J Neurosci.* 1989;9(3):1073–1081.
5. Hatcher NG, Atkins Jr N, Annangudi SP, et al. Mass spectrometry-based discovery of circadian peptides. *Proc Natl Acad Sci USA.* 2008;105(34):12527–12532.
6. Ralph MR, Foster RG, Davis FC, Menaker M. Transplanted suprachiasmatic nucleus determines circadian period. *Science.* 1990;247(4945):975–978.
7. Silver R, LeSauter J, Tresco PA, Lehman MN. A diffusible coupling signal from the transplanted suprachiasmatic nucleus controlling circadian locomotor rhythms. *Nature.* 1996;382(6594):810–813.
8. Roenneberg T, Merrow M. The network of time: understanding the molecular circadian system. *Curr Biol.* 2003;13(5):R198–R207.
9. Wager-Smith K, Kay SA. Circadian rhythm genetics: from flies to mice to humans. *Nat Genet.* 2000;26(1):23–27.
10. King DP, Takahashi JS. Molecular genetics of circadian rhythms in mammals. *Annu Rev Neurosci.* 2000;23:713–742.
11. Wang TA, Yu YV, Govindaiah G, et al. Circadian rhythm of redox state regulates excitability in suprachiasmatic nucleus neurons. *Science.* 2012;337(6096):839–842.

12. Yoo SH, Yamazaki S, Lowrey PL, et al. PERIOD2::LUCIFERASE real-time reporting of circadian dynamics reveals persistent circadian oscillations in mouse peripheral tissues. *Proc Natl Acad Sci USA*. 2004;101(15):5339–5346.
13. Moore RY, Eichler VB. Loss of a circadian adrenal corticosterone rhythm following suprachiasmatic lesions in the rat. *Brain Res*. 1972;42(1):201–206.
14. Moore RY, Lenn NJ. A retinohypothalamic projection in the rat. *J Comp Neurol*. 1972;146(1):1–14.

CHAPTER ONE

The Suprachiasmatic Nucleus and the Circadian Timing System

Robert Y. Moore
Department of Neurology, University of Pittsburgh, Pittsburgh, Pennsylvania, USA

Contents

Abstract

The circadian timing system (CTS) in mammals may be defined as a network of interconnected diencephalic structures that regulate the timing of physiological processes and behavioral state. The central feature of the CTS is the suprachiasmatic nucleus (SCN) of the hypothalamus, a self-sustaining circadian oscillator entrained by visual afferents, input from other brain and peripheral oscillators. The SCN was first noted as a distinct component of the hypothalamus during the late nineteenth century and recognized soon after as a uniform feature of the mammalian and lower vertebrate brain. But, as was true for so many brain components identified in that era, its function was unknown and remained so for nearly a century. In the latter half of the twentieth century, numerous tools for studying the brain were developed including neuroanatomical tracing methods, electrophysiological methods including long-term recording *in vivo* and *in vitro*, precise methods for producing localized lesions in the brain, and molecular neurobiology. Application of these methods provided a body of data strongly supporting the view that the SCN is a circadian pacemaker in the mammalian brain. This chapter

Progress in Molecular Biology and Translational Science, Volume 119
ISSN 1877-1173
http://dx.doi.org/10.1016/B978-0-12-396971-2.00001-4

presents an analysis of the functional organization of the SCN as a component of a neural network, the CTS. This network functions as a coordinator of hypothalamic regulatory systems imposing a temporal organization of physiological processes and behavioral state to promote environmental adaptation.

1. INTRODUCTION

"Even when current knowledge emphasizes the importance of a single program-pacemaker, like the suprachiasmatic nucleus in mammals, it seems likely its control of the overall program reduces to entrainment of slave oscillations inherent in those systems it times" (Ref. 1, p 48).

Circadian timing is a fundamental adaptation of living organisms.[1] In mammals, a circadian timing system (CTS) regulates the timing of behavioral state coordinated with the timing of physiological and cellular processes throughout the organism. In this chapter, I focus on the functional organization of the suprachiasmatic nucleus (SCN) of the hypothalamus, a principal feature of the CTS. The SCN is a prominent component of the anterior hypothalamus lying in the ventral periventricular zone, dorsal to the optic chiasm, lateral to the third ventricle, and medial to the anterior hypothalamic area. In most mammals, it lies caudal to the medial preoptic area and extends to the retrochiasmatic area. The SCN was described as a distinct nucleus in the 1880s and subsequently was recognized in a number of species. Comprehensive comparative studies of the hypothalamus by Crosby and Woodburne[2] found the SCN present in a wide range of metatherian and eutherian mammalian brains. Despite the anatomical prominence of the SCN, it was nearly 100 years from its discovery to recognition of a function.

An analysis of the functional organization of the SCN can best be evaluated in three contexts: (1) the concept of localization of function; (2) the neuronal organization of the SCN as a mini network; and (3) the SCN as a component of a functional network, the CTS. Localization of function in the brain was first presented as a coherent theory by Gall[3] who posited it in three questions: (1) what are the functions of the nervous system? (2) where are these functions localized? and (3) how do we understand the relationship? These were reasonable, even prescient, questions particularly for the time. However, Gall's interest in complex human behaviors influenced him to put forth four additional premises: (1) the brain is the organ of the mind; (2) the mind is comprised of a set of definable faculties or functions; (3) the size of brain components mediating these faculties corresponds with the efficiency of each; and (4) the development of these brain areas is reflected in the size and shape of the overlying skull. The latter was his scientific downfall as he applied it to the development of phrenology that was then used to make predictions of individual cognitive abilities based on palpation of the skull. This was quite popular during the nineteenth century but became, as characterized by the famous British neurologist McDonald Critchley, "a theory of brain function that began as a heresy and ended as a superstition."[4]

Nonetheless, Critchley and others noted that the clarity of Gall's formulation of localization of function provided a framework for investigating functional localization clinically and experimentally. Stated in a more modern form, it is "the doctrine that various parts of the brain have relatively distinct mental, behavioral and/or physiological functions" (Ref. 5, p 10). To discuss this further, it is important to review brain organization briefly.

The adult brain is made up of individual neurons that are specialized during development for their function in brain areas in which they are generated. Brain neurons are grouped in two fundamental patterns, laminated groups (cortices) and nonlaminated groups (nuclei and fields). Nuclei may be made up of a homogeneous group of neurons or have divisions. The designation of a neuronal group as a nucleus, or a division of a nucleus, was originally based on appearance in gross anatomical preparations or in stained tissue sections to determine cytoarchitecture. More recently, the precision of such designations has been improved by the addition of other information from analysis of chemoarchitecture, patterns of gene expression, connections, and physiology.

We also recognize that the definition of function must be made in the context of levels of organization. For the nervous system, these are generally considered to be as shown in Table 1.1. Reductionism is a dominant theme in modern biology, and much of the work in circadian neurobiology in the present era is at the cellular and molecular levels. Nevertheless, this chapter is directed to the structural level with functional correlations as they are understood and begins with a general description of brain organization. The brain has five subdivisions from rostral to caudal: telencephalon, diencephalon, mesencephalon, metencephalon, and myelencephalon (cf. Ref. 7, for review). The hypothalamus is the ventral longitudinal cell group of the diencephalon. The hypothalamus began to be recognized as a distinct subdivision of the diencephalon in the mid-nineteenth century. The fundamental organization of the diencephalon as comprising four longitudinal cell columns was established early in the twentieth century.[8,9] The hypothalamus has three rostrocaudal subdivisions: the anterior, or chiasmal, hypothalamus lying predominantly above the optic chiasm; the tuberal hypothalamus situated above the pituitary stalk; and, behind this, the posterior hypothalamus including the mammillary bodies. The structure and functions of the hypothalamus have

Table 1.1 Levels of Organization

Ecological/evolutionary
Organism
Specialized organs/tissues
Nervous system
Brain
Functional systems
Networks
Neurons/Glia
Molecular Mechanisms

been studied intensively for more than a century. Between the late 1880s and the early 1930s, comprehensive descriptions of hypothalamic nuclear structure were made for a number of mammalian and nonmammalian species (Ref. 2, for reviews). The modern nomenclature for hypothalamic nuclei was first used in a detailed description of the rat hypothalamus by Gurdjian.[10] The nuclear descriptions and nomenclature were subsequently refined by LeGros Clark[11] and accepted by a group of hypothalamic morphologists at a meeting of the Association for Research on Nervous and Mental Disease.[12] Research on hypothalamic anatomy and function advanced, particularly stimulated by the development of neuroendocrinology, following the description and discovery of the pathological basis for Fröhlich's syndrome (cf. Ref. 13, for historical review) and Harvey Cushing's clinical studies establishing the basic phenomenology of hypothalamic–pituitary interaction.[14] In the 1920s and 1930s, hypothalamic control of osmoregulation, autonomic function, feeding, drinking, temperature regulation, metabolism, behavioral state regulation, reproductive function and behavior, and emotional behavior were discovered (cf. Refs. 11,13,15, for reviews from that era). All of this work established the function of the hypothalamus as the highest brain level of regulatory function.

The mechanisms of hypothalamic control of endocrine function emerged through the work of Harris[16] and other pioneering neuroendocrinologists, particularly Ernst and Berta Scharrer (cf. Ref. 17, for review), establishing that neurons could produce and release small peptides. Harris was the first to establish that the control of anterior pituitary function was through peptide release into the portal circulation to stimulate or inhibit anterior pituitary hormone release. Guillemin and Schally[18] independently elucidated the amino acid sequences of such peptides. Further data accumulated over a number of years established that these and other peptides are produced by virtually all neurons and resulted in the expansion of the current views of synaptic transmission from an emphasis on small-molecule neurotransmitters to include peptides and, hence, to extend what was referred to as "Dale's principle"[19] to be "defined as stating that at all the axonal branches of a neuron, there was liberation of the same transmitter substance or substances."[20] The discovery of the structure of small peptides produced and released by hypothalamic neurons in the 1960s was the crucial validation of Harris' hypotheses on control of the anterior pituitary (cf. Refs. 21–26, for reviews). The 1960s and 1970s witnessed the discovery of many new peptides and, with rapid developments in immunohistochemistry, led to new insights into the nature of chemical neurotransmission. Colocalization of neuroactive substances in neurons was recognized as the rule, rather than the exception, and then included findings that a single neuron can produce and release as many as two small-molecule transmitters and multiple peptides (cf. Refs. 27–29, for reviews). These data are important to understanding the function of the SCN. Peptides also brought forth a new problem. Peptides are released from axon terminals but not necessarily into synaptic clefts so that their effects depend upon the localization of appropriate receptors. These data have stimulated development of the concept of "volume transmission" (cf. Refs. 30–32, for reviews). The importance of peptides as complex and important signaling molecules in the CTS will be discussed below.

2. THE EMERGENCE OF A CIRCADIAN NEUROBIOLOGY

Over the first half of the twentieth century, the field of circadian biology was established by a small group of pioneering scientists working on a diverse set of organisms from prokaryotes to humans. Daily rhythms in behavioral state in humans and other animals have been recorded since antiquity, but the critical contribution of these scientists was recognition that circadian rhythms are a fundamental property of living systems playing a critical role in adaptation (cf. Refs. 1,33–37,144 for reviews). The term circadian was first applied by Franz Halberg in 1959 (from *circa* and *diem*; cf. Ref. 37). The fundamental features of circadian rhythms were summarized by Colin Pittendrigh as "Biological activities that characteristically occur once per day in nature continue in laboratory conditions of constant darkness and temperature as a persistent rhythm with a period (τ) that is close to but not exactly 24 h: the period is said to be circadian (L. circa, dies). Such circadian rhythmicity has been observed at all levels of organization, from the behavior of mammals, flies, and single cells, to the specific activity of enzymes, the activity of ribosomes, and the transcription of identified genes... they are driven by some self-sustaining cellular oscillation as pacemaker of the system".[1] From its beginnings, our understanding of circadian biology has expanded rapidly at all levels, from molecular to behavioral, over the past 40 years with publications in the field increasing rapidly. The essential localization-of-function questions for circadian neurobiology is summarized in Fig. 1.1. As the diagram indicates, the essential features are intrinsic self-sustaining oscillators with a period (τ) approximating 24 h, entrainment pathways, and output from the oscillators to systems under circadian regulation. The experimental task was identification and characterization of each.

My interest in the field began in 1964 when I was introduced to Julius Axelrod at a Federated American Societies for Experimental Biology (FASEB) meeting. Axelrod had made important contributions to understanding catecholamine neurotransmission leading to the Nobel Prize in Physiology or Medicine[145] and, in the course of this, discovered *O*-methylating enzymes including the pineal melatonin-forming enzyme, hydroxyindole-*O*-methyltransferase (HIOMT). One of his postdoctoral fellows, Richard Wurtman, was interested in the pineal gland and had published data with Axelrod showing a diurnal rhythm in pineal HIOMT content that is dependent on an intact sympathetic innervation.[143] We had been studying the effects of hypothalamic and brainstem lesions on monoamines in the brain,[39] and Axelrod asked the question of whether central monoamine pathways participated in the regulation of pineal function. This interaction resulted in a collaborative study between Axelrod and colleagues at the NIH and Alfred Heller and I at the University of Chicago. In these, we found a loss of HIOMT responses with bilateral medial forebrain bundle lesions.[146] These data indicated that the information

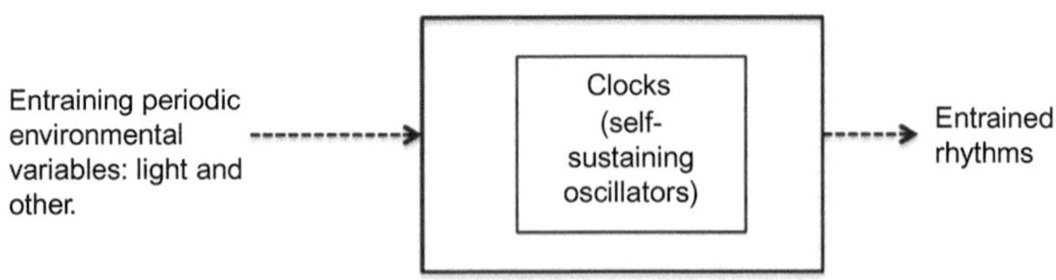

Figure 1.1 Diagram showing the necessary components of a circadian timing system: entrainment pathways, endogenous self-sustaining oscillators (ESSO's), output pathways. *Modified from Ref. 38.*

driving rhythmicity originated in the forebrain and raised the question in my mind of whether we could approach the issue of localizing the control of mammalian circadian function. A number of studies had shown that the lateral eyes are necessary for entrainment in mammals. The general outline for a circadian system in Fig. 1.1 requires that the brain mechanisms for mammalian circadian rhythm generation and regulation must have three parts: photoreceptors, entrainment pathways that are a component of central retinal projections, and brain pacemakers with output to systems under circadian control. As no other optic photoreceptors were known at the time, we assumed that rod and/or cone photoreception was the mechanism for converting light to neural information. The localization question, then, was identification of the retinal projections mediating entrainment. It seemed likely that the retinal projection would be to hypothalamus as circadian timing is clearly a regulatory function. The most recent study of retinal projections in the rat[40] available at that time had used the Nauta stain (Refs. 41,147). This was the most powerful method for tract tracing, and it had not revealed evidence for retinohypothalamic projections. Indeed, there was general consensus among experts on the projections of retinal ganglion cells that there were no projections to hypothalamus.[42–44] We employed a new variant of the Nauta stain, the Fink-Heimer method,[45] and found tantalizing hints but no definitive evidence for a projection. It was at this time, however, that axoplasmic transport was discovered using the technique of applying tritiated amino acids near or in a neuronal population and following their transport, after incorporation into protein, to axon terminals by measuring either radioactivity or autoradiography.[46,47] Grafstein's work using intraocular injections of a tritiated amino acid to trace retinal projections in the frog, in particular, suggested that the existence of retinohypothalamic projections could be explored with this methodology in mammals. With Nicholas Lenn, then a graduate student, we carried out the autoradiographic experiment and found evidence for a retinal projection to the ventral portion of the SCN, but not to other medial hypothalamic nuclei, and confirmed the projection by showing degenerating axon terminals by electron microscopy in the same area after eye removal.[48] Similar findings were published at the same time by Hendrickson.[49] During this period, I also demonstrated that the retinohypothalamic tract (RHT) was a consistent feature of the mammalian visual system finding it in six other mammalian species from metatherians to primates.[50] This did not prove definitively that the RHT was the entrainment pathway but a selective transection of the RHT was not possible technically at the time, and it seemed to me that the next obvious experiment was to ablate the SCN and study the effects on circadian function. I had extensive experience making localized stereotactic lesions in the rat brain but did not have the animal facilities to study rest-activity rhythms. However, I was performing biochemical assays in my laboratory and decided to analyze the effects of SCN ablation on the adrenal corticosterone rhythm in the rat. This is a robust rhythm and the only disadvantage was that we could show only group effects as animals had to be sacrificed at four time periods around the clock. Control groups included a blinded group, a group with electrolytic lesions of the SCN, a group with a Halasz knife[51] coronal transection of the medial hypothalamus rostral to the SCN, one with a transection just caudal to the optic chiasm and a sham-operated group. The blinded group showed a shift in the peak of the rhythm compared to controls indicating a free running rhythm. The SCN lesion group and the group with knife cuts just caudal to the SCN showed a loss of the adrenal rhythm whereas the group with knife cuts rostral to the SCN had no rhythm impairment. Thus, these data are consistent with the view that the SCN is a circadian pacemaker. SCN lesions abolish the rhythm, and a knife cut rostral to the SCN does not affect rhythmicity indicating that SCN lesions do not alter rhythm generation by transecting pathways running near or through the SCN. That knife cuts caudal to the SCN also abolish the rhythm indicating that SCN projections caudal to the nucleus are crucial to rhythm control (Ref. 52; Fig. 1.2). Loss of the rest-activity rhythm after SCN lesions was reported in the same year.[53] Over the next 20 years, the effects of SCN ablation were studied many times (cf. Refs. 54–56, for detailed reviews). With some exceptions that will be discussed later, SCN lesions reliably abolished behavioral and physiological rhythms. Three other sets of data were extremely important. First, Schwartz and colleagues reported an *in vivo* rhythm in glucose utilization employing a new method.[57–59] Second, Inouye and Kawamura[60] demonstrated circadian rhythms in multiunit firing rate recorded *in vivo* from the SCN in the intact brain and in a hypothalamic island isolated from the remaining brain. In the latter case, rhythms in other brain areas were lost indicating that the SCN was the source of rhythmicity.[61] Subsequently, three groups reported single-unit rhythms in the SCN in hypothalamic slices *in vitro*.[62–64] No other area sampled showed neuronal rhythms, and the peak of the firing rate rhythms was in the middle of what would have been subjective day in the intact animal, and the peak of the rhythm in glucose utilization.[58] Second, transplants containing fetal SCN placed in the third ventricle of arrhythmic hosts with SCN lesions restored locomotor activity rhythms (Refs. 65–67,136). A crucial study demonstrated that the transplant, not the host, determined the period of the restored

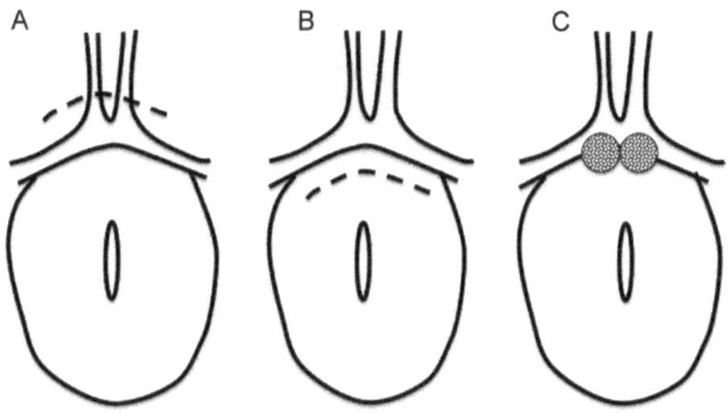

Serum corticosterone rhythm

Experimental group	Rhythm status
Control	Present, entrained
Blinded	Present, free-running
Primary optic tract transection	Present, entrained
Accessory optic tract transection	Present, entrained
Prechiasmal knife-cut (A)	Present, entrained
Postchiasmal knife-cut (B)	Absent
Suprachiasmatic nucleus lesions (C)	Absent

Figure 1.2 Diagrams showing the location of lesions. The data interpretations are shown in the table below the diagrams. *Ref. 52; reprinted with permission, Elsevier.*

rhythm.[68] An intriguing aspect of this is that the transplants are effective without reestablishing connections to the host brain.[69] Taken together, these studies clearly established SCN function as a circadian pacemaker. What was not entirely clear was the definition of "pacemaker" in this situation.

Nevertheless, these data formed the basis for continuing investigation of the neural basis of circadian timing. I will review aspects of this in four contexts: (1) anatomical organization of the SCN including intrinsic organization and the issue of subdivisions, (2) visual and nonvisual SCN afferents, (3) efferent projections including issues of nonsynaptic communication, and (4) functional organization of the SCN as a component of a brain CTS. The review is not intended to be comprehensive but to present an overview of the information that gradually has come available over the past 40 years as this is interpreted by one who was "in at the beginning."

3. ANATOMICAL ORGANIZATION OF THE SCN

3.1. General organization

In Nissl-stained material, the SCN is a compact nucleus of small cells situated just above the medial optic chiasm and lateral to the periventricular nucleus adjacent to the third ventricle. Dorsally and laterally, the SCN is bordered by the anterior hypothalamic area. Caudally it extends into the rostral retrochiasmatic area. The compact nature of the nucleus suggests relatively less neuropil than adjacent nuclei, and there are no apparent SCN subdivisions in Nissl-stained material. The general shape of the SCN in most species is that of a triaxial ellipsoid extending from the rostral optic chiasm to the caudal chiasm. A detailed quantitative analysis of the rat SCN was reported by Güldner[70,71] Each SCN has dorsoventral, mediolateral, and rostrocaudal dimensions not exceeding 360, 450, and 750 μm, respectively, with an astroglia to neuron ratio of approximately 1:3. The total neuron number is about 8000[72] to 12,000.[6,71] SCN volume is about 0.064 mm^3.[71,72] Although SCN subdivisions are not evident in Nissl-stained material, they are with other methods; "the present study finds striking subpopulations of cells within the SCN... Although other minor subpopulations exist, two predominant ones are designated the dorsomedial group of cells and the ventrolateral group of cells."[72] In an extensive Golgi analysis and morphometry on thick, plastic-embedded sections of the rat SCN, van den Pol[72] reported that neurons in the dorsal and medial part of the nucleus had an area and mean diameter of 84 ± 4 and 7.8 ± 0.9 μm, respectively, whereas those of the ventrolateral SCN were 102 ± 6 and 9.6 ± 1.5 μm. Neurons in the ventrolateral region also had more nuclear invaginations and more complex dendritic ramifications than those in the dorsomedial region. This division of the SCN also is supported by developmental studies showing that neurons of the ventrolateral region show final cell division and migration prior to those of the dorsomedial division.[73–75]

By the early 1980s, there were data from immunohistochemical studies supporting this view of SCN organization (Refs. 76,77,85). As available information increased the nomenclature for the subdivisions was based on their position in the nucleus (e.g., "ventrolateral" and "dorsomedial" for the rat) became difficult to apply as it is not uniform among mammals, and in 1996, I proposed a nomenclature for the divisions not based on relative position of the subdivisions. In this, the subdivision lying above the optic chiasm and predominantly populated by vasoactive intestinal polypeptide (VIP+) and gastrin-releasing peptide (GRP+) immunoreactive neurons is designated "core" and the subdivision lying above it containing arginine vasopressin (AVP+) immunoreactive neurons is designated "shell".[78] In most species, core is largely surrounded by shell. This nomenclature has been used quite extensively but has been criticized as being excessively simplistic (Ref. 79). In the next section, a detailed account of SCN organization and summary of the argument that the SCN in mammals has two major functional subdivisions will be presented.

3.2. Neurochemical organization

Neurochemical organization as discussed here refers to the presence of biological molecules involved in intercellular communication. This is a topic that has grown increasingly complex over the past three decades and is very important to understanding SCN function. Although Cajal[80,133] definitively proved that neurons are discontinuous elements with functional polarity and Ref. 142 introduced the word "synapse," identification of synaptic transmitters developed slowly between the 1920s and the 1960s with acetylcholine and glutamate as excitatory transmitters and GABA and glycine as inhibitory transmitters. Approximately 66% of all brain neurons produce glutamate and 33% GABA with acetylcholine produced by all motor neurons and small populations of pontine and basal forebrain neurons. Glycine is primarily found in spinal cord. The other groups of small molecule transmitters are the catecholamines dopamine, found in the substantia nigra and ventral tegmental area, posterior hypothalamus, the medial hypothalamus, olfactory bulb and retina, and norepinephrine found in pontine cell groups (Ref. 140). Neurons producing the indoleamine serotonin are found in brainstem raphe nuclei.[81] It has been evident for a number of years from immunocytochemistry, or glutamic acid decarboxylase *in situ* hybridization, that all SCN neurons in the species studied contain GABA (Refs. 82, 83,84,141). GABA is colocalized in SCN neurons with one or more neuroactive peptides (cf. Refs. 6,85,88, for reviews). In the rat, eight peptides have been found in SCN neurons by immunocytochemistry, totaling nearly the same number as GABA neurons (Ref. 6; see Table 1.2). The postsynaptic effects of peptides are complex (cf. Refs. 32,86,87, for reviews). In general, colocalized peptides are released from terminals concomitantly with small-molecule transmitters, interact with receptors, and are metabolized by peptidases with postsynaptic effects similar to the small-molecule transmitter, but more complex and prolonged.

3.3. SCN organization in the rat brain

The pattern of organization in which the SCN is comprised of two subdivisions evident in Nissl and Golgi material is quite evident as well in material prepared with antisera against a variety of peptides. The major peptide phenotypes are AVP, VIP, GRP, and calretinin (CAR). Smaller numbers of SCN neurons produce enkephalin (ENK), somatostatin (SS), substance P (SP), and neurotensin (NT; Table 1.1). As we have previously noted most, if not all SCN, neurons produce the inhibitory small molecule transmitter, GABA. The total number of neurons characterized by peptide phenotype is 79% of the total number of SCN neurons

Table 1.2 Neurotransmitter/peptide phenotype of SCN neurons

Transmitter/ peptide phenotype	Number of neurons ± SD	Perikaryal area (μm²±SD)	Perikaryal diameter (μm²±SD)	Percentage of total SCN population
AVP	3176 ± 43	44.0 ± 9.8	10.2 ± 1.4	37
VIP	2081 ± 81	41.7 ± 10.4	9.7 ± 1.3	24
CAR	1237 ± 105	41.4 ± 9.3	10.1 ± 1.3	14
GRP	1169 ± 77	29.6 ± 7.7	8.9 ± 1.2	14
NT	316 ± 60	43.8 ± 10.0	10.5 ± 1.5	4
ENK	530 ± 129	45.7 ± 11.9	10.5 ± 1.4	3[a]
SS	277 ± 28	51.1 ± 13.6	11.4 ± 1.6	3[a]
SP	164 ± 36	57.6 ± 13.9	12.1 ± 1.6	2
Total peptide[b]	8630	–	–	100
				(79)[e]
Total GABA[c]	7945 ± 115	–	–	–
Total Nissl neurons[d]	10,972 ± 1179	–	–	–

[a]Approximately 50% of ENK+ neurons colocalize AVP, and this is accounted for in the total percent of characterized neurons. Similarly, 20% of SS+ neurons colocalize AVP, and this is also corrected for the final count.
[b]The total number of SCN neurons characterized by peptide phenotype is corrected for a 50% colocalization of AVP with ENK and a 20% colocalization of AVP with SS. Similarly, quantitative data for angiotensin II are not shown because all angiotensin II neurons colocalize AVP.
[c]Total number of GABA neurons is taken from Moore and Speh (1993).
[d]Total neurons is from two brains (4 SCNs) which were embedded in paraffin, 20μm coronal sections were stained with cresyl violet and neuronal number in the SCN was estimated using an unbiased stereological method.
[e]The total number of neurons characterized by peptide phenotype is 79% of the total number of SCN neurons determined without correction from Nissl material. The lower number of peptide-containing neurons most likely represents the lower sensitivity of the immunohistochemical method compared to the Nissl stain.
Table from Ref. 6. Reproduced with permission (Elsevier license ID 3058031028613).

estimated from Nissl material (Table 1.2). This difference between total neuron number determined from Nissl material as opposed to from immunocytochemical material is almost certainly a result of an underestimate due to limitations of immunocytochemical sensitivity. The total number from Nissl material is consistent with prior studies (see earlier). Thus, the immunocytochemical method can be viewed as 80% as sensitive as the Nissl method. It is noteworthy that the perikaryal size of the individual groups is quite comparable except for the GRP neurons which are significantly smaller ($p < 0.05$, t test) than the AVP, VIP, and ENK groups.

The location of neurons immunoreactive for AVP, VIP, GRP, CAR, ENK, SS, SP, and NT is shown diagrammatically in Fig. 1.3. AVP neurons occupy the rostral pole of the SCN and extend to the caudal pole. Throughout most of the rostrocaudal extent of the nucleus, AVP perikarya lie in the shell. The AVP axonal plexus is evident throughout the shell and extends dorsally into the PSCN and SPVZ. Within the core, VIP perikarya are ventral, in part, to the GRP perikarya, and some VIP perikarya extend into the dorsal chiasm as small pockets of neurons and neuropil. The GRP perikarya are generally situated more dorsal and lateral than the VIP perikarya, but there is substantial overlap. The VIP and GRP populations make up 25% and 14% of the SCN, respectively. CAR-containing perikarya are an unusual group. First described by Ref. 137, these neurons are located in the dorsal and lateral portion of the SCN but extend beyond the borders of the SCN into the adjacent PSCN-anterior hypothalamic area. CAR neurons make up 14% of the nucleus, and we interpret the entire group to be part of the shell. All of the remaining peptide phenotypes are included in the shell. The shell contains 57% of SCN neurons and the core 43%. A calculation of subdivision volumes indicates that the shell volume is $0.0022+0.0004\ mm^3$ and the core $0.0014+0.0002\ mm^3$ (R.Y. Moore and J.C. Speh, unpublished).

3.4. SCN organization in the human brain

The human exhibits interesting differences from the rat and mouse.[88] The AVP+ and VIP+ neurons are located in exactly the same areas as in the mouse and rat. Mai[89] reports GRP+ neurons in the core but does not note their relative number compared to VIP+ neurons. In addition to the VIP+ and AVP+ populations, there is a substantial number of neuropeptide Y (NPY+) neurons in the core with a sparse-associated axonal plexus and a large population of NT+ neurons throughout the nucleus (Fig. 1.4) and also with a dense plexus.[90] Thus, although the core–shell configuration of AVP and VIP neurons is maintained, the human SCN is more complex than that of rodents and the macaque monkey.[90]

3.5. The subdivision view of SCN organization

The hypothesis that the SCN is organized into two anatomical and functional subdivisions has been a subject of some controversy since it was offered as a general principle.[78] This seems somewhat unusual as hypothalamic nuclei frequently have subdivisions (e.g., the paraventricular nucleus and the ventromedial nucleus; Ref. 91). The arguments against the

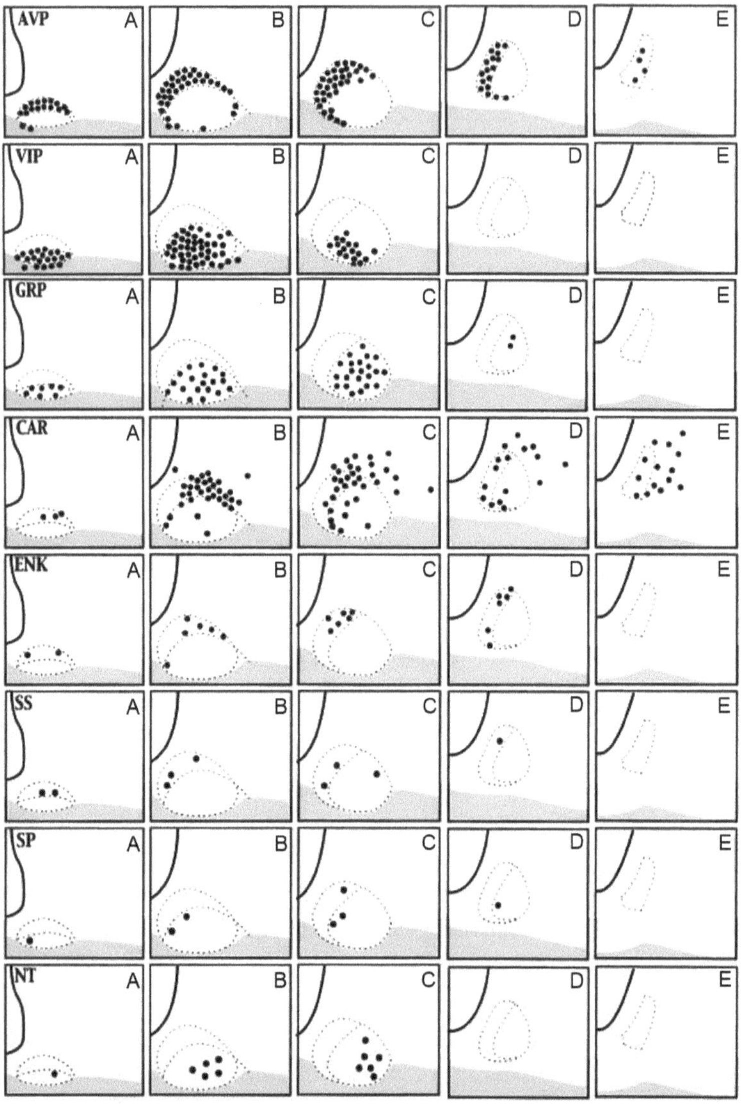

Figure 1.3 Drawings of successive rostral to caudal levels (A–E) depicting the distribution of peptide phenotype of SCN neurons (AVP, arginine vasopressin; VIP, vasoactive intestinal polypeptide; CAR, calretinin; SS, somatostatin; GRP, gastrin-releasing peptide, ENK, enkephalin; SP, substance P; NT, neurotensin). *From Ref. 6 with permission (Elsevier).*

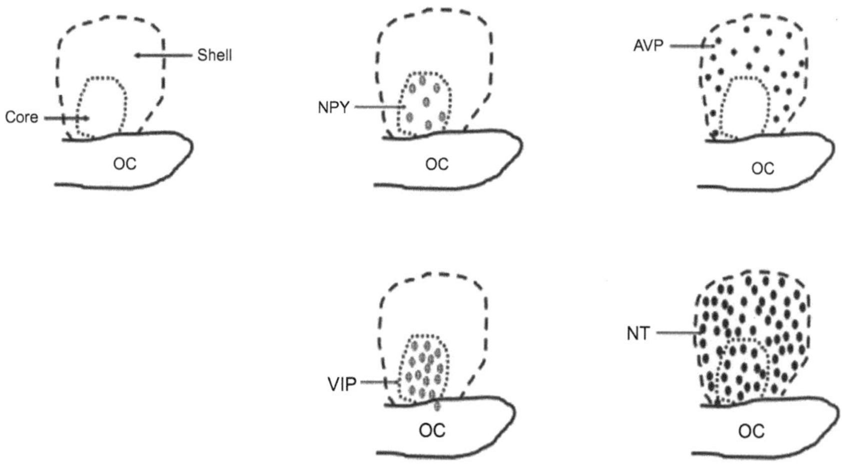

Figure 1.4 Drawing the localization of peptide-containing neurons in the human SCN. AVP, arginine vasopressin; NPY, neuropeptide Y; NT, neurotensin; OC, optic chiasm; VIP, vasoactive intestinal polypeptide.

subdivisions come in a number of forms that have one of two bases. The first is that the SCN is too complex to fit into such a rigid package, and the second is that the subdivisions do not have clear functional correlates.[79,92] The first argument is reminiscent of the debates in the early to mid-twentieth century about cortical organization. Cytoarchitectonics, the study of the laminar organization of the cerebral cortex, was initiated by Campbell[93] who described 14 cortical areas for which he proposed distinct functions. Brodmann[94] proposed 44 areas, and the undertaking reached its acme with[95] describing 107 areas. No generally accepted map was accomplished by mid-century, and no further attempts were reported. But the argument was between "lumpers" and "splitters," those espousing few areas and those with many. Recent studies using functional imaging indicate that Brodmann's areas are a good approximation of real functional areas. The conclusion is that anatomical descriptions are of value if they lead to functional correlations.

The arguments for the core/shell are summarized in Table 1.3. They can be placed in several categories. Core and shell connections differ. Expressions of function such as light responses, intrinsic activity, responses to forced desynchrony, development of photoperiodic responses versus intrinsic rhythmicity, and clock gene expression all differ between core and shell. Some individuals object to the terms "core" and "shell." The intent in introducing this designation was twofold. First, anatomically core lies adjacent to the optic chiasm and is largely surrounded by the shell in most mammalian

Table 1.3 Evidence for cell and core subdivisions in the SCN

- Rhythms in AVP and VIP release in organotypic culture show phase dissociation.
- Primary and secondary visual projections terminate predominantly in core.[78]
- Nonvisual projections terminate predominantly in the shell.[96]
- Commissural connections between SCNs of each cerebral hemisphere maintain a shell–shell, core–core organization.[97]
- Light-induced Fos expression occurs in core.
- SCN efferent projections are predominantly core- or shell specific.
- Per 1 rhythms in LD occur only in the shell.
- Behavioral forced desynchronization produces loss of synchrony of core–shell Per 1 and BMAL 1 expression.
- Circadian rhythmicity appears to develop sooner in the shell than in the core, whereas the photoperiodic response may develop sooner in the core.
- Core VIP neurons are necessary for shell synchronization and neuronal coherence.
- Physiological responses to light differ in core and shell.[98]
- Circadian rhythms in intracellular calcium differ in timing between core and shell in SCN and are dissociated by TTX.

brains. Second, the core, and particularly the VIP neuron group, is critical to SCN function (Refs. 86,99,148,149).

4. AFFERENT AND EFFERENT CONNECTIONS

4.1. Intrinsic and commissural connections

The organization of SCN connections is summarized in Fig. 1.5. Three methods have been useful in studying intrinsic and commissural connections. The first is the Golgi method. In his Golgi material, van den Pol[72] observed axons emerging from both perikarya and dendrites of SCN neurons. Occasional axons extended across to the opposite SCN. Within the SCN of origin, axons had numerous branches and varicosities. We have made similar observations in our material (J.P. Card and R.Y. Moore, unpublished). In one instance, an axon originating from the cell body of a ventral SCN neuron formed 64 collateral branches within 100 μm of the cell of origin. Some of the collaterals that could be followed for at least short distances had numerous axonal varicosities. We would conclude from the Golgi material that there is an extensive intrinsic plexus derived from SCN neurons.

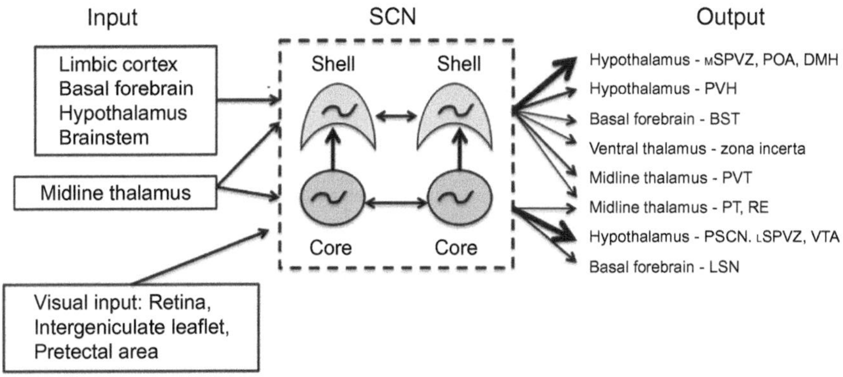

Figure 1.5 Diagram showing major connections of the SCN core and shell. BST, bed nucleus of the stria terminalis; DMH, dorsomedial hypothalamic nucleus; LSN, lateral septal nucleus; POA, preoptic area; PT, parataenial nucleus; PSCN, perisuprachiasmatic nucleus; PVH, paraventricular hypothalamic nucleus; PVT, paraventricular thalamic nucleus; RE, nucleus reuniens; LSPVZ, lateral subparaventricular area; MSPVZ, medial subparaventricular area; VTA, ventral tuberal area; ZI, zona incerta.

The second method is immunocytochemistry. In material stained with antisera for peptides, AVP+ axons are densely distributed over the medial and dorsal portions of the SCN, extending laterally but largely sparing the ventral region occupied by the VIP and GRP neurons. In contrast, the axonal plexus produced by VIP and GRP neurons extends densely over the entire SCN.[100–102] Commissural fibers were first noted by van den Pol.[72,103] Using transynaptic transport of the swine herpes virus (pseudorabies virus; Ref. 97), after injection of virus into multiple areas outside the SCN, we found evidence for a precise pattern of intrinsic and commissural connections that confirms and extends the observations from normal material. First, neurons of the shell project to contralateral shell and neurons of the core to contralateral core. Second, neurons of the core project to shell on the ipsilateral side, but there is no evidence for a commissural core to shell projection.

4.2. Afferent connections

4.2.1 Visual system projections

4.2.1.1 Retina

The first studies of RHT projections demonstrated a preponderance of innervation to the ventral half of the SCN.[48–50] This has been confirmed with increasingly sensitive techniques (Refs. 104,150), and it is clear that the RHT termination in the SCN is almost exclusively to the core, overlapping the distribution of VIP+ and GRP+ perikarya. The RHT projection is about equally dense to the contralateral and ipsilateral SCN in all species studied thus far. Because the only photosensitive elements known in the retina at the time were rods and cones, and many mammals have predominantly rod retinas, it was assumed that rods were the necessary receptors for entrainment. This conclusion was rendered untenable by data showing normal entrainment in animals lacking both rods and cones, leading to an elegant elucidation of the mechanisms of photoentrainment over the past 15 years,[105] identified and characterized a novel photopigment, melanopsin, derived from the gene, *Opn4*. As noted earlier, studies on mice without rod and cone photoreceptors had shown a loss of vision but retention of visual reflexes and light-mediated entrainment of circadian rhythms.[106] This was followed by the discovery of photosensitive retinal ganglion cells containing the photopigment, melanopsin, in the inner retina and that this mediated circadian entrainment and all other nonimage forming visual functions.[107–110,134,135] The full pathway of the RHT can be shown by immunocytochemistry with antisera to melanopsin including the RHT extension beyond the SCN into the SPVZ and anterior hypothalamic area and nucleus and caudally into the area of the dorsomedial nucleus and ventral tuberal area (Ref. 111, for review of RHT). Proof that the RHT is the entrainment pathway was shown with selective transection of the tract that led to a selective loss of entrainment with preservation of other visual functions.[112]

4.2.1.2 Lateral geniculate nucleus and the pretectal area

The pattern of RHT innervation of the SCN core is largely mimicked by a projection from the intergeniculate leaflet (IGL) of the lateral geniculate complex through the geniculohypothalamic tract (GHT), a projection from neurons that contain NPY colocalized with GABA (Ref. 141). This projection is also coextensive with another secondary visual projection from the pretectal area to the SCN in the rat (Ref. 138). The function of these visual projections to the SCN has remained elusive. The components of the "visual" portions of the CTS are shown in Fig. 1.6. The other major input to the SCN core is a serotonin projection from the midbrain raphe nuclei, particularly the median raphe.[96,113,114] The origin of afferents to the shell is complex with dense afferents arising from other hypothalamic nuclei, basal forebrain, limbic cortex, septal area, and brainstem.[96] There also are afferents from the paraventricular nucleus of the midline thalamus that distribute to both shell and core (Ref. 139). All afferents to the core terminate over the entire subdivision.

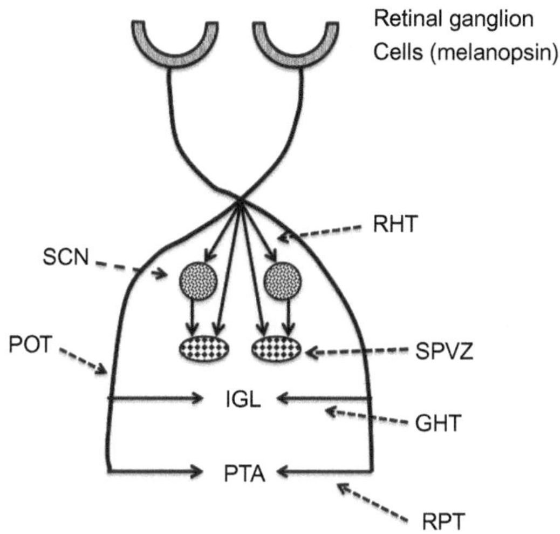

Figure 1.6 Diagram showing retinal projections to the components of the circadian timing system. IGL, intergeniculate leaflet; GHT, geniculohypothalamic tract; PTA, pretectal area; POT, primary optic tract; RHT, retinohypothalamic tract; RPT, retinopretectal tract; SCN, suprachiasmatic nucleus; SPVZ, subparaventricular zone.

4.3. Efferent organization

SCN efferents were first described in normal material[115] but were not easily distinguishable from other systems. Early studies used an autoradiographic tracing method,[116] but it was difficult in such material to differentiate labeling of fibers of passage from terminal plexuses. The most extensive data come from studies on the rat using anterograde transport using *Phaseolus* leucoagglutinin (Refs. 117–119,150). These showed a pattern of efferents distributing predominantly to adjacent hypothalamus, preoptic area, anterior hypothalamic area, particularly the subparavenricular zone, retrochiasmatic area, tuberal and posterior hypothalamic areas, as well as more limited projections to basal forebrain, midline thalamus, IGL, and periaqueductal gray (Fig. 1.6). With retrograde analysis, it is apparent that projections to these areas arise differentially from core and shell (Refs. 119,151). In addition to the peptides reviewed earlier, SCN neurons contain prokineticin 2 (PKC2), a member of a family of secreted proteins with multiple functions.[120] All SCN neurons appear to produce PKC2, the protein is found in SCN neurons and projections and appears necessary for transmission of the circadian signal (Refs. 121,122).

5. CONCLUSIONS

The SCN is comprised of individual neuronal oscillators that are coupled by neural connections to function as a pacemaker controlling a series of effector systems including those regulating the rest–activity cycle, core body temperature, autonomic nervous system, neuroendocrine function, and a psychomotor performance.[98] The SCN is made up of two anatomical and functional subdivisions, core and shell. Core contains VIP+ and GRP+ neurons; receives primary and secondary visual afferents and input from the median raphe; and projects upon shell, contralateral core, and a select set of effector areas. Shell contains populations of AVP+ and CAR+ neurons; receives input from hypothalamus, basal forebrain, limbic cortical areas, thalamus, and brainstem; and projects to a wider set of effector areas than core. Information processing in the SCN involves four steps: (1) integration of visual and related entraining input in the core; (2) intrinsic connections among core neurons, from core to shell neurons and among shell neurons; (3) commissural projections from core to core and from shell to shell; (4) integration in the shell of input from the core and from a wide set of nonvisual inputs. The conclusion from *in vitro* data, both explant cultures and cultures containing individual neurons, is that most, if not all, SCN neurons are circadian oscillators.[98,123] If this is indeed true, the further conclusion is that the circadian information relayed from shell and core to effector regions must differ. That is, output from core is the circadian output of core neurons modified primarily by photic entrainment influences (from RHT, GHT, and PHT), whereas the output from shell reflects the circadian output of shell neurons modified both by entraining stimuli from core and nonvisual modulating inputs from a wide set of areas (Fig. 1.7). Thus, the circadian signal from the SCN is likely to vary with respect to the subdivision of origin. The complexity of the output signal is further enhanced by the number and variety of neuroactive substances that have been advanced as candidate transmitters/modulators. In addition, the output to the preoptic and anterior hypothalamic areas, SPVZ, and tuberal hypothalamus is likely to be further affected by overlapping direct retinal projections (Refs. 104,111,150). A further level of complexity is introduced by humoral and behavioral events that are components of feedback input to the SCN.

At this point, a comment on experimental models and methods is necessary. The criteria for selecting an experimental paradigm include utility in approaching a variety of problems, ease of use, replication of data both within and between laboratories, minimal expense, and wide acceptance

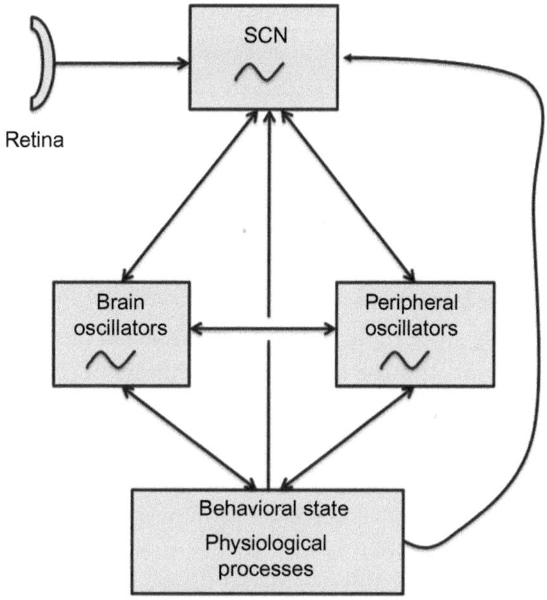

Figure 1.7 Schematic representation of a mammalian circadian timing system (CTS). Retinal input to the SCN provides a primary entraining stimulus. Secondary entraining stimuli are contributed by brain oscillators, peripheral oscillators, physiological processes of the internal milieu, and behavioral state. Specific behaviors such as activity also contribute.

of data obtained. In Section 1, I described the historical localization of function formulation for understanding brain function. Early studies relied on lesion effects, and it took many years for the full realization to emerge that localized destruction of the brain, either experimental or with disease or injury, informed the investigator about what the brain could do in the absence of the part but not necessarily what the part did. Neurophysiology was often more informative but was limited in that recordings were most often made from anesthetized animals with evident limitations. Invasive studies are not possible with the potentially most informative species, *Homo sapiens*. In circadian neurobiology, a predominant paradigm has been use of the brain slice preparation, either acutely or in organotypic culture. These *in vitro* preparations can provide much insight into molecular and cellular events and intrinsic organization, but they have limitations,[124] and we need to investigate the SCN as a component of the CTS, and the CTS as a component of regulatory systems in the brain *in vivo*. Perhaps, a model for this can be derived from recent developments in cognitive neuroscience. The recent evolution of functional imaging, particularly functional magnetic resonance imaging, has required that localization of function be viewed on a network, or system, level.[125,126] Advances in imaging technology are bringing the resolution of functional imaging to a level that would permit studying the human CTS (Refs. 127,128), so we have much to look forward to.

6. REPRISE

In the early years after discovery of SCN involvement in circadian regulation, the simplest hypothesis was to assign it a role as "master pacemaker." After the unraveling of the molecular basis of circadian function, however, it became evident rapidly that the molecular machinery of circadian function is widely distributed among living organisms and in cells and tissues throughout mammalians (cf. Ref. 129). Further, the entrainment pathway was found to have specialized photoreceptor neurons that are the beginning of the RHT. These advances have required revision of ideas about the neurobiology of circadian timing. Gone is the "master pacemaker." In its place is a network of brain structures and peripheral tissues with the SCN, a specialized brain oscillator, sitting at the interface between the circadian photic world and the brain to transduce light information into neural information (Fig. 1.7). This process is clearly complex, and "understanding" the SCN will be more difficult than we imagine. The problems will be centered not only at the molecular level, or even the cellular level, but will involve explanations at the network and systems levels. As the scientific focus of circadian neurobiology moves increasingly to cellular and molecular studies, how will elegant explanations of the parts lead to a comprehension of CTS? This is an old problem in biology, discussed often but less often resolved (cf. Ref. 130, for review). It was stated well by the great systems biologist Ernst Mayr in a paper on emergent properties in biology, "When two entities are combined at a higher level of integration, not all the properties of the new entity are necessarily a logical or predictable consequence of the properties of the components." (Ref. 131, p 1505). This has been noted as a problem in the study of circadian function,[132] and one we should remember as circadian neurobiology moves on.

REFERENCES

1. Pittendrigh CS. Temporal organization: reflections of a Darwinian clock-watcher. *Annu Rev Physiol.* 1993;55:17–54.
2. Crosby EC, Woodburne RT. The comparative anatomy of the preoptic area and the hypothalamus. In: Fulton JF, ed. *The Hypothalamus.* Baltimore: Williams and Wilkins; 1939:52–169.

3. Gall FJ. *Anatomie et physiologie du système nerveux en général et sur celui du cerveau en particulier.* Vol. III-IV. Paris: Schoell; 1818–1819, 1819.
4. Critchley M. Neurology's debt to FJ Gall (1758-1828). *Br Med J.* 1965;2:775–781.
5. Young RM. *Mind, Brain and Adaptation in the Nineteenth Century.* Oxford: Clarendon Press; 1970.
6. Moore RY, Speh JC, Leak RK. Suprachiasmatic nucleus organization. *Cell Tissue Res.* 2002;309:89–98.
7. Swanson LW. What is the brain? *Trends Neurosci.* 2000;23:519–527.
8. Herrick CJ. *An Introduction to Neurology.* Philadelphia: Saunders; 1915.
9. Kühlenbeck H. *The Central Nervous System of Vertebrates.* Vol. 3. In: Karger S, Basel AG, eds. Baltimore, MD: Springer; 1978.
10. Gurdjian ES. The diencephalon of the albino rat. *J Comp Neurol.* 1927;43:1–114.
11. Le Gros Clark WE. The topography and homologies of the hypothalamic nuclei in man. *J Anat.* 1936;70:203–214.
12. Rioch DM, Wislocki GB, O'Leary JL. A precis of preoptic, hypothalamic and hypophysial terminology with atlas. In: Fulton JF, ed. *The Hypothalamus.* Baltimore: Williams and Wilkins; 1939:3–30.
13. Fulton JF. Introduction: historical resume. In: Fulton JF, ed. *The Hypothalamus.* Baltimore: Williams and Wilkins; 1939:xiii–xvi.
14. Cushing H. *The Pituitary Body and Its Disorders: Clinical States.* Philadelphia: JB Lippincott; 1912.
15. Hess WR. *The Functional Organization of the Diencephalon.* New York: Grune and Stratton; 1957.
16. Harris GW. *Neural Control of the Pituitary Gland.* London: E.J. Arnold; 1955.
17. Scharrer B. Neurosecretion: beginnings and new directions in neuropeptide research. *Annu Rev Neurosci.* 1987;10:1–17.
18. Guillemin R. Neuroendocrinology: a short historical review. *Ann N Y Acad Sci.* 2011;1220:1–5.
19. Dale HH. Pharmacology and nerve-endings. *Proc R Soc Med.* 1934;28:319–330.
20. Eccles JC. From electrical to chemical transmission in the central nervous system. *Notes Rec R Soc London.* 1976;30:219–230.
21. Klavdieva MM. The history of neuropeptides I. *Front Neuroendocrinol.* 1995;16:293–321.
22. Klavdieva MM. The history of neuropeptides II. *Front Neuroendocrinol.* 1996;17:126–153.
23. Klavdieva MM. The history of neuropeptides III. *Front Neuroendocrinol.* 1996;17: 155–179.
24. Klavdieva MM. The history of neuropeptides IV. *Front Neuroendocrinol.* 1996;17:247–280.
25. Meites J. The 1977 nobel prize in physiology or medicine. *Science.* 1977;198:594–596.
26. Raisman G. An urge to explain the incomprehensible: Geoffrey Harris and the discovery of the neural control of the pituitary gland. *Annu Rev Neurosci.* 1997;20:533–566.
27. Gundersen V. Co-localization of excitatory and inhibitory transmitters in the brain. *Acta Neurol Scand.* 2008;17(suppl 188):29–33.
28. Hökfelt T, Broberger C, Xu ZQ, Sergeyev V, Ubink R, Diez M. Neuropeptides—an overview. *Neuropharmacology.* 2000;39:1337–1356.
29. Sulzer D, Rayport S. Dale's principle and glutamate corelease from ventral midbrain dopamine neurons. *Amino Acids.* 2000;19:45–52.
30. Agnati LF, Guidolin D, Guescini M, Genedani S, Fuxe K. Understanding wiring and volume transmission. *Brain Res Rev.* 2010;64:137–159.
31. Fuxe K, Dahlström AB, Jonsson G, et al. The discovery of central monoamine neurons gave volume transmission to the wired brain. *Prog Neurobiol.* 2010;90:82–100.
32. Van den Pol AN. Neuropeptide transmission in brain circuits. *Neuron.* 2012;76:99–115.
33. Aschoff J. Exogenous and endogenous components in circadian rhythms. *Cold Spring Harb Symp Quant Biol.* 1960;25:11–28.

34. Aschoff J. Circadian rhythms in man. *Science*. 1965;148:1427–1432.
35. Richter CP. *Biological Clocks in Medicine and Psychiatry*. Springfield, IL: C.C. Thomas; 1965.
36. Daan S. A history of chronobiological concepts. In: Albrecht U, ed. *The Circadian Clock*. New York: Springer; 2010:1–36.
37. Halberg F, Cornélissen G, Katinas G, et al. Transdisciplinary unifying implications of circadian findings in the 1950s. *J Circadian Rhythms*. 2003;1:2.
38. Pittendrigh C, Bruce VG. An oscillator model for biological clocks. In: Rudnick D, ed. *Rhythmic and Synthetic Processes in Growth*. Princeton: Princeton University Press; 1957:75–109.
39. Heller A, Moore RY. Effect of central nervous system lesions on brain monoamines in the rat. *J Pharmacol Exp Ther*. 1965;150:1–9.
40. Hayhow WR, Webb C, Jervie A. The accessory optic fiber system in the rat. *J Comp Neurol*. 1960;115:187–215.
41. Nauta WJH, Gygax PA. Silver impregnation of degenerating axons in the central nervous system: a modified technique. *Stain Technol*. 1954;29:91–93.
42. Ingram WR. Nuclear organization and chief connections of the primate hypothalamus. *Res Publ Assoc Res Nerv Ment Dis*. 1940;20:195–244.
43. Kiernan JA. On the probable absense of retino-hypothalamic connections in five mammals and an amphibian. *J Comp Neurol*. 1967;131:405–408.
44. Nauta JHW, Haymaker W. Hypothalamic nuclei and fiber connections. In: Haymaker W, Anderson E, Nauta WJH, eds. *The Hypothalamus*. Springfield, IL: Charles C. Thomas; 1969:136–209.
45. Fink RP, Heimer L. Two methods for selective silver impregnation of degenerating axons and their synaptic endings in the central nervous system. *Brain Res*. 1967;4:369–374.
46. Droz B, Leblond CP. Axonal migration of proteins in the central nervous system and peripheral nerves as shown by radioautogaphy. *J Comp Neurol*. 1963;121:325–346.
47. Grafstein B. Transport of protein by goldfish optic nerve fibers. *Science*. 1967;157:1986–1988.
48. Moore RY, Lenn NJ. A retinohypothalamic projection in the rat. *J Comp Neurol*. 1972;146:1–14.
49. Hendrickson AE, Wagoner N, Cowan WM. An autoradiographic and electron microscopic study of retinohypothalamic connections. *Z Zellforsch*. 1972;135:1–26.
50. Moore RY. Retinohypothalamic projection in mammals: a comparative study. *Brain Res*. 1973;51:403–409.
51. Halasz B, Pupp L. Hormone secretion of the anterior pituitary gland after physical interruption of all nervous pathways to the hypophysiotrophic area. *Endocrinology*. 1965;77:553–562.
52. Moore RY, Eichler VB. Loss of a circadian adrenal corticosterone rhythm following suprachiasmatic lesions in the rat. *Brain Res*. 1972;42:201–206.
53. Stephan FK, Zucker I. Circadian rhythms in drinking behavior and locomotor activity of rats are eliminated by hypothalamic lesions. *Proc Natl Acad Sci USA*. 1972;69:1583–1586.
54. Rusak B, Zucker I. Neural regulation of circadian rhythms. *Physiol Rev*. 1979;59:449–526.
55. van Esseveldt KE, Lehman MN, Boer GJ. The suprachiasmatic nucleus and the circadian time-keeping system revisited. *Brain Res Rev*. 2000;33:34–77.
56. Van den Pol AN, Powley T. A fine-grained anatomical analysis of the role ofthe rat suprachiasmatic nucleus in circadian rhythms of feeding and drinking. *Brain Res*. 1979;160:307–326.
57. Schwartz WJ, Gainer H. Suprachiasmatic nucleus: use of 14C-labeled deoxyglucose uptake as a functional marker. *Science*. 1977;197:1089–1091.
58. Schwartz W, Davidsen LC, Smith CB. In vivo metabolic activity of a putative circadian oscillator, the rat suprachiasmatic nucleus. *J Comp Neurol*. 1980;189:157–167.

59. Sokoloff L, Reivich M, Kennedy C, et al. The [14C]deoxyglucose method for the measurement of local cerebral glucose utilization: theory, procedure, and normal values in the conscious and anesthetized albino rat. *J Neurochem*. 1977;28:897–916.
60. Inouye ST, Kawamura H. Persistence of circadian rhythmicity in a mammalian hypothalamic "island" containing the suprachiasmatic nucleus. *Proc Natl Acad Sci USA*. 1979;76:5962–5966.
61. Kubota A, Inouye ST, Kawamura H. Reversal of multiunit activity within and outside the suprachiasmatic nucleus in the rat. *Neurosci Lett*. 1981;27:303–308.
62. Green DJ, Gillette R. Circadian rhythm of firing rate recorded from single cells in the rat suprachiasmatic nucleus. *Brain Res*. 1982;245:198–200.
63. Groos GA, Hendriks J. Circadian rhythms in electrical discharge of rat suprachiasmatic neurons recorded in vitro. *Neurosci Lett*. 1982;34:283–288.
64. Shibata S, Oomura Y, Kita H, Hattori K. Circadian rhythmic changes of neuronal activity in the suprachiasmatic slice. *Brain Res*. 1982;247:154–158.
65. Lehman MN, Silver R, Gladstone WR, Kahn RM, Gibson M, Bittman EL. Circadian rhythmicity restored by neural transplant. Immunocytochemical characterization of the graft and its integration with the host brain. *J Neurosci*. 1987;7:1626–1638.
66. Saitoh Y, Nihonmatsu I, Kawamura H. Location of the suprachiasmatic nucleus grafts in rats which restored circadian rhythmicity after transplantation. *Neurosci Lett*. 1990;118:45–48.
67. Sawaki Y, Nihonmatsu I, Kawamura H. Transplantation of the neonatal suprachiasmatic nuclei into rats with complete bilateral suprachiasmatic lesions. *Neurosci Res*. 1984;1:67–72.
68. Ralph MR, Foster RG, Davis FC, Menaker M. Transplanted suprachiasmatic nucleus determines circadian period. *Science*. 1990;247:975–978.
69. Silver R, LeSauter J, Tresco PA, Lehman MN. A diffusible coupling signal from the transplanted suprachiasmatic nucleus controlling circadian locomotor rhythms. *Nature*. 1996;382:810–813.
70. Güldner FH. Synaptology of the rat suprachiasmatic nucleus. *Cell Tissue Res*. 1976;165:509–544.
71. Güldner F-H. Numbers of neurons and astroglial cells in the suprachiasmatic nucleus of male and female rats. *Exp Brain Res*. 1983;50:373–376.
72. Van den Pol AN. The hypothalamic suprachiasmatic nucleus of rat: intrinsic anatomy. *J Comp Neurol*. 1980;19:661–702.
73. Altman J, Bayer SA. The development of the rat hypothalamus. *Adv Anat Embryol Cell Biol*. 1986;100:1–178.
74. Davis FC, Boada R, LeDeaux J. Neurogenesis of the hamster suprachiasmatic nucleus. *Brain Res*. 1990;519:192–199.
75. Kabrita CS, Davis FC. Development of the mouse suprachiasmatic nucleus: determination of time of cell origin and spatial arrangements within the nucleus. *Brain Res*. 2008;1195:20–27.
76. Sofroniew MV, Weindl A. Projections from the parvocellular vasopressin- and neurophysin-containing neurons of the suprachiasmatic nucleus. *Am J Anat*. 1978;153:391–430.
77. Vandesande F, Dierickx K, DeMey J. Identification of the vasopressin-neurophysin producing neurons of the rat suprachiasmatic nuclei. *Cell Tissue Res*. 1975;156:377–380.
78. Moore RY. Entrainment pathways and the functional organization of the circadian system. *Prog Brain Res*. 1996;111:103–119.
79. Morin LP. SCN organization reconsidered. *J Biol Rhythms*. 2007;22:3–13.
80. Cajal SR. *Textura del Sistema Nervioso del Hombre y de los Vertebrados*. Madrid: Moya; 1899.
81. Hornung JP. The human raphe nuclei and the serotonergic system. *J Chem Neuroanat*. 2003;26:331–343.

82. Okamura H, Bérod A, Julien JF, et al. Demonstration of GABAergic cell bodies in the suprachiasmatic nucleus: in situ hybridization of glutamic acid decarboxylase (GAD) mRNA and immunocytochemistry of GAD and GABA. *Neurosci Lett.* 1989;102:131–136.
83. Gao B, Fritschy JM, Moore RY. GABAA-receptor subunit composition in the circadian timing system. *Brain Res.* 1995;700:142–156.
84. Gao B, Moore RY. Glutamic acid decarboxylase message isoforms in human suprachiasmatic nucleus. *J Biol Rhythms.* 1996;11:172–179.
85. Van den Pol AN, Tsujimoto KL. Neurotransmitters of the hypothalamic suprachiasmatic nucleus: immunocytochemical analysis of 25 neuronal antigens. *Neuroscience.* 1985;15:1049–1086.
86. Maywood ES, Chesham JE, O'Brien JA, Hastings MH. A diversity of paracrine signals sustains molecular circadian cycling in suprachiasmatic nucleus circuits. *Proc Natl Acad Sci USA.* 2011;108:14306–14311.
87. Nässel DR. Neuropeptide signaling near and far: how localized and timed is the action of neuropeptides in brain circuits? *Invert Neurosci.* 2009;9:57–75.
88. Abrahamson EE, Moore RY. Suprachiasmatic nucleus in the mouse: retinal innervation, intrinsic organization and efferent projections. *Brain Res.* 2001;916:172–191.
89. Mai JK, Kedziora O, Teckhaus L, Sofroniew MV. Evidence for subdivisions in the human suprachiasmatic nucleus. *J Comp Neurol.* 1991;305:508–525.
90. Moore RY. Organization of the primate circadian system. *J Biol Rhythms.* 1992;8: S3–S9.
91. Thompson RH, Swanson LW. Structural characterization of a hypothalamic visceromotor pattern generator network. *Brain Res Rev.* 2003;41:153–202.
92. Morin LP, Allen CN. The circadian visual system, 2005. *Brain Res Rev.* 2006;51:1–60.
93. Campbell AW. Histological studies on cerebral localisation. *Proc Royal Soc (London).* 1903;72:488–492.
94. Brodmann K. Vergleichende Lokalisationslehre der Grosshirnrinde in ihren Prinzipien dargestellt auf Grund des Zellenbaues. Leipzig: Johann Ambrosius Barth; 1909.
95. von Economo C, Koskinas GN. *Die Cytoarchitektonik der Hirnrinde des erwachsenen Menschen.* Vienna: Julius Springer; 1925.
96. Moga MM, Moore RY. Organization of neural inputs to the suprachiasmatic nucleus in the rat. *J Comp Neurol.* 1997;389:508–534.
97. Leak RK, Card JP, Moore RY. Suprachiasmatic pacemaker organization analyzed by viral transynaptic transport. *Brain Res.* 1999;819:23–27.
98. Welsh DK, Takahashi JS, Kay SA. Suprachiasmatic nucleus: cell autonomy and network properties. *Annu Rev Physiol.* 2010;72:551–577.
99. Tanaka M, Matsuda T, Shigeyoshi Y, Ibata Y, Okamura H. Peptide expression in GABAergic neurons in rat suprachiasmatic nucleus in comparison with other forebrain structures: a double labeling in situ hybridization study. *J Histochem Cytochem.* 1990;45:1231–1237.
100. Daikoku S, Hisato S, Kagotani Y. Neuronal associations in the rat suprachiasmatic nucleus demonstrated by immunoelectron microscopy. *J Comp Neurol.* 1992;325:559–571.
101. Ibata Y, Tanaka M, Ichitani Y, Takahashi Y, Okamura H. Neuronal interaction between VIP and vasopressin neurones in the rat suprachiasmatic nucleus. *Neuroreport.* 1993;4:128–130.
102. Ibata Y, Okamura H, Tanaka M, et al. Functional morphology of the suprachiasmatic nucleus. *Front Neuroendocrinol.* 1999;20:241–268.
103. Card JP, Brecha N, Karten HJ, Moore RY. Immunocytochemical localization of vasoactive intestinal polypeptide-containing cells and processes in the suprachiasmatic nucleus of the rat: light and electron microscopic analysis. *J Neurosci.* 1981;11:1289–1303.

104. Johnson RF, Moore RY, Morin LP. Loss of entrainment and anatomical plasticity after lesions of the hamster retinohypothalamic tract. *Brain Res*. 1988;460:297–313.
105. Provencio I, Jiang G, De Grip WJ, Hayes WP, Rollag MD. Melanopsin: an opsin in melanophores, brain, and eye. *Proc Natl Acad Sci USA*. 1998;95:340–345.
106. Lucas RJ, Freedman MS, Lupi D, Munoz M, David-Gray ZK, Foster RG. Identifying the photoreceptive inputs to the mammalian circadian system using transgenic and retinally degenerate mice. *Behav Brain Res*. 2001;125:97–102.
107. Güler AD, Ecker JL, Lall GS, et al. Melanopsin cells are the principal conduits for rod-cone input to non-image-forming vision. *Nature*. 2008;453:102–105.
108. Hattar S, Lucas RJ, Mrosovsky N, et al. Melanopsin and rod-cone photoreceptive systems account for all major accessory visual functions in mice. *Nature*. 2003;424: 76–81.
109. Lall GS, Revell VL, Momiji H, et al. Distinct contributions of rod, cone, and melanopsin photoreceptors to encoding irradiance. *Neuron*. 2010;66:417–428.
110. Lucas RJ, Lall GS, Allen AE, Brown TM. How rod, cone, and melanopsin photoreceptors come together to enlighten the mammalian circadian clock. *Prog Brain Res*. 2012;199:1–18.
111. Hattar S, Kumar M, Park A, et al. Central projections of melanopsin-expressing retinal ganglion cells in the mouse. *J Comp Neurol*. 2006;497:326–349.
112. Johnson RF, Morin LP, Moore RY. Retinohypothalamic projections in the hamster and rat demonstrated using cholera toxin. *Brain Res*. 1988;462:301–312.
113. Moore RY, Halaris AE, Jones BE. Serotonin neurons of the midbrain raphe: ascending projections. *J Comp Neurol*. 1978;180:417–438.
114. Morin LP, Meyer-Bernstein EL. The ascending serotonergic system in the hamster: comparison with projections of the dorsal and median raphe nuclei. *Neuroscience*. 1999;91:81–105.
115. Krieg WJS. The hypothalamus of the albino rat. *J Comp Neurol*. 1932;55:19–89.
116. Stephan FK, Berkley KJ, Moss RL. Efferent connections of the rat suprachiasmatic nucleus. *Neuroscience*. 1981;6:2625–2641.
117. Kriegsfeld LJ, Leak RK, Yackulic CB, LeSauter J, Silver R. Organization of suprachiasmatic nucleus projections in Syrian hamsters (*Mesocricetus auratus*): an anterograde and retrograde analysis. *J Comp Neurol*. 2004;468:361–379.
118. Watts AG, Swanson LW, Sanchez-Watts G. Efferent projections of the suprachiasmatic nucleus: I. Studies using anterograde transport of *Phaseolus vulgaris* leucoagglutinin in the rat. *J Comp Neurol*. 1987;258:204–229.
119. Watts AG, Swanson LW. Efferent projections of the suprachiasmatic nucleus: II. Studies using retrograde transport of fluorescent dyes and simultaneous peptide immunohistochemistry in the rat. *J Comp Neurol*. 1987;258:230–252.
120. Negri L, Lattanzi R, Giannini E, Melchiorri P. Bv8/Prokineticin proteins and their receptors. *Life Sci*. 2007;81:1103–1116.
121. Zhou QY, Cheng MY. Prokineticin 2 and circadian clock output. *FEBS J*. 2005;272:5703–5709.
122. Zhang C, Truong KK, Zhou QY. Efferent projections of prokineticin 2 expressing neurons in the mouse suprachiasmatic nucleus. *PLoS One*. 2009;4:e7151.
123. Mohawk JA, Green CB, Takahashi JS. Central and peripheral circadian clocks in mammals. *Annu Rev Neurosci*. 2012;35:445–462.
124. Beaule C, Granados-Fuentes D, Marpegan L, Herzog E. In vitro circadian rhythms: imaging and electrophysiology. *Essays Biochem*. 2011;49:103–117.
125. Friston KJ. Modalities, modes, and models in functional neuroimaging. *Science*. 2009;326:399–403.
126. Ross ED. Cerebral localization of functions and the neurology of language: fact versus fiction or is it something else? *Neuroscientist*. 2010;16:222–243.

127. Lang EW, Tomé AM, Keck IR, Górriz-Sáez JM, Puntonet CG. Brain connectivity analysis: a short survey. *Comput Intell Neurosci*. 2012;2012:412512.
128. Mantini D, Gerits A, Nelissen K, et al. Default mode of brain function in monkeys. *J Neurosci*. 2011;31:12954–12962.
129. Dibner C, Schibler U, Albrecht U. The mammalian circadian timing system: organization and coordination of central and peripheral clocks. *Annu Rev Physiol*. 2010;72:517–549.
130. MacDougall-Shackleton SA. The levels of analysis revisited. *Philos Trans R Soc Lond B*. 2011;366:2076–2085.
131. Mayr E. Cause and effect in biology. *Science*. 1961;134:1501–1506.
132. Beersma DG, van Bunnik BA, Hut RA, Daan S. Emergence of circadian and photoperiodic system level properties from interactions among pacemaker cells. *J Biol Rhythms*. 2008;23:362–373.
133. Cajal SR. The structure and connexions of neurons. *Nobel Lectures, Physiology or Medicine 1901-1921*. Amsterdam: Elsevier Publishing Company; 1967, 220–253.
134. Schmidt TM, Do MT, Dacey D, Lucas R, Hattar S, Matynia A. Melanopsin-positive intrinsically photosensitive retinal ganglion cells: from form to function. *J Neurosci*. 2011;31:16094–16101.
135. Schmidt TM, Chen SK, Hattar S. Intrinsically photosensitive retinal ganglion cells: many subtypes, diverse functions. *Trends Neurosci*. 2011;34:572–580.
136. Aguilar-Roblero R, Shibata S, Speh JC, Drucker-Colin R, Moore RY. Morphological and functional development of the suprachiasmatic nucleus in transplanted fetal hypothalamus. *Brain Res*. 1992;580:288–296.
137. Jacobowitz DM, Winsky L. Immunocytochemical localization of calretinin in the forebrain of the rat. *J Comp Neurol*. 1991;304:198–218.
138. Mikkelsen JD, Vrang N. A direct pretectosuprachiasmatic projection in the rat. *Neuroscience*. 1994;62:497–505.
139. Moga MM, Weis RP, Moore RY. Efferent projections of the paraventricular thalamic nucleus in the rat. *J Comp Neurol*. 1995;359:221–238.
140. Moore RY. Catecholamine neuron systems in brain. *Ann Neurol*. 1982;12:321–327.
141. Moore RY, Speh JC. GABA is the principal neurotransmitter in the mammalian circadian system. *Neurosci. Letters*. 1993;150:112–116.
142. Sherrington CS. *The Integrative Action of the Nervous System*. New York: Scribners; 1906.
143. Wurtman RJ, Axelrod J. The pineal gland. *Sci Am*. 1965;213:50–60.
144. Bünning E. *The physiological clock*. 3rd ed. New York: Springer-Verlag; 1973.
145. Axelrod J. Noradrenaline: fate and control of its biosynthesis. *Science*. 1971;173:598–606.
146. Axelrod J, Snyder SH, Heller A, Moore RY. Light-induced changes in pineal hydroxyindole-O-methyltransferase: abolition by lateral hypothalamic lesions. *Science*. 1966;154:898–899.
147. Nauta WJH. Some early travails of tracing axonal pathways in the brain. *J Neurosci*. 1993;13:1337–1345.
148. Maywood ES, Reddy AB, Wong GK, O'Neill JS, O'Brien JA, McMahon DG, Harmar AJ, Okamura H, Hastings MH. Synchronization and maintenance of timekeeping in suprachiasmatic circadian clock cells by europeptidergic signaling. *Curr Biol*. 2006;16:599–605.
149. Vosko AM, Schroeder A, Loh DH, Colwell CS. Vasoactive intestinal peptide and the mammalian circadian system. *Gen Comp Endocrinol*. 2007;152:165–175.
150. Canteras NS, Ribeiro-Barbosa ER, Goto M, Cipolla-Neto J, Swanson LW. The retinohypothalamic tract: comparison of axonal projection patterns from four major targets. *Brain Res Rev*. 2011;65:150–183.
151. Leak RK, Moore RY. Suprachiasmatic nucleus projections in the rat: retrograde analysis. *J Comp Neurol*. 2001;433:312–334.

CHAPTER TWO

Epigenetic Regulation of the Molecular Clockwork

Kristin Eckel-Mahan, Paolo Sassone-Corsi
Center for Epigenetics and Metabolism, University of California, Irvine, California, USA

Contents

Abstract

Circadian rhythms control a wide range of physiological events in all organisms. Typical of our modern lifestyles is the flexibility to rest, exercise, eat, or socialize at any time of the circadian day or night; yet, these allowances correlate with rising disorders of a metabolic nature, which are thought to be mediated by changes in the molecular events underlying metabolic gene expression. Because the clock confers on gene expression changes in activity that are not necessarily related to changes in DNA sequence, the study of circadian rhythms is inseparable from epigenetics. Increasingly evident is that energy balance at the systems level relies on precise and collaborative circadian timing of epigenetic events within individual cells and tissues of the body. At the center of these rhythms resides the circadian clock machinery, a remarkably well-orchestrated transcription–translation feedback system that incorporates a fluctuating landscape of mRNA expression, protein stability, chromatin state, and metabolite abundance to keep correct time. Understanding more fully the ties that exist between cellular metabolism and the circadian clock at the epigenetic level will produce not only needed insights about circadian physiology but also novel strategies for the pharmacological and nonpharmacological treatment of metabolic disorders.

Progress in Molecular Biology and Translational Science, Volume 119
ISSN 1877-1173
http://dx.doi.org/10.1016/B978-0-12-396971-2.00002-6

1. INTRODUCTION TO EPIGENETICS AND THE CIRCADIAN CLOCK

Epigenetics, which involves changes in gene expression or activity that are not due to fundamental alterations in DNA sequence that are passed down, encompasses several levels of regulation. These include histone modifications, methylation of DNA, histone composition, and small RNA presence.[1–8] In most cases, epigenetic processes are mediated not by one but several of these mechanisms at once.

Rhythmicity in gene expression falls under the umbrella of epigenetics in the sense that changes in the temporal aspects of gene expression can greatly alter a cellular response while the genetic material being expressed might be identical to another cell or organism (Fig. 2.1). At least 10% of genes in a given cell are oscillatory and are controlled directly by cyclical changes in the chromatin structure and function at relevant gene promoters.[9–12] Post-translational modifications at histone proteins account for much of the plasticity in chromatin remodeling processes.[4] Circadian changes in chromatin modifications have even recently been demonstrated as important for a

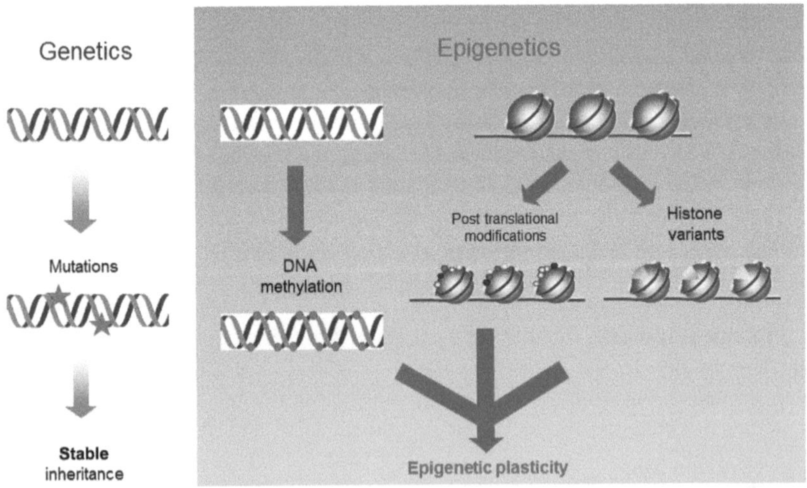

Figure 2.1 Genetics versus epigenetics. The mechanisms underlying genetic and epigenetic inheritance involve the passing on of different information, such as mutations, found within the coding region of the DNA itself (genetics) and changes in the expression patterns of the same DNA code (epigenetics). Circadian fluctuations in DNA expression are considered to be epigenetic in nature. (See Color Insert.)

number of antisense RNAs, lincRNAs, and microRNAs, underscoring the pervasive nature of these modifications in overall cellular circadian operations.[13]

Higher order chromatin structure involves the wrapping of DNA around histone proteins. Due to the enormous length of DNA strands relative to the size of a cell, DNA must be packaged extremely tightly. DNA wrapped around the histone octamer (two copies of each of the histone proteins H2A, H2B, H3, and H4) is referred to as a nucleosome. Many nucleosomes packaged together become part of the condensed structure knows as chromatin. As chromosomes must be replicated prior to cell division, it is essential that genes necessary for the development and survival of offspring cells be replicated. In addition, RNA synthesis requires the binding of additional polymerase structures which are necessary for initiation and progression of gene transcription (reviewed in Ref. 14). These processes depend on priming of the chromatin, a step which involves relaxing of this higher order structure and unwinding of specific DNA at nucleosomes. This is accomplished in part by the modification of histone protein tails, which are essential components of chromatin structure and function.[15] Modification of histone tails is essential for providing permissive states for gene transcription. The ways in which these modifications mediate a permissive state are complex but reveal the importance of generating specific chromatin structures that allow transcription to occur. During interphase, chromosomes reside in precise territories which are enveloped by the interchromatin compartment. It is generally thought that gene-dense chromosome territories reside in the interior regions of the nucleus while regions of the chromosome that harbor few genes get localized at the nuclear periphery. Specific marks that occur at histone tails can lead to chromatin decondensation which ultimately repositions actively transcribed genes out from regions of compact chromatin structures (reviewed in Ref. 16). Some of these histone tail modifications and their circadian profiles will be addressed below.

Histones contain a number of amino acid residues within their N-terminal tails that can be modified. These modifications include acetylation, phosphorylation, methylation, sumoylation, ADP-ribosylation, biotinylation, and ubiquitylation. The type and combination of these modifications are thought to form a type of "epigenetic code," a process originally defined by Strahl and Allis.[8] Alterations at chromatin associated with gene activation typically include phosphorylation, acetylation, and methylation. Specifically, at the histone 3 tail, phosphorylation at serine 10 (S10), acetylation at lysine 9/14 (K9/14), and methylation at lysine 4/27 (Me; K4 and K27) often involve robust changes in gene expression at relevant loci. Histone 3 (H3) K4 trimethylation particularly is associated with activation of gene transcription and is essential for circadian gene transcription and the recruitment of chromatin remodeling complexes.

Recent high-throughput approaches used in studying the role of the circadian clock in chromatin structure and function have revealed that modulation of chromatin structure throughout the circadian period occurs in a genome-wide way and on a much larger scale than originally appreciated.[17] Specifically, wide scale changes in the circadian regulation of histone modifications have been observed via RNA-seq experiments with antibodies for H3K4me3, H3K9Ac, and H3K27Ac. These marks show robust circadian rhythmicity in occupancy near the transcription start sites of many circadian genes. Interestingly, H3K27Ac is enriched at both intragenic and intergenic enhancer sites.[17] This and other recent observations underscore the remarkable ways in which the circadian clock participates in rhythmic gene expression through the epigenome.

1.1. Histone proteins under circadian control

While the expression of a number of histone modifiers is under circadian control,[9] some histone modifying proteins directly interact with the circadian clock proteins themselves. For example, enhancer of zeste homology 2 (EZH2), JumonjiC and ARID domain-containing histone lysine demethylase 1A (JARID1A), histone deacetylase 3 (HDAC3), WD repeat domain-containing protein 5, mixed-lineage leukemia histone methyltransferase 1 (MLL1), and p300/cAMP response element-binding protein (CREB), p300/CBP all have been reported to bind directly to clock proteins.[18–24] In addition, current studies that are ongoing have begun to address the rhythmic modification of proteins throughout the circadian cycle. Specifically, the acetylation of many proteins across the circadian cycle is oscillatory, allowing rhythmicity in activity even when the expression of the protein itself remains constant over time.[25] The role of some of these modifiers and their interaction with the clock machinery will be discussed further in Sections 1.2-1.4.

1.2. Histone acetylases under circadian control

As acetylation is a known circadian event at both histone- and nonchromatin-associated proteins,[25] the expression or activity of several enzymes that regulate protein acetylation has been studied in a circadian context. A list of several enzymes can be seen in Table 2.1 and includes the circadian transcription factor CLOCK itself. CLOCK functions dually as a transcription factor and a histone acetyltransferase, modifying its own binding partner, BMAL1 by direct acetylation.[26] It also functions as a HAT at the H3K9/K14 positions. The role of CLOCK as a HAT appears to extend beyond BMAL1 and H3. CLOCK HAT activity influences gene transcription via direct acetylation of nuclear receptors.[27] The discovery that CLOCK protein functions as a histone acetyltransferase opened up new directions in our understanding of how this transcriptional activator contributes to such robust oscillations in gene expression via its binding partner, BMAL1.

Other transcription factor complexes have also been observed to be potent activators of transcription via both their direct binding to DNA and their ability to modify chromatin structure. Sometimes, this is

Table 2.1 Chromatin-modifying enzymes with known association with the circadian clock

Enzyme	Type	Targets
p300	Histone acetyltransferase	H3K9/K14
CREB-binding protein, CBP	Histone acetyltransferase	H3K4
CLOCK	Histone acetyltransferase	H3K9/K14, BMAL1 (ARNTL)
p300/CBP-associated factor, PCAF	Histone acetyltransferase	H3K4
SIRT1	Histone deacetylase	H3K9, BMAL1, PER2
HDAC3	Histone deacetylase	H3K9
MLL1/MLL3	Histone methyltransferase	H3K4
EZH2	Histone methyltransferase	H3K27
JARID1A/KDM5	Histone demethylase	H3K4

Several factors associated with cAMP-responsive element-binding protein (CREB)-mediated gene transcription as well as the circadian protein CLOCK exhibit acetyltransferase activity. The sirtuin 1 protein (SIRT1), which associates directly with the clock machinery, and histone deacetylase 3 (HDAC3) both contain deacetylase activity and are potent regulators of rhythmic transcription. Among the histone methylases best known for their role in circadian control of gene expression are mixed-lineage leukemia 1 and 3 (MLL1 and MLL3), enhancer of zeste homolog 2 (EZH2), and JumonjiC and ARID domain-containing histone lysine demethylase 1A (JARID1A).

attributable to their recruitment of a histone acetyltransferase, but other times, the factors themselves contain HAT activity. For example, p300, a binding partner of CREB, has been demonstrated to contain histone acetyltransferase activity and is involved in H3K9/14 acetylation.[28] Also associated with CREB-mediated gene transcription is the protein CBP (CREB-binding protein). This protein contains a HAT domain and is predominantly associated with H3K4 acetylation. This is generally associated positively with gene transcription.[29] p300/CBP is also thought to be important for H3K27 acetylation, a modification which takes place in both enhancer and promoter regions.[24] The p300/CBP complex mediates these rhythmic modifications in part by being rhythmically recruited to the CLOCK:BMAL1 complex in an oscillatory fashion.[17,30] The role of CBP in rhythmic gene transcription appears to be complex, however, as CBP also interacts with the period 2 protein (PER2), thereby participating in the negative limb of the negative transcriptional loop of the circadian clock.[99] p300/CBP-associated factor (PCAF) is yet another H3K4 modifier, inducing acetylation and subsequent gene activation at relevant gene sites.[29]

1.3. Histone deacetylases under circadian control

There are a number of histone deacetylase (HDAC) enzymes that have been demonstrated as circadian in expression or activity. The circadian expression or activity of these proteins is equally important in circadian gene expression as proteins with HAT activity as HDACs remove acetyl groups from histones and nonhistone proteins, ultimately affecting activity that is dependent on acetylation state. Among the HDACs associated with circadian control, the sirtuin family members may be among the most famous HDACs. The sirtuin family of proteins has been recognized for its peripheral and central roles in both circadian rhythmicity and metabolism. The sirtuin family is composed of seven family members, which display unique subcellular localization. Specifically, some sirtuins are mitochondrial (SIRT3, SIRT4, and SIRT5) and others are principally cytoplasmic (SIRT2) nuclear (SIRT1, SIRT6, and SIRT7) or expressed in several compartments within the cell.[31] Sir2 (silent information regulator 2, the homolog of the mammalian SIRT1) is a nicotinamide adenine dinucleotide (NAD^+)-dependent HDAC, and well known for its role in longevity,[32,33] although it is not clear that it contributes to a long lifespan in all situations and species.[34] Consequently, the mammalian ortholog of Sir2, SIRT1, has become a well-recognized component linking the circadian clock to metabolism. SIRT1 is a class III HDAC, which differs from the class I and II deacetylases in that it requires NAD^+. NAD^+ serves as a cofactor for SIRT1 and directly modulates its activity. SIRT1 breaks down NAD^+ during lysine deacetylation and produces *O*-acetyl-ADP-ribose as a by-product. In the fasting state, levels of NAD^+ rise, and the activity of SIRT1 is elevated as a result.[35] However, when energy is in excess, NAD^+ becomes depleted because of the rampant flux through the glycolytic cycle which promotes the conversion of NAD^+ to NADH. Recent studies show that SIRT1 directly interacts with circadian clock machinery and that its rhythmic enzymatic activity is necessary for robust oscillations in gene expression *in vivo*.[36,37] The expression of SIRT1 has been reported as oscillatory by some,[38] but the activity of SIRT1 has been demonstrated to be oscillatory based on the oscillatory abundance of its cofactor, nucleotide adenine dinucleotide (NAD^+) as depicted in Fig. 2.2.[37,38] SIRT1 is an HDAC for the circadian protein BMAL1 and it leads to the both the acetylation and degradation of PER2.[38] In addition, SIRT1 is a chromatin modifier, deacetylating H3K9/14 and thereby influencing subsequent gene transcription.[39,40]

Other HDACs associated with the clock include the protein HDAC3 and its associated nuclear receptor corepressor 1 protein (NCoR1).[41] This complex is involved in the deacetylation of H3K9 and can dramatically influence the expression levels of associated genes. HDAC3 has been demonstrated as oscillatory in activity and modulates much of the rhythmic gene expression that occurs throughout the circadian cycle.[42] Loss of HDAC3

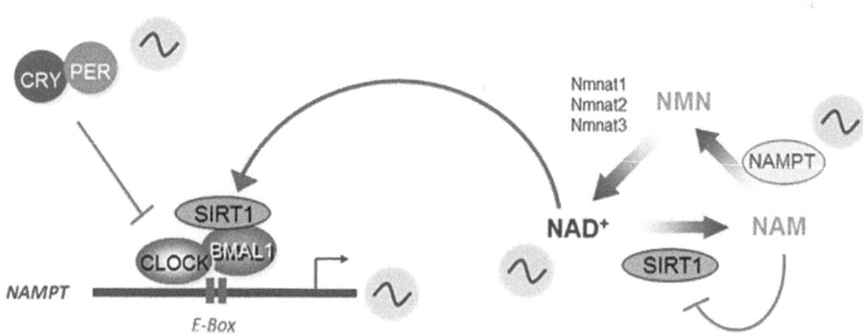

Figure 2.2 The role of SIRT1 in the circadian clock is maintained by NAD^+. The HDAC SIRT1 associates with CLOCK and BMAL1 proteins at the promoter of some genes, including those involved in the abundance of its own cofactor, NAD^+. Oscillations in the recycling of NAD^+ by *Nampt* are perpetuated by rhythmic activation of the *Nampt* promoter by CLOCK, BMAL1, and SIRT1. (See Color Insert.)

interaction with NCoR1 eliminates circadian rhythmicity of many clock genes as well as clock output genes important for cellular metabolism.

1.4. Other histone protein modifiers associated with the circadian clock

In addition to acetylation as a modifier of histone activity and chromatin structure, histone protein methylation has been associated with changes in gene expression. Studies looking at single locus methylation at histone tails have now been expanded to studies looking at the global changes in this mark across the genome. These studies reveal not only the importance of this mark in rhythmic gene transcription but also the complexity of the histone code during the process of cellular timekeeping.

During a repressive state of circadian gene expression, there is typically di- and trimethylation of H3K27 at the *Per* gene locus, a modification that is dependent on the polycomb group protein, EZH2.[23] At the D-element-binding protein (*Dbp*) locus, H3K9 acetylation turns into methylation in a rhythmic fashion, a process that is necessary for the robust changes in *Dbp* expression throughout the circadian cycle.[21] H3K4 methylation is strongly rhythmic and occurs in a circadian manner at numerous rhythmic genes. Among the histone methyltransferases that participate in the rhythmic methylation of target genes, the MLL1 protein has been demonstrated as a protein with potent influence on cellular timekeeping. Specifically, H3K4 methylation by the histone methyltransferase MLL1 is important for circadian gene expression.[20] MLL1, a homolog of Drosophila trithorax, is essential for cyclic H3K4 trimethylation and functions in a complex with the CLOCK–BMAL1 heterodimer. MLL1 appears necessary for the development of a permissive chromatin state during the process of circadian transcription. In addition to MLL1, the methyltransferase MLL3 also contributes to circadian transcription.[43] MLL3 is expressed in a circadian manner, and it is also associated with H3K4me3. Unlike MLL1, it does not appear to require CLOCK or BMAL1 binding for circadian occupancy, although it does dramatically affect the expression of numerous core clock genes.[43] Similar rhythms in H3K4me3 have been observed in *Arabidopsis thaliana*, underscoring the role of methyltransferases in the maintenance of the circadian clock across organisms.[44]

The JumonjiC and ARID domain-containing histon elysine demethylase 1A (JARID1A, also known as KMD5A) is another histone methyltransferase implicated in circadian gene expression. JARID1A binds directly to CLOCK–BMAL1. In addition to its role as a histone demethylase, however,

JARID1A appears to regulate rhythmic gene transcription in part via its ability to inhibit histone deacetylation.[19] Finally, the EZH2 protein also affects histone methylation in a rhythmic manner, acting predominately at H3K27 to alter gene transcription at relevant promoter sites.[23,45] Other jumonji domain-containing proteins are also implicated in circadian transcription. Specifically, *Jmj5B* (*Kdm8*) is oscillatory in expression and contributes to circadian timekeeping.[46] Loss of this protein in both mammalian systems and *Arabidopsis* causes a phase advance in the circadian cycle.

Finally, a number of other enzymes involved in histone tail methylation have also been observed to be important in the maintenance of the clock in cells. The lysine methyltransferase SET1 in *Neurospora* plays an interesting role in *frq* expression and changing not only trimethylation at H3K4 but via this modification, allowing methylation of the DNA itself.[47]

2. CIRCADIAN EPIGENETICS AND SYNAPTIC PLASTICITY IN THE BRAIN

The role of transcription in synaptic plasticity has long been known but only recently has the role of chromatin modification in driving synaptic plasticity been appreciated.[6] The role of histone modifications in chromatin structure appears to be as important in neuronal communication as it is for cells of the periphery to perform their functions. The role of chromatin structure in driving synaptic plasticity is an essential component of circadian physiology as the brain houses the "master" circadian clock for the body.[48] Specifically, in mammals, the suprachiasmatic nucleus (composed of two small lobes in the anterior hypothalamus) drives many circadian rhythms in the body and lesion of this region causes arrhythmicity of an organism.[49] As an essential driver of circadian rhythmicity, the role of gene transcription in this region is essential for normal clock ticking across the body. The discovery of epigenetic processes at work in the SCN as well as in other regions within the central nervous system has underscored the importance of chromatin structure in driving additional physiological processes.

2.1. Light and epigenetics in the suprachiasmatic nucleus

The mammalian SCN responds to light via neurons extending from the retina via the retinohypothalamic tract, providing synchronization to the surrounding light/dark cycles. The circadian clock must undergo reentrainment in order for an organism to synchronize with its environment. This process is extremely important due to the fact that humans are often subject to changes in their environment such as during travel across time zones. The entrainment process takes days to week, temporarily involving desynchronization between the central clock and the peripheral organs. For example, the disruption of the circadian clock during travel across time zones produces jet lag, as homeostatic and circadian cues temporarily are out of sync. The endogenous circadian clock must reentrain, a process that relies on distinct signaling transduction pathways that lie upstream of the transcriptional translational loops of the clock machinery already described.

The clock proteins are required to maintain circadian oscillations in the SCN (i.e., *Bmal1* knockout animals and *Clock* mutant mice have altered circadian rhythms in SCN tissue). The clock system uses both photic and nonphotic inputs to transduce the signals necessary for alterations in chromatin structure. Specifically, photic and nonphotic inputs are integrated in the SCN by various signal transduction pathways which result in epigenetic changes which alter gene expression. For example, neurons of the SCN respond to pituitary adenylyl cyclase-activating peptide during the subjective day,[50] while at night they respond to acetylcholine as well as other cGMP-activating analogs.[51] While nonphotic SCN resetting can occur via serotonergic innervation,[52] light exposure during the subjective night can also reset the SCN clock by inducing the release of glutamate from retinal ganglion cells, leading to activation of NMDA receptors on SCN neurons. NMDA-induced depolarization of SCN neurons causes an activation of calcium-sensitive adenylyl cyclases in the SCN and the production of cAMP. In turn, cAMP activates proteins such as the guanine nucleotide exchange factors (EPAC proteins) as well as the mitogen-activated protein kinase (MAPK) signaling cascade, which in neuronal cells couples depolarization to transcription via activation of CREB. Indeed, organisms exposed to a light pulse at night show a rapid and robust MAPK phosphorylation in the SCN as well as phosphorylation of CREB and activation of cAMP response element (CRE)-mediated gene transcription.[53,54] cAMP oscillations are necessary for circadian rhythmicity in the SCN.[55] Inhibition of cAMP-producing adenylyl cyclase enzymes prolongs period length, an event which is abrogated by mutations in the central clock machinery. Understanding of the central pathways involved in SCN plasticity provides an essential framework for understand how chromatin remodeling in the brain participates in cellular and systems-wide level timekeeping.

Initial studies looking at the role of epigenetic processes in circadian gene expression revealed that a pulse of light administered to a nocturnal rodent during its subjective night period can induce phosphorylation as H3S10.[56]

Light uses the signal transduction pathways described above to induce phosphorylation of H3 and subsequently drive induction of the *c-Fos* and *Per1* genes. This observation led to the idea that chromatin modification at circadianly expressed genes might be important for driving their rhythmic expression. Subsequently, numerous studies now reveal that changes in histone protein posttranslational modifications are a driving force in the circadian expression of CCGs.[22,26,30,57] This is true of mammalian and nonmammalian systems. For example, dynamic chromatin modifications occur in the *Arabidopsis* CCA1/LHY and TOC1 regulatory regions and are an essential part of circadian rhythms in this organism.[58]

It is interesting to speculate whether histone acetylation at promoters of genes necessary for long-term memory is rhythmic. There is accumulating evidence that synaptic plasticity, the basis of all cognitive processes, involves the circadian clock.[59] The basic MAPK pathway, which is necessary for both circadian rhythms in the SCN and for long-term memory formation in the hippocampus, is under circadian control and involves rhythmicity in CREB activation and subsequent gene transcription.[60] Interestingly, histone acetylation at promoters of genes in the central nervous is an important process which is required for long-term memory formation.[61] To date, little has been done to address whether this activity is a rhythmic event in regions of the central nervous system associated with learning and memory; but based on the circadian oscillation of upstream events, it is quite likely to be so. What is known is that both acetylation and methylation occur in the central nervous system and both types of modifications have been demonstrated to produce effects on cognition. After a learning event, the transcription factor CREB is activated by phosphorylation, allowing its binding to CREs.[62] This event is an essential step for the formation of long-term memory. Interestingly, HDAC inhibitors have been shown to enhance synaptic plasticity and memory, a process that is in part mediated by the HAT activity of CREB and CBP.[63] Among other tasks dependent on CREB-mediated gene transcription, the object recognition task (a hippocampus-dependent learning paradigm) has been demonstrated to be influenced by HDAC inhibitors in a CBP-dependent manner.[64]

3. LINKING THE CIRCADIAN METABOLOME AND METABOLISM TO EPIGENETICS

While genetics are essential for the maintenance of energy balance,[65–67] numerous studies have revealed that environmental cues including those affecting circadian rhythms also play pivotal roles in metabolic homeostasis.[68,69] Because today's technology provides exogenous lighting throughout the 24-h cycle, night shift and rotating shift work have become increasingly common. These extensions of the conventional workday influence not only temporal aspects of work and social activities but also physiological and molecular rhythms in the body. Studies on night shift and rotating shift workers present compelling evidence that circadian rhythm disruptions contribute to metabolic disorders. For example, an association between cardiovascular disease, increased body mass, and elevated plasma glucose and lipid levels has been observed in circadian disturbance and humans which engage in nighttime shift work.[70–73] Small rodent studies also support this link and simply changing their normal light/dark cycle to one in which dim light replaces the normal dark period causes profound changes in metabolism that are observed on a physiological level.[74]

Integral to cellular metabolism are small metabolites. Metabolites are gaining attention based on their potent zeitgeber effects on some tissues. Many small metabolites are used for epigenetic processes. Specifically, biochemicals such as *S*-adenosyl methionine, citrate, acetyl-CoA, NAD^+, GlcNAc, glucose, glutamine, methionine, and ATP are all direct or indirect modulators of chromatin structure (see Fig. 2.3). In part because of this, recent studies have begun to focus on more global oscillations within the

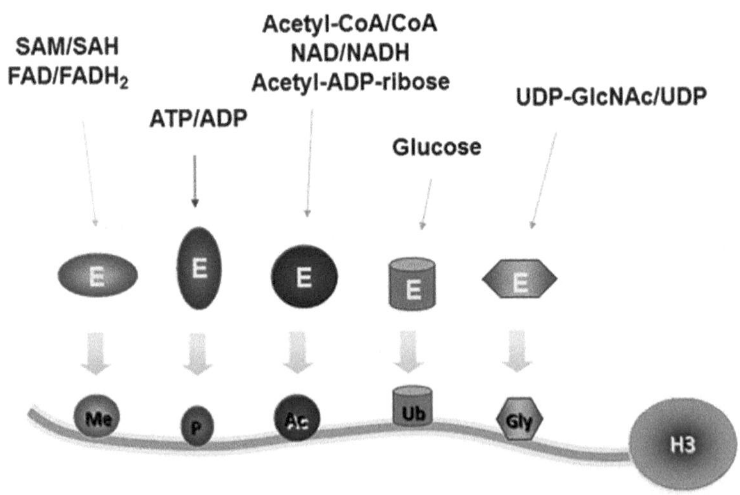

Figure 2.3 Metabolites affect modifications at histone tails and act as substrates, cofactors, or upstream metabolic precursors for the many modifications at histone tails. Me, methylation; P, phosphorylation; Ac, acetylation; Ub, ubiquitination; Gly, glycosylation. (See Color Insert.)

metabolome. This approach is important for several reasons. An obvious provider of many biochemicals in the cell is food. Thereby, the rhythmicity in energy intake and likely its quality too are critical for the maintenance of circadian oscillations in chromatin modifications. Food functions as a zeitgeber for peripheral tissues, specifically, and if food can entrain peripheral clocks, it is likely that its timekeeping properties have much to do with its ultimate metabolic effects on chromatin. Studies looking at the role of the circadian clock in controlling metabolite levels demonstrate that much of the liver metabolome is under circadian control. Numerous metabolites oscillate in a circadian fashion, and via studies in *Clock* knockout animals, many of these metabolite oscillations have been demonstrated to be dependent directly on the expression of *Clock*.[75] Bioinformatics techniques which involve large-scale mining of the literature have been instrumental in mapping these circadian metabolites within the context of their cellular surroundings. For example, circadian (and noncircadian) controlled metabolites are now depicted in interactive maps which incorporate data from multiple sources to display data on circadian gene oscillation of metabolite-producing and/or-degrading enzymes,[11] transcription factor binding sites for these metabolic enzymes,[76,77] and the metabolite's own interacting metabolites.[75,78–80] These networks can be visualized on an interactive database, http://circadiomics.igb.uci.edu/. The circadian oscillation of metabolites includes metabolites representing numerous metabolic pathways. Amino acid and xenobiotic metabolites tend to be higher at night in nocturnal rodents, while carbohydrate, lipid, and nucleotide metabolites peak during the rest phase.[75] While much of the hepatic metabolome is under circadian influence, studies looking at the metabolome of serum from humans in which circadian activities (such as sleeping, diurnal eating patterns, and activity) have been removed reveal that a remarkable 15% of metabolites still oscillate in a circadian fashion.[81] The oscillations of many amino acid-related metabolites as well as urea cycle metabolites appear to be tightly linked to the circadian clock and have been reported in several studies.[75,81–83]

The number of oscillatory metabolites that are known to influence various aspects of clock machinery is growing. For example, a recently discovered function of the CRY protein depends on rhythmic metabolic factors. Specifically, in *Drosophila*, CRY controls neuronal firing rate. ILNv neurons, which are light-sensing neurons of *Drosophila* pacemaker cells, require CRY for a rapid light response.[84] This is accomplished via a redox-based flavin mechanism, which induces the CRY-dependent neuronal response to blue light.

This response can be attenuated by the administration of an antagonist, which blocks the flavin binding site in CRY. Interestingly, flavin adenine dinucleotide (FAD^+) is robustly oscillatory in hepatic tissue, and dependent on the circadian protein CLOCK.[75] Whether flavin oscillates in the *Drosophila* pacemaker cells has not been discovered to date; however, light-induced CRY activity does oscillate in a metabolite-dependent way, and it suggests that there may be an oscillatory pattern of FAD^+ itself. The CLOCK-dependent oscillation of FAD^+ in mice suggests that its oscillation may be present in other tissues and organisms.[75]

Another mode of coupling between metabolite availability and the circadian clock centers on the ability of NPAS2 to function as a gas sensor.[85] Specifically, the PAS domain of the NPAS2 protein can bind heme directly, allowing it to function as a carbon dioxide-sensitive transcription factor. Interestingly, there are two enzymes involved in heme and CO biosynthesis, which are differentially expressed in regions of the brain that highly express NPAS2, an observation which is likely more than mere coincidence. Aminolevalinic acid synthases (*Alas*), a rate-limiting enzyme in heme biosynthesis, expression is oscillatory in the brain as is the heme oxygenase 2 protein, which is involved in CO generation. These proteins are expressed robustly in the same regions that express high levels of NPAS2 as well, particularly the forebrain. NPAS2 is not solely a heme sensor but appears to be directly regulated by other metabolites as well. Both heme bound- and nonheme-bound NPAS2 PAS domains depend on a high NADPH/NADP ratio for efficient DNA binding. Heme is also a ligand for the circadian protein REV-ERBα,[86,87] the interaction of which promotes the recruitment of the NCoR/HDAC3 corepressor complex that leads to the repression of PGC-1α expression among other genes. Itself a powerful inducer of heme synthesis, PGC-1α can negatively feedback to control the levels of cellular heme, thereby regulating cellular metabolism.[88] NCoR and the HDAC3 complex get recruited to the promoter of *Bmal1* by REV-ERBα, where they control the expression of *Bmal1*. The formation of this complex is essential for circadian rhythmicity as well as for the oscillation in the expression of many metabolic genes.[42,89]

3.1. The role of specific metabolites in chromatin structure

Studies focused on the oscillatory nature of specific metabolites and their link to the circadian clock have been revealing in demonstrating how small biochemicals can affect chromatin structure and function. For example, the oscillation of NAD^+ has numerous effects in the cell which includes its ability to feed back into the clock system via CLOCK:BMAL1-mediated gene transcription. Mitochondrial NAD^+ is used as a carrier molecule for oxidation–reduction reactions and as such is essential for cellular energy balance. Electrons and a hydrogen atom removed from a substrate molecule are picked up by NAD^+ and ultimately used to generate additional ATP, driving ATP-dependent reactions. NAD^+ is generated from niacin and participates as a coenzyme in many cellular dehydrogenase reactions, including β-oxidation of fatty acids and the Krebs cycle. Adding to its role in ATP generation, NAD^+ also provides the ADP-ribose necessary for ADP-ribosylation of some proteins. Increases in the activity of the NAD^+-dependent ribosylating enzyme, poly(ADP-ribose) polymerase-1 (PARP-1), can deplete cellular NAD^+ pools, ultimately causing damage and cell death.[90–92] PARP-1 may be an important player in the circadian influence of transcription via chromatin modification. PARP-1 has been shown to have a direct role in the regulation of chromatin structure through the modulation of the histone demethylase KDM5B.[93] PARP-1 is thought to allow a permissive chromatin state such that proper loading of the RNA Pol II machinery can take place. This is accomplished by PARP-1-mediated inhibition of the histone demethylase KDM5B. As a modifier of PARP-1 but also as an activator of SIRT1, the contribution of NAD^+ to cellular metabolic homeostasis is mediated in large part through its role in gene expression. Figure 2.4 summarizes the pleotropic roles of NAD^+ in the cell.

The small molecule cyclic ADP-ribose (cADPR) is made from NAD^+ by ADP-ribosyl cyclases, and it has also been reported to oscillate in a circadian fashion. cADPR is a ligand for type-3 ryanodine receptors, which are receptors important for generating cytosolic calcium in plants and animal cells.[94] The dependence of this metabolite on the levels of cellular NAD^+ levels provides another example of how oscillations in metabolites participate in the circadian clock system and thereby integrate cellular metabolism into cellular timekeeping.

In addition to NAD^+, a number of other metabolites are important for epigenetic changes at chromatin, particularly when it comes to modification of histone tails. Figure 2.3 depicts the many modifications that can take place on the tail of H3 and shows just a small number of the many metabolites that assist in these marks.

The importance of epigenetics in gene expression extends to that of nuclear hormone receptors. While the expression of numerous nuclear hormone receptors oscillates, the expression of some does not. Regardless of

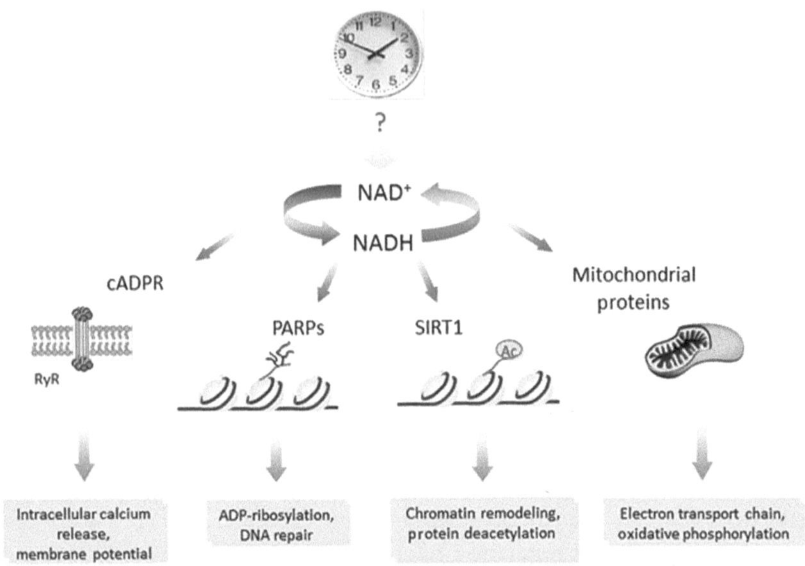

Figure 2.4 The many roles of NAD^+ in the cell. As a cofactor for SIRT1 and PARP proteins, NAD^+ contributes to rhythmicity in chromatin structure and rhythmic gene expression. In addition, the mitochondrial role of NAD^+ as an electron acceptor is essential for cellular metabolic homeostasis. Its role as a substrate for cyclic ADP-ribose (cADPR) modulates intracellular levels of calcium. (See Color Insert.)

receptor oscillation, a number of ligands for these receptors do oscillate in a circadian fashion and therefore must coordinate with the expression levels of their corresponding nuclear receptors to correctly time gene expression.[95–97] Heme is an example of the many nuclear receptor ligands that appear to be under circadian regulation. Fatty acids, bile acids, prostaglandins, leukotrienes, vitamins, and hormones are also part of the dynamic circadian metabolome that contributes to circadian physiology via changes in nuclear receptor activity and expression.[75]

In addition to the metabolites mentioned above, several other metabolites have been studies in the context of their role in chromatin remodeling. Specifically, adenosine triphosphate (ATP) is a potent modulator of chromatin structure as many chromatin remodeling enzymes are ATP dependent. In *Neurospora*, the ATP-dependent enzymes Clockswitch (CSW-1/CRF10) and chromo-domain helicase DNA-binding protein are both required for oscillations in *frq* expression.[98]

How these metabolites influence circadian function is distinct and probably tissue specific. A better understanding of how the circadian metabolome changes in distinct physiological states promises to be important for our future understanding of how the energy state of the cell couples epigenetics to the circadian clock.

4. CONCLUSIONS

The elaborate mechanisms underlying the role of the circadian clock in epigenetics are providing a whole new arena in which to study circadian gene expression. The advancement of technology has assisted in this endeavor. Specifically, high-throughput techniques have allowed the circadian epigenome to be better elucidated across species and in a variety of experimental conditions which have been enlightening for the field. ChIP-seq and RNA-seq among other techniques have revealed the global state of specific chromatin marks and binding proteins at specific chromosomal locations throughout the circadian cycle. In addition, an understanding of the dynamic nature of the circadian metabolome has built on our knowledge of the mechanisms underlying changes in chromatin structure and function. Becoming increasingly evident is that the metabolic state of the cell has a profound influence on circadian gene transcription not only through the modification of transcription factors themselves but also via the role they have in the activation of chromatin-modifying enzymes. In addition, cellular metabolites play a central role in chromatin modification by themselves serving as substrates or upstream intermediates that ultimately drive chromatin marks in a rhythmic fashion. Considering this, it is important to remember that energy intake is a rhythmic activity for most organisms, and thus, it drives the circadian changes in overall cellular energy levels. Recent studies showing that abnormal temporal energy intake can dramatically influence metabolic homeostasis ultimately underscore the importance of cellular metabolism and metabolites in driving rhythmic gene transcription for metabolic health.

REFERENCES

1. Bernstein E, Allis CD. RNA meets chromatin. *Genes Dev.* 2005;19(14):1635–1655.
2. Berger SL. The complex language of chromatin regulation during transcription. *Nature.* 2007;447(7143):407–412.
3. Borrelli E, Nestler EJ, Allis CD, Sassone-Corsi P. Decoding the epigenetic language of neuronal plasticity. *Neuron.* 2008;60(6):961–974.
4. Cheung P, Allis CD, Sassone-Corsi P. Signaling to chromatin through histone modifications. *Cell.* 2000;103(2):263–271.
5. Katada S, Imhof A, Sassone-Corsi P. Connecting threads: epigenetics and metabolism. *Cell.* 2012;148(1–2):24–28.

6. Masri S, Sassone-Corsi P. Plasticity and specificity of the circadian epigenome. *Nat Neurosci.* 2012;13(11):1324–1329.
7. Miranda TB, Jones PA. DNA methylation: the nuts and bolts of repression. *J Cell Physiol.* 2007;213(2):384–390.
8. Strahl BD, Allis CD. The language of covalent histone modifications. *Nature.* 2000;403 (6765):41–45.
9. Akhtar RA, Reddy AB, Maywood ES, et al. Circadian cycling of the mouse liver transcriptome, as revealed by cDNA microarray, is driven by the suprachiasmatic nucleus. *Curr Biol.* 2002;12(7):540–550.
10. Duffield GE, Best JD, Meurers BH, Bittner A, Loros JJ, Dunlap JC. Circadian programs of transcriptional activation, signaling, and protein turnover revealed by microarray analysis of mammalian cells. *Curr Biol.* 2002;12(7):551–557.
11. Hughes M, Deharo L, Pulivarthy SR, et al. High-resolution time course analysis of gene expression from pituitary. *Cold Spring Harb Symp Quant Biol.* 2007;72:381–386.
12. Panda S, Antoch MP, Miller BH, et al. Coordinated transcription of key pathways in the mouse by the circadian clock. *Cell.* 2002;109(3):307–320.
13. Vollmers C, Schmitz RJ, Nathanson J, Yeo G, Ecker JR, Panda S. Circadian oscillations of protein-coding and regulatory RNAs in a highly dynamic mammalian liver epigenome. *Cell Metab.* 2012;16(6):833–845.
14. Barton MC, Crowe AJ. Chromatin alteration, transcription and replication: what's the opening line to the story? *Oncogene.* 2001;20(24):3094–3099.
15. Luger K, Richmond TJ. The histone tails of the nucleosome. *Curr Opin Genet Dev.* 1998;8(2):140–146.
16. Bartova E, Krejci J, Harnicarova A, Galiova G, Kozubek S. Histone modifications and nuclear architecture: a review. *J Histochem Cytochem.* 2008;56(8):711–721.
17. Koike N, Yoo SH, Huang HC, et al. Transcriptional architecture and chromatin landscape of the core circadian clock in mammals. *Science.* 2012;338(6105):349–354.
18. Brown SA, Ripperger J, Kadener S, et al. PERIOD1-associated proteins modulate the negative limb of the mammalian circadian oscillator. *Science.* 2005;308(5722):693–696.
19. DiTacchio L, Le HD, Vollmers C, et al. Histone lysine demethylase JARID1a activates CLOCK-BMAL1 and influences the circadian clock. *Science.* 2011;333(6051):1881–1885.
20. Katada S, Sassone-Corsi P. The histone methyltransferase MLL1 permits the oscillation of circadian gene expression. *Nat Struct Mol Biol.* 2010;17(12):1414–1421.
21. Ripperger JA, Schibler U. Rhythmic CLOCK-BMAL1 binding to multiple E-box motifs drives circadian Dbp transcription and chromatin transitions. *Nat Genet.* 2006;38(3):369–374.
22. Naruse Y, Oh-hashi K, Iijima N, Naruse M, Yoshioka H, Tanaka M. Circadian and light-induced transcription of clock gene Per1 depends on histone acetylation and deacetylation. *Mol Cell Biol.* 2004;24(14):6278–6287.
23. Etchegaray JP, Yang X, DeBruyne JP, et al. The polycomb group protein EZH2 is required for mammalian circadian clock function. *J Biol Chem.* 2006;281 (30):21209–21215.
24. Ramos YF, Hestand MS, Verlaan M, et al. Genome-wide assessment of differential roles for p300 and CBP in transcription regulation. *Nucleic Acids Res.* 2010;38(16):5396–5408.
25. Masri S, Patel VR, Eckel-Mahan KL, et al. Circadian acetylome reveals regulation of mitochondrial metabolic pathways. *Proc Natl Acad Sci USA.* 2013;110(9):3339–3344.
26. Doi M, Hirayama J, Sassone-Corsi P. Circadian regulator CLOCK is a histone acetyltransferase. *Cell.* 2006;125(3):497–508.
27. Nader N, Chrousos GP, Kino T. Circadian rhythm transcription factor CLOCK regulates the transcriptional activity of the glucocorticoid receptor by acetylating its hinge region lysine cluster: potential physiological implications. *FASEB J.* 2009;23(5):1572–1583.

28. McManus KJ, Hendzel MJ. Quantitative analysis of CBP- and P300-induced histone acetylations in vivo using native chromatin. *Mol Cell Biol.* 2003;23(21):7611–7627.
29. Wang Z, Zang C, Cui K, et al. Genome-wide mapping of HATs and HDACs reveals distinct functions in active and inactive genes. *Cell.* 2009;138(5):1019–1031.
30. Etchegaray JP, Lee C, Wade PA, Reppert SM. Rhythmic histone acetylation underlies transcription in the mammalian circadian clock. *Nature.* 2003;421(6919):177–182.
31. Haigis MC, Sinclair DA. Mammalian sirtuins: biological insights and disease relevance. *Annu Rev Pathol.* 2010;5:253–295.
32. Landry J, Sutton A, Tafrov ST, et al. The silencing protein SIR2 and its homologs are NAD-dependent protein deacetylases. *Proc Natl Acad Sci USA.* 2000;97(11):5807–5811.
33. Lin SJ, Kaeberlein M, Andalis AA, et al. Calorie restriction extends Saccharomyces cerevisiae lifespan by increasing respiration. *Nature.* 2002;418(6895):344–348.
34. Burnett C, Valentini S, Cabreiro F, et al. Absence of effects of Sir2 overexpression on lifespan in C. elegans and Drosophila. *Nature.* 2011;477(7365):482–485.
35. Rodgers JT, Lerin C, Haas W, Gygi SP, Spiegelman BM, Puigserver P. Nutrient control of glucose homeostasis through a complex of PGC-1alpha and SIRT1. *Nature.* 2005;434(7029):113–118.
36. Belden WJ, Dunlap JC. SIRT1 is a circadian deacetylase for core clock components. *Cell.* 2008;134(2):212–214.
37. Nakahata Y, Kaluzova M, Grimaldi B, et al. The NAD+-dependent deacetylase SIRT1 modulates CLOCK-mediated chromatin remodeling and circadian control. *Cell.* 2008;134(2):329–340.
38. Asher G, Gatfield D, Stratmann M, et al. SIRT1 regulates circadian clock gene expression through PER2 deacetylation. *Cell.* 2008;134(2):317–328.
39. Prozorovski T, Schulze-Topphoff U, Glumm R, et al. Sirt1 contributes critically to the redox-dependent fate of neural progenitors. *Nat Cell Biol.* 2008;10(4):385–394.
40. Bellet MM, Sassone-Corsi P. Mammalian circadian clock and metabolism—the epigenetic link. *J Cell Sci.* 2010;123(Pt 22):3837–3848.
41. Zhang J, Kalkum M, Chait BT, Roeder RG. The N-CoR-HDAC3 nuclear receptor corepressor complex inhibits the JNK pathway through the integral subunit GPS2. *Mol Cell.* 2002;9(3):611–623.
42. Alenghat T, Meyers K, Mullican SE, et al. Nuclear receptor corepressor and histone deacetylase 3 govern circadian metabolic physiology. *Nature.* 2008;456(7224):997–1000.
43. Valekunja UK, Edgar RS, Oklejewicz M, et al. Histone methyltransferase MLL3 contributes to genome-scale circadian transcription. *Proc Natl Acad Sci USA.* 2012;110(4):1554–1559.
44. Malapeira J, Khaitova LC, Mas P. Ordered changes in histone modifications at the core of the Arabidopsis circadian clock. *Proc Natl Acad Sci USA.* 2012;109(52):21540–21545.
45. Lu SX, Tobin EM. Chromatin remodeling and the circadian clock: Jumonji C-domain containing proteins. *Plant Signal Behav.* 2011;6(6):810–814.
46. Jones MA, Covington MF, DiTacchio L, Vollmers C, Panda S, Harmer SL. Jumonji domain protein JMJD5 functions in both the plant and human circadian systems. *Proc Natl Acad Sci USA.* 2010;107(50):21623–21628.
47. Raduwan H, Isola AL, Belden WJ. Methylation of histone H3 on lysine 4 by the lysine methyltransferase SET1 protein is needed for normal clock gene expression. *J Biol Chem.* 2013;288(12):8380–8390.
48. Schibler U, Sassone-Corsi P. A web of circadian pacemakers. *Cell.* 2002;111(7):919–922.
49. Stephan FK, Swann JM, Sisk CL. Anticipation of 24-hr feeding schedules in rats with lesions of the suprachiasmatic nucleus. *Behav Neural Biol.* 1979;25(3):346–363.
50. Hannibal J, Ding JM, Chen D, et al. Pituitary adenylate cyclase activating peptide (PACAP) in the retinohypothalamic tract: a daytime regulator of the biological clock. *Ann N Y Acad Sci.* 1998;865:197–206.

51. Gillette MU, Mitchell JW. Signaling in the suprachiasmatic nucleus: selectively responsive and integrative. *Cell Tissue Res*. 2002;309(1):99–107.
52. Smith VM, Sterniczuk R, Phillips CI, Antle MC. Altered photic and non-photic phase shifts in 5-HT(1A) receptor knockout mice. *Neuroscience*. 2008;157(3):513–523.
53. Obrietan K, Impey S, Smith D, Athos J, Storm DR. Circadian regulation of cAMP response element-mediated gene expression in the suprachiasmatic nuclei. *J Biol Chem*. 1999;274(25):17748–17756.
54. Obrietan K, Impey S, Storm DR. Light and circadian rhythmicity regulate MAP kinase activation in the suprachiasmatic nuclei. *Nat Neurosci*. 1998;1(8):693–700.
55. O'Neill JS, Maywood ES, Chesham JE, Takahashi JS, Hastings MH. cAMP-dependent signaling as a core component of the mammalian circadian pacemaker. *Science*. 2008;320 (5878):949–953.
56. Crosio C, Cermakian N, Allis CD, Sassone-Corsi P. Light induces chromatin modification in cells of the mammalian circadian clock. *Nat Neurosci*. 2000;3 (12):1241–1247.
57. Curtis AM, Seo SB, Westgate EJ, et al. Histone acetyltransferase-dependent chromatin remodeling and the vascular clock. *J Biol Chem*. 2004;279(8):7091–7097.
58. Hemmes H, Henriques R, Jang IC, Kim S, Chua NH. Circadian clock regulates dynamic chromatin modifications associated with Arabidopsis CCA1/LHY and TOC1 transcriptional rhythms. *Plant Cell Physiol*. 2012;53(12):2016–2029.
59. Gerstner JR, Lyons LC, Wright Jr KP, et al. Cycling behavior and memory formation. *J Neurosci*. 2009;29(41):12824–12830.
60. Eckel-Mahan KL, Phan T, Han S, et al. Circadian oscillation of hippocampal MAPK activity and cAMP: implications for memory persistence. *Nat Neurosci*. 2008;11:1074–1082.
61. Levenson JM, Sweatt JD. Epigenetic mechanisms in memory formation. *Nat Rev Neurosci*. 2005;6(2):108–118.
62. Impey S, Smith DM, Obrietan K, Donahue R, Wade C, Storm DR. Stimulation of cAMP response element (CRE)-mediated transcription during contextual learning. *Nat Neurosci*. 1998;1(7):595–601.
63. Vecsey CG, Hawk JD, Lattal KM, et al. Histone deacetylase inhibitors enhance memory and synaptic plasticity via CREB:CBP-dependent transcriptional activation. *J Neurosci*. 2007;27(23):6128–6140.
64. Haettig J, Stefanko DP, Multani ML, Figueroa DX, McQuown SC, Wood MA. HDAC inhibition modulates hippocampus-dependent long-term memory for object location in a CBP-dependent manner. *Learn Mem*. 2011;18(2):71–79.
65. Hegele RA, Pollex RL. Genetic and physiological insights into the metabolic syndrome. *Am J Physiol Regul Integr Comp Physiol*. 2005;289(3):R663–R669.
66. Eckel RH, Grundy SM, Zimmet PZ. The metabolic syndrome. *Lancet*. 2005;365 (9468):1415–1428.
67. Groop L. Genetics of the metabolic syndrome. *Br J Nutr*. 2000;83(Suppl 1):S39–S48.
68. Heerwagen MJ, Miller MR, Barbour LA, Friedman JE. Maternal obesity and fetal metabolic programming: a fertile epigenetic soil. *Am J Physiol Regul Integr Comp Physiol*. 2010;299(3):R711–R722.
69. Fonken LK, Workman JL, Walton JC, et al. Light at night increases body mass by shifting the time of food intake. *Proc Natl Acad Sci USA*. 2010;107(43):18664–18669.
70. Kawachi I, Colditz GA, Stampfer MJ, et al. Prospective study of shift work and risk of coronary heart disease in women. *Circulation*. 1995;92(11):3178–3182.
71. van Amelsvoort LG, Schouten EG, Kok FJ. Duration of shiftwork related to body mass index and waist to hip ratio. *Int J Obes Relat Metab Disord*. 1999;23(9):973–978.
72. Parkes KR. Shift work and age as interactive predictors of body mass index among offshore workers. *Scand J Work Environ Health*. 2002;28(1):64–71.

73. Karlsson BH, Knutsson AK, Lindahl BO, Alfredsson LS. Metabolic disturbances in male workers with rotating three-shift work. Results of the WOLF study. *Int Arch Occup Environ Health*. 2003;76(6):424–430.
74. Fonken LK, Finy MS, Walton JC, et al. Influence of light at night on murine anxiety- and depressive-like responses. *Behav Brain Res*. 2009;205(2):349–354.
75. Eckel-Mahan KL, Patel VR, Mohney RP, Vignola KS, Baldi P, Sassone-Corsi P. Coordination of the transcriptome and metabolome by the circadian clock. *Proc Natl Acad Sci USA*. 2012;109(14):5541–5546.
76. Xie X, Rigor P, Baldi P. MotifMap: a human genome-wide map of candidate regulatory motif sites. *Bioinformatics*. 2009;25(2):167–174.
77. Daily K, Patel VR, Rigor P, Xie X, Baldi P. MotifMap: integrative genome-wide maps of regulatory motif sites for model species. *BMC Bioinformatics*. 2011;12(1):495.
78. Kanehisa M, Goto S, Sato Y, Furumichi M, Tanabe M. KEGG for integration and interpretation of large-scale molecular data sets. *Nucleic Acids Res*. 2011;40(Database issue): D109–D114.
79. Kanehisa M, Goto S, Hattori M, et al. From genomics to chemical genomics: new developments in KEGG. *Nucleic Acids Res*. 2006;34(Database issue):D354–D357.
80. Kanehisa M, Goto S. KEGG: Kyoto encyclopedia of genes and genomes. *Nucleic Acids Res*. 2000;28(1):27–30.
81. Dallmann R, Viola AU, Tarokh L, Cajochen C, Brown SA. The human circadian metabolome. *Proc Natl Acad Sci USA*. 2012;109(7):2625–2629.
82. Minami Y, Kasukawa T, Kakazu Y, et al. Measurement of internal body time by blood metabolomics. *Proc Natl Acad Sci USA*. 2009;106(24):9890–9895.
83. Jeyaraj D, Scheer FA, Ripperger JA, et al. Klf15 orchestrates circadian nitrogen homeostasis. *Cell Metab*. 2012;15(3):311–323.
84. Fogle KJ, Parson KG, Dahm NA, Holmes TC. CRYPTOCHROME is a blue-light sensor that regulates neuronal firing rate. *Science*. 2011;331(6023):1409–1413.
85. Dioum EM, Rutter J, Tuckerman JR, Gonzalez G, Gilles-Gonzalez MA, McKnight SL. NPAS2: a gas-responsive transcription factor. *Science*. 2002;298 (5602):2385–2387.
86. Raghuram S, Stayrook KR, Huang P, et al. Identification of heme as the ligand for the orphan nuclear receptors REV-ERBalpha and REV-ERBbeta. *Nat Struct Mol Biol*. 2007;14(12):1207–1213.
87. Yin L, Wu N, Curtin JC, et al. Rev-erbalpha, a heme sensor that coordinates metabolic and circadian pathways. *Science*. 2007;318(5857):1786–1789.
88. Wu N, Yin L, Hanniman EA, Joshi S, Lazar MA. Negative feedback maintenance of heme homeostasis by its receptor, Rev-erbalpha. *Genes Dev*. 2009;23 (18):2201–2209.
89. Yin L, Wu N, Lazar MA. Nuclear receptor Rev-erbalpha: a heme receptor that coordinates circadian rhythm and metabolism. *Nucl Recept Signal*. 2010;8:e001.
90. Cipriani G, Rapizzi E, Vannacci A, Rizzuto R, Moroni F, Chiarugi A. Nuclear poly(-ADP-ribose) polymerase-1 rapidly triggers mitochondrial dysfunction. *J Biol Chem*. 2005;280(17):17227–17234.
91. Ha HC, Snyder SH. Poly(ADP-ribose) polymerase is a mediator of necrotic cell death by ATP depletion. *Proc Natl Acad Sci USA*. 1999;96(24):13978–13982.
92. Kauppinen TM, Swanson RA. The role of poly(ADP-ribose) polymerase-1 in CNS disease. *Neuroscience*. 2007;145(4):1267–1272.
93. Krishnakumar R, Kraus WL. PARP-1 regulates chromatin structure and transcription through a KDM5B-dependent pathway. *Mol Cell*. 2010;39(5):736–749.
94. Guse AH, da Silva CP, Berg I, et al. Regulation of calcium signalling in T lymphocytes by the second messenger cyclic ADP-ribose. *Nature*. 1999;398(6722):70–73.
95. Sladek FM. What are nuclear receptor ligands? *Mol Cell Endocrinol*. 2011;334(1–2):3–13.

96. Yang X, Lamia KA, Evans RM. Nuclear receptors, metabolism, and the circadian clock. *Cold Spring Harb Symp Quant Biol.* 2007;72:387–394.
97. Yang X, Downes M, Yu RT, et al. Nuclear receptor expression links the circadian clock to metabolism. *Cell.* 2006;126(4):801–810.
98. Belden WJ, Loros JJ, Dunlap JC. Execution of the circadian negative feedback loop in Neurospora requires the ATP-dependent chromatin-remodeling enzyme CLOCKSWITCH. *Mol Cell.* 2007;25(4):587–600.
99. Visel A, Blow MJ, Li Z, et al. ChIP-seq accurately predicts tissue-specific activity of enhancer. *Nature.* 2009;457:854–858.

CHAPTER THREE

Diversity of Human Clock Genotypes and Consequences

Luoying Zhang*, Louis J. Ptáček*,†, Ying-Hui Fu*
*Department of Neurology, University of California, San Francisco, California, USA
†Howard Hughes Medical Institute, University of California, San Francisco, California, USA

Contents

Abstract

The molecular clock consists of a number of genes that form transcriptional and post-transcriptional feedback loops, which function together to generate circadian oscillations that give rise to circadian rhythms of our behavioral and physiological processes. Genetic variations in these clock genes have been shown to be associated with phenotypic effects in a repertoire of biological processes, such as diurnal preference, sleep, metabolism, mood regulation, addiction, and fertility. Consistently, rodent models carrying mutations in clock genes also demonstrate similar phenotypes. Taken together, these studies suggest that human clock-gene variants contribute to the phenotypic differences observed in various behavioral and physiological processes,

Progress in Molecular Biology and Translational Science, Volume 119
ISSN 1877-1173
http://dx.doi.org/10.1016/B978-0-12-396971-2.00003-8

although to validate this requires further characterization of the molecular consequences of these polymorphisms. Investigating the diversity of human genotypes and the phenotypic effects of these genetic variations shall advance our understanding of the function of the circadian clock and how we can employ the clock to improve our overall health.

1. INTRODUCTION

The circadian clock regulates daily rhythms of behavior and physiology in organisms ranging from bacteria to human,[1] with the daily sleep and wake cycle in animals being one of the most prominent functions regulated by the clock. An intact clock enables the organism to adjust its biological processes to anticipate daily changes in the environment, whereas a disrupted clock underlies various disorders and/or diseases.[2]

Our understanding of the human molecular clock is largely based on studies in rodents and *in vitro*. The molecular clock consists of a series of transcriptional/posttranscriptional feedback loops with *Clock* and *Bmal1* at the center of the loops.[3] CLOCK/BMAL1 dimers activate the transcription of three *Period* genes (*Per1*, *2*, and *3*) and two *Cryptochrome* genes (*Cry1* and *Cry2*). PER and CRY heterodimerize and translocate into the nucleus, inhibiting the transcriptional activity of CLOCK/BMAL1. In a second loop, CLOCK/BMAL1 activates the transcription of retinoic acid-related orphan receptors, *Rev-erbα* and *Rorα*. The former inhibits, whereas the latter activates transcription of *Bmal1*. In certain tissues, neuronal PAS domain protein 2 (NPAS2) functions as a CLOCK analog.[4] CLOCK and BMAL1 are also believed to drive the expression of *Dec1* and *Dec2*, which function to repress the transactivation of CLOCK/BMAL1 at clock-gene promoters.[5,6] In addition, DBP and E4BP4 are clock-controlled positive and negative regulators, respectively, of D-boxes in the promoter regions of clock genes.[7–9] TIMELESS may also function in the clock by associating with PER/CRY and inhibiting CLOCK/BMAL1-stimulated transcription of *Per*.[10]

Besides transcriptional control, posttranslational modifications also play a critical role in setting the speed of clock. Casein kinase 1 epsilon (CK1ε) and casein kinase 1 delta (CK1δ) impinge on the negative limb of the feedback loop by phosphorylating PERs, resulting in enhanced protein turnover and nuclear translocation, which in turn affects the transactivation by CLOCK/BMAL1.[11] Mutation in *Ck1ε* dramatically shortens the period of circadian rhythms in both hamster and mouse.[12,13] Consistently, a mutation in *CK1δ* results in familial advanced sleep phase (FASP) in humans and shorter period in a transgenic mouse model.[14] One route that phosphorylation impinges on protein turnover is to target the substrate for ubiquitylation and degradation by the 26S proteasome. CK1-mediated phosphorylation of PER leads to recruitment of Skp1-Cul1-F-box protein ubiquitin ligase and a ubiquitin ligase adaptor protein, β-transducin repeat protein (β-TrCP), leading to ubiquitylation, and degradation of PER.[15–17] Similarly, an F-box protein FBXL3 regulates the degradation of CRY.[18–20]

Genetic variations of these clock genes can contribute to physiological changes, which ultimately lead, in some cases, to alterations in disease susceptibility. In this chapter, we bring together findings from studies that examine the effects of human clock-gene variations on diverse aspects of behavior and physiology such as sleep, mood, metabolism, and cancer.

2. BMAL1

BMAL1 variants may play a causative role in type 2 diabetes (T2D) and hypertension. A genetic association study that examined 1304 individuals from 424 families primarily selected for T2D demonstrates that two *BMAL1* haplotypes are associated with T2D and hypertension.[21] Similarly in rodents, *Bmal1* is located within hypertension susceptibility loci and maps closely to a region that is genetically divergent between normotensive and spontaneously hypertensive rat.[21] Cell culture experiments revealed that a polymorphism in *Bmal1* promoter significantly affects transcriptional activation by GATA-4, which is a transcription factor known to be expressed in the cardiovascular system.[21] Therefore, this polymorphism could potentially affect *Bmal1* expression in tissues that are critical for regulating blood pressure. Moreover, *Bmal1* mutant mice show defects in glucose tolerance, reduced islet size, islet proliferation, and insulin secretion that worsen with age, consistent with genetic association studies in human.[22] Conditional knockout mice with *Bmal1* deficiency specifically in the pancreas exhibit diabetes mellitus due to impaired beta-cell function at the latest stage of stimulus-secretion coupling.[22] Notably, one of the single-nucleotide polymorphisms (SNPs) identified in the human haplotype associated with T2D is also significantly associated with susceptibility to prostate cancer.[23]

BMAL1 has been implicated in the pathogenesis of seasonal affective disorder (SAD). SNP analysis in 189 patients and 189 matched controls found an intronic variation in *BMAL1* to be associated with winter depression. Based on *in silico* studies, this site may affect the binding of transcription factors.[24]

Furthermore, this variation correlates with differences in experiencing seasonal variation of energy levels.[25]

BMAL1 may contribute to fertility. An intronic polymorphism has been shown to link to the number of pregnancies and miscarriages.[25] This is in agreement with studies in *Bmal1*-null mutant mice, which demonstrates that *Bmal1* is necessary for fertility.[26,27] Loss of *Bmal1* in male mice results in reduced testosterone production,[26] while in female mice, *Bmal1* deficiency leads to impaired growth and development of the reproductive system, reduced ovulation rate, and failure of fertilized oocytes to implant.[27]

Lastly, a study investigating whether clock-gene polymorphisms predispose to alcohol use identified an intronic variant in *BMAL1* to be associated with alcohol consumption in socially drinking controls but not in individuals with alcohol dependence or abuse.[28]

3. CLOCK

The first polymorphism identified in clock genes to be associated with human phenotypes is a SNP located in the 3′-untranslated region (UTR) of *CLOCK*, rs1801260 A/G. Subjects carrying the G allele have significantly lower scores on the Horne–Östeberg (HÖ) questionnaire, which assays morningness/eveningness preference and a lower score means eveningness is preferred.[29] The G allele carriers show 10- to 44-min delay in preferred timing for active and sleep phases. This finding of rs1801260 G allele associating with evening preference was further validated by independent investigations.[30–32] However, there are also several studies that were not able to observe this association between the G allele and eveningness, which may be due to differences in ethnic heritages and/or linkages to other polymorphisms (reviewed in Refs. 33 and 34).

Apart from playing a central role in the circadian clock, CLOCK is believed to participate in the regulation of sleep as well. Based on HÖ questionnaire, rs1801260 G/G homozygotes show significantly shorter sleep duration and increased daytime sleepiness compared to individuals carrying the A allele.[30,32,35] The association of rs1801260 and sleep has also been observed in patients with psychiatric disorders. Among patients with major depressive disorder (MDD) or bipolar disorder (BP), G/G homozygotes exhibit significantly increased occurrence of sleep disturbance and BP patients that are homozygous of the G allele show decreased need for sleep.[36] rs1801260 G/G homozygotes also demonstrate higher presence of insomnia during antidepressant treatment.[37] Moreover, in patients with major psychosis (mainly schizophrenia), rs1801260 G/G correlates with daytime sleepiness induced by clozapine treatment, suggesting an interaction between clozapine and *CLOCK* rs1801260 polymorphism.[38] Besides the much studied rs1801260 SNP, two variants in the intronic regions of *CLOCK* have also been shown to be associated with sleep duration based on assessment using Munich ChronoType Questionnaire.[39] Consistently, mutation in the *Clock* gene alters sleep homeostasis in mice.[40] Heterozygous and homozygous *Clock* mutant mice sleep approximately 1 and 2 h less, respectively, than wild type. The heterozygous and homozygous mutants also show 25% and 51% smaller increase of rapid eye movement (REM) sleep, respectively, during 24 h recovery sleep relative to wild-type mice.

Given the reciprocal connections between circadian rhythms/sleep and psychiatric disorders, a number of studies searched for association of *CLOCK* gene polymorphisms with mood. The much studied rs1801260 G allele exhibits significant association with BP,[41] and in patients with over 5 years of BP history, recurrence rate for bipolar depression is significantly higher in rs1801260 G/G homozygotes.[42] Two variants downstream of the *CLOCK* gene are also significantly linked to BP.[43,44] In BP and unipolar patients undergoing a depressive episode, rs1801260 is related to neuropsychological performance and neural responses in the cingulate cortex to stimuli with moral valence.[31] In addition, the rs1801260 G allele has been shown to be associated with schizophrenia.[45] Interestingly, the rs1801260 A allele significantly correlates with attention deficit hyperactivity disorder (ADHD), implicating a protective role of the G allele in this disorder.[46,47] A synonymous polymorphism in exon 20 of the *CLOCK* gene is linked to fluvoxamine therapeutic response in MDD patients as well as remission with fluvoxamine,[48] implying interaction between the *CLOCK* polymorphism and the mechanistic actions of fluvoxamine. Again these findings are consistent with studies in *Clock* mutant mice. These mice exhibit overall behavioral profile similar to human mania, including hyperactivity, decreased sleep, reduced depression-like and anxiety-like behaviors, as well as an increase in the reward value for cocaine, sucrose, and medial forebrain bundle stimulation.[49,50] Chronic administration of the mood stabilizer lithium can bring many of these behavioral phenotypes back to wild-type levels.[49] The mutant animals exhibit increased dopaminergic activity in the ventral tegmental area, a key reward region in the brain, which could lead to the phenotypes.[49,50] Taken together, these findings in mice are in agreement with the human studies and corroborate the notion that CLOCK is involved in mood regulation.

Similar to BMAL1, CLOCK has also been suggested to play a role in metabolic processes. A number of *CLOCK* polymorphisms are related to body mass index (BMI). Two of these variants located in intron 12 (rs1554483) and the promoter region of the *CLOCK* gene (rs4864548) form a haplotype associated with BMI, while two additional variants, rs1801260 and rs3749474 (located in 3′-UTR), are individually associated with BMI.[35,51,52] Both rs1801260 and rs3749474 are significantly associated with weight, and the latter with waist circumference as well.[35] Under weight-reduction programs, rs1801260 G allele carriers display greater difficulty losing weight, higher plasma ghrelin levels, altered eating behavior, and dietary habits compared to the noncarriers.[32,35] Both rs1801260 and rs3749474, along with an additional SNP in intron 9, rs4580704, are significantly linked to changes in serum cholesterol at the end of dietary treatment.[35] In contrast to overweight/obese individuals, patients with anorexia nervosa or bulimia nervosa carrying the rs1801260 G allele have a lifetime body weight significantly lower than those carrying the A/A genotype, implying a rather complex mechanism of how this rs1801260 variant interacts with metabolism.[53] Notably, rs1801260 G carriers exhibit significantly less small dense low-density lipoprotein, an abnormal lipid metabolite and one of the risk parameters for cardiometabolic disorders, compared to individuals with rs1801260 A/A.[54] Several SNPs in the *CLOCK* gene are significantly associated with energy intake, including the aforementioned 3′-UTR SNP rs3749474, intron 9 SNP rs4580704, promoter SNP rs4864548, and an SNP in intron 11.[35] Moreover, rs3749474, rs4580704, and rs1801260 are related to plasma cytokine levels, particularly those that highly correlated with energy intake.[35] These energy intake-associated SNPs, including rs1801260, are also linked to the monounsaturated fatty acid content of red blood cell membrane, which plays a critical metabolic role.[55] rs4580704 and rs1801260 exhibit dietary fatty acid-dependent associations with metabolic syndrome traits including glucose and insulin resistance as well as waist circumference.[55] This suggests that *CLOCK* polymorphisms interact with fatty acid to modulate metabolic processes. In addition, rs4580704 is significantly associated with the risk of hypertension.[55] A number of SNPs in the promoter and intronic regions of *CLOCK* show significant associations with susceptibility to and severity of nonalcoholic fatty liver disease, which is one of the most common abnormalities observed in obese people.[56] Two of these SNPs, promoter SNP rs4864548 and intron 12 SNP rs1554483, have been reported to be linked with BMI and energy intake as described earlier, adding further evidence suggesting a role for CLOCK in metabolic pathways. *Clock* mutant mice nicely recapitulate many of the metabolic phenotypes associated with human *CLOCK* gene polymorphisms. These animals are hyperphagic and obese and develop hyperleptinemia, hyperlipidemia, hepatic steatosis, hyperglycemia, and hypoinsulinemia.[57] This supports the idea that the polymorphisms in the human *CLOCK* gene are causatively linked to metabolic alterations observed in human subjects.

Lastly, *CLOCK* variants correlate with the risk and survival rate of cancer. Several SNPs located in intronic regions and 3′-UTR of *CLOCK*, including rs1801260, are significantly associated with susceptibility to prostate cancer or breast cancer.[23,58,59] Both rs1801260 and rs3749474, which have been implicated in various metabolic traits as described earlier, exhibit significant association with survival of colorectal cancer.[60]

4. NPAS2

As a paralogue of CLOCK, NPAS2 has also been implicated in circadian timing and sleep. A SNP in intron 3 of the *NPAS2* gene is associated with timing of sleep in nurses on shift-work schedule, while another SNP in intron 3 correlates with sleepiness during shift work and self-reported adaptation levels to shift-work schedule.[61] Notably, this latter SNP is also significantly linked to alcohol consumption.[61] Consistently, *Npas2*-deficient mice show reduction in sleep during the active phase and enhanced adaptability to phase advance of light–dark schedule.[62]

Like the other two circadian activators, BMAL1 and CLOCK, NPAS2 may be involved in mood regulation as well. In patients with SAD, the frequency of *NPAS2* 471 Leu/Leu genotype is significantly higher than in controls, suggesting that *NPAS2* 471 Leu/Leu contributes to disease susceptibility.[24,63] Furthermore, *NPAS2* 394 Thr correlates with lack of experiencing seasonal variation, assayed by Global Seasonal Scores which measures six items, including seasonal variation of sleep length, social activity, mood, weight, appetite, and energy level, whereas an intronic variant of *NPAS2* is associated with seasonal variation of weight.[25] Another intronic SNP is related to the number of miscarriages, implying that NPAS2 influences fertility.[25] Intronic polymorphisms in *NPAS2* have also been linked to unipolar major mood depression, autistic disorder, and chronic fatigue syndrome.[44,64,65] Notably, *NPAS2* expression is increased in patients with chronic fatigue syndrome.[65]

A missense polymorphism in *NPAS2*, 394 Ala/Thr, is linked to risks of human tumors. *NPAS2* 394 Thr is associated with reduced risk for non-Hodgkin's lymphoma[66] and prostate cancer[67] but increased risk for breast cancer.[68] In terms of effects on physiology, *NPAS2* 394 Thr correlates with lower and bioavailable testosterone, providing support for a role for NPAS2 in hormone-related cancers.[69] Another intronic SNP in *NPAS2* has been shown to be significantly associated with susceptibility to prostate cancer as well.[23]

5. PER1

PER1 may be involved in circadian timing in human. A silent polymorphism in *PER1*, 2434 T/C located in exon 18, is associated with extreme diurnal preference.[70] The C allele is more frequent in individuals with extreme morning preference than in individuals with extreme evening preference.

PER1 is believed to regulate alcohol consumption under psychosocial stress. A SNP in the promoter region of *PER1* is associated with frequency of heavy drinking in adolescents, and significant interaction is observed between this SNP and social adversity on drinking measures.[71] Consistently, this SNP is associated with alcohol dependence in adults as well. Molecular analysis revealed that cortisol-induced transcriptional activation of *PER1* is reduced in human cell lines carrying the risk allele of this SNP. Binding affinity of the transcription factor SNAIL1 to *PER1* promoter containing the risk allele is also reduced. Furthermore, m*Per1* mutant mice show increased alcohol consumption relative to wild type in response to social defeat, supporting a role for PER1 in regulating alcohol drinking induced by psychosocial stress.

Two intronic variants of *PER1* significantly correlate with susceptibility to prostate cancer.[23] One of these SNPs is also significantly associated with autistic disorder, along with a couple additional intronic SNPs.[64] *PER1* 962 Ala/Pro variant is linked to serum levels of sex steroid and insulin-like growth factor-binding protein 3, providing physiological support for a role of PER1 in hormone-related cancer.[69]

6. PER2

PER2 is the first gene found to carry mutation that causes FASP. FASP is formerly known as familial advanced sleep phase syndrome and currently referred to as familial advanced sleep phase disorder. However, according to American Academy of Sleep Medicine's classification of sleep disorders, advanced and delayed sleep phase (DSP) is only a disorder when it is problematic for the individual.[72] Therefore, in this chapter, advanced and DSP phenotypes will not be called disorders. FASP is originally identified as a highly penetrant autosomal dominant trait in three families in which affected individuals exhibit very early sleep onset and offset time.[73] HÖ questionnaire was performed on family members, and FASP subjects scored significantly higher than unaffected relatives. FASP is early onset: the youngest affected individual was 8 years old, and most FASP subjects knew they were obligate "morning larks" by 30 years of age, which is distinctly different from ASPD caused by aging.[74,75] FASP subjects from the first identified family demonstrate a 4-h phase advance of the time of sleep onset, sleep offset, first slow-wave sleep, and REM sleep compared to that of the controls, although sleep quality and quantity are not significantly different between the two groups. Narcolepsy, obstructive sleep apnea, "restless legs" syndrome, and depression were ruled out as possible causes of early sleep onset in these FASP subjects. Consistent with the sleep–wake cycle, dim-light melatonin onset, a reliable marker of circadian phase, and core body temperature rhythms are also advanced by approximately 4 h in FASP subjects

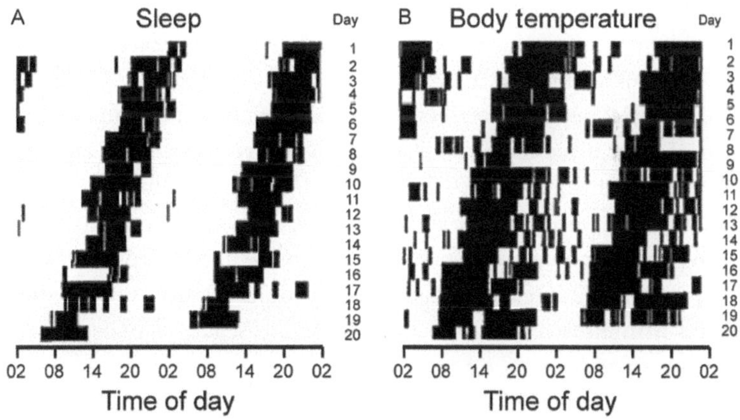

Figure 3.1 Free-running period of sleep/wake and body temperature cycles in a FASP subject. Sleep/wake (A) and body temperature (B) rhythms of a 69-year-old female monitored in time isolation for 18 days. The data are double plotted. (A) Filled bars indicate periods of sleep derived from polygraphically-recorded sleep scored using "standard" criteria. (B) Filled bars indicate periods when body temperature is below the daily mean. The free-running period of both variables are 23.3 h based on chi-squared periodogram. *Adapted from Ref. 73.*

from this family. Sleep–wake and temperature rhythms of one FASP subject were monitored in time isolation and show a circadian period of 23.3 h (Fig. 3.1), which is substantially shorter than that of control subjects (24.2 h) and is consistent with the advanced phase of sleep–wake cycle.

In order to identify the mutation that leads to FASP in the subjects in this family, linkage analysis was performed, which mapped the allele to chromosome 2qter.[76] Further physical mapping was carried out and led to identification of ~40 cDNAs localized to this region. The only coding mutation identified is in the *PER2* cDNA at position 2106 (A–G), which results in substitution of a serine at amino acid 662 with a glycine (S662G). Functional characterization was subsequently carried out to establish whether the S662G mutation causes FASP. *In vitro* study using PER2 truncation mutants demonstrates that S662 is located within CK1-binding region and the S662G mutation causes hypophosphorylation by CK1. Sequence analysis of PER2 reveals four additional serine residues that are C-terminal to S662 and each with two amino acids in between (i.e., S665, S668, S671, and S674), consistent with the CK1 recognition consensus motif. Furthermore, mutating S662 to aspartate (S662D), which mimics a phosphoserine, restores CK1-dependent phosphorylation, suggesting that S662 is a phosphorylation site on PER2. Similarly, *in vitro* phosphorylation assays using PER2 peptides that encompass residues from 660 to 674 demonstrate that PER2 peptide with a phosphate covalently linked to S662 is phosphorylated at the other residues by CK1, whereas PER2 peptide without a phosphate at S662 is not phosphorylated by CK1.[77] A quantitative assay using PER2 peptides shows that an additional 4 mol of phosphate were incorporated per mole of the PER2 peptide, corresponding to the four serine residues C-terminal to S662. Subsequent phosphoamino acid analysis revealed that the threonine and tyrosine residues on the peptide are not phosphorylated, implying that phosphorylation occurs at the serine residues. Taken together, these results suggest that phosphorylation at S662 of PER2 serves as a priming event that is critical for a cascade of phosphorylations downstream of S662 by CK1.

To investigate the functional consequences of the S662G mutation *in vivo*, transgenic mice carrying wild-type h*PER2* and h*PER2* with S662G or S662D mutations were generated using a human bacterial artificial chromosome (BAC) which carries the *cis*-acting genomic regulatory elements that can faithfully recapitulate endogenous *PER2* expression.[77] Behavior analysis shows that the S662G transgenic mice exhibit ~2 h shorter free-running period, whereas the S662D mice exhibit 0.5 h lengthening of period versus wild type. Under 12 h light:12 h dark (12L:12D) conditions, the S662G mice show ~4 h phase advance of locomotor activity rhythms which is almost identical to that of human FASP subjects carrying this mutation. The S662G mutation does not significantly affect PER2 degradation or nuclear localization, but it affects *PER2* transcript levels. In the transgenic mice, both h*PER2* and the endogenous mouse *Per2* (m*Per2*) mRNA levels peak earlier for S662G and later for S662D relative to wild type, corresponding to the shorter and longer behavioral periods, respectively. Moreover, the mRNA levels are lower in S662G mice and higher in S662D mice compared to wild type. Because both mutant h*PER2* and the endogenous wild-type m*Per2* transcript levels are reduced in the S662G mice, this argues for reduced transcriptional activity rather than reduced *PER2* mRNA stability as a result of the mutation.

Consistently, association studies have linked *PER2* to diurnal preference as well. The allele frequency of a SNP in the 5′-UTR 12 bases upstream of the translational start codon of *PER2*, -111G, is significantly higher in individuals with extreme morning preference than individuals with extreme evening preference.[78] Based on computer prediction, this polymorphism may alter the secondary structure of *PER2* mRNA. A missense variant 1244 Gly/Glu is also associated with morningness: carriers of 1244 Gly show significantly higher morning scores based on composite scale for morningness.[79] This 1244 Gly/Glu SNP is also part of a haplotype in *PER2* linked to depression vulnerability.[80] In addition, *PER2* has been implicated in sleep regulation. A synonymous SNP in *PER2*, 2229 G/A, correlates with the duration of sleep for nurses on day shift but not night shift.[61]

The *PER2* -111G allele is also linked to reduced activity in adolescents in the key neural component of the reward circuitry (medial frontal cortex).[81] Supporting the idea of a role for PER2 in reward function, m*Per2* mutant mice exhibit hypersensitized response to cocaine and strong cocaine-induced place preference.[82] Collectively, these results strongly suggest that PER2 modulates reward.

The *PER2* -111G/C SNP correlates with metabolic and eating behavior-related phenotypes, including abdominal obesity, probability of withdrawing from weight-reduction program, extreme snacking, stress with dieting, eating when bored, and skipping breakfast.[83] Among individuals with metabolic syndromes and high levels of saturated fatty acid (SFA), -111G carriers have higher plasma lipid concentrations,[84] suggesting that the -111G/C allele interacts with plasma SFA to modify lipid levels.

PER2 participates in modulating alcohol consumption, similar to its counterpart PER1. A SNP located in intron 3 of *PER2*, 10,870 A/G, is associated with the quantity of alcohol intake.[85,86] This SNP resides in a CAT-TTT motif, which is conserved in human, chimpanzee, and rat.[85] It is also in an enhancer-like structure, which contains several transcriptional factor-binding site motifs. This SNP alters the binding motifs for Sp1, c-myb, and NF-κB, possibly resulting in altered transactivation of *PER2*. m*Per2* mutant mice exhibit increased alcohol consumption, accompanied by enhanced glutamate levels in the extracellular space in the brain. This is believed to be a result of reduced expression of the glutamate transporter gene, *Eaat1*, and thus reduced uptake of glutamate by astrocytes. Acamprosate, a drug used to prevent craving and relapse in alcoholic patients, reduced the enhanced glutamate levels and normalized the increased alcohol intake in m*Per2* mutant mice. Collectively, these data suggest that PER2 acts to suppress glutamatergic signaling, which in turn influences alcohol drinking. Besides being involved in modulating alcohol consumption, the *PER2* 10870 A/G SNP is also associated with SAD.[24]

PER2 is linked to the risk of cancer. An intronic SNP in *PER2* is significantly associated with susceptibility to prostate cancer,[23] whereas the aforementioned 1244 Gly/Glu associated with morningness and depression vulnerability also correlates with the risk of breast cancer in combination with an SNP in *CLOCK*.[59]

7. PER3

Several polymorphisms in *PER3* have been suggested to contribute to determination of diurnal preference and DSP.[63,87–91] The most well-studied polymorphism among these is a polymorphic repeat region with four or five copies of a 54-bp repetitive sequence (4-repeat vs. 5-repeat) in exon 18. However, this association with morningness/eveningness attenuates with age.[90,92] Human subjects homozygous for the long allele are particularly sensitive to blue-enriched light, as such light significantly suppresses evening rise of endogenous melatonin in homozygotes for the long allele but not the short allele.[93] Likewise, individuals homozygous for the long allele exhibit more pronounced response to the alerting effects of light compared to homozygotes for the short allele. Waking electroencephalographic (EEG) activity in the theta range (5–7 Hz), which is a putative correlate of sleepiness, is substantially attenuated during exposure to blue-enriched light in subjects homozygous for the long allele but not the short allele. This length polymorphism has also been shown to be one of the alleles associated with self-reported adaptation levels to shift-work schedules and sleep phase in nurses working on shifts.[61] Another SNP reported to be associated with morning–evening scores is *PER3* 647 Val/Gly.[63] A few polymorphisms in the promoter region of *PER3* are linked to DSP as well.[91] *In vitro* studies demonstrate that these promoter polymorphisms may modify the transcription of *PER3*.

PER3 may exert effects on sleep homeostasis. Individuals homozygous for the 5-repeat allele exhibit increase in markers of sleep homeostasis, including slow-wave sleep, EEG slow-wave activity (0.75–4.5 Hz) in non-REM sleep, as well as theta and alpha activity (8–12 Hz) during REM sleep and wakefulness.[94] The decrement in cognitive performance as a result of sleep deprivation is significantly larger in subjects homozygous for the long allele. Individuals of this genotype also perform worse on tests of executive function at early morning during sleep deprivation relative to homozygotes for the short allele.[95] Further study employing functional magnetic resonance imaging indicates that both genotypes recruit brain regions typically involved in working memory, but individuals homozygous for the short allele recruit supplemental anterior frontal, temporal, and subcortical regions in addition.[96] In contrast, widespread reductions of activation in prefrontal, temporal, parietal, and occipital areas were observed in homozygotes for the long allele. Accompanying increased slow-wave sleep in subjects homozygous for the long allele is an elevated sympathetic predominance and a reduction of parasympathetic predominance in the autonomic nervous system.[97] Both homozygosity for the long allele and a SNP in exon 18, 1148 Arg, are associated with reduced daytime sleepiness, and also sleepiness in nurses working during shifts.[61,98] On the other hand, homozygosity of the 4-repeat allele correlates with insomnia in alcohol-dependent patients.[99] Consistent with the idea of a role for PER3 in modulating sleep homeostasis in human, mice deficient for *Per3* exhibit altered patterns of sleep both under baseline condition and after sleep deprivation.[100]

A role for PER3 has been implicated in metabolic processes. *PER3* 639 Val is associated with T2D, while the much studied *PER3* length polymorphism modifies the effects of the timing and duration of sleep on BMI.[98,101] This is supported by *in vitro* study that demonstrates PER3 functions to inhibit adipogenesis, and *Per3* knockout mice display increased adipose tissue and decreased muscle tissue relative to wild type.[102]

An intronic SNP in *PER3* is significantly associated with susceptibility to prostate cancer.[23] At a physiological level, the aforementioned 5-repeat allele correlates with higher levels of serum insulin-like growth factor-I (IGF) and the ratio of IGF-I to IGF-binding protein 3, which may contribute to hormone-related cancer.[69] Furthermore, inflammation is an established cancer risk factor and carriers of the 5-repeat allele show elevated levels of the cytokine interleukin-6.[103]

Lastly, the 4-repeat allele of *PER3* is significantly linked to heroin dependence and postpartum onset of BP.[104,105]

8. CRY1

Two SNPs located in the promoter region of *CRY1* are associated with susceptibility to and mortality from prostate cancer, respectively.[23,106] On the other hand, a SNP within 3′-UTR of *CRY1* correlates with risk of breast cancer.[59] In addition, a SNP located 3′ downstream of *CRY1* is significantly associated with MDD.

9. CRY2

CRY2 has been suggested by various studies to act as a modulator of cancer development. Two intronic SNPs in *CRY2* are significantly associated with susceptibility to prostate cancer.[23,67] For one of these SNPs located in intron 2, rs1401417 G/C, carriers of the C allele exhibit 1.7-fold increased risk of prostate cancer.[67] This risk is increased to 4.1-fold in the C allele carriers with higher insulin resistance. This allele is also linked to breast cancer risk, along with two additional SNPs, and all three of these SNPs are significantly associated with the risk of non-Hodgkin's lymphoma.[59,107,108] Breast cancer patients have significantly higher levels of *CRY2* promoter methylation relative to controls, consistent with lower levels of *CRY2* in tumor tissues compared to adjacent normal tissues.[108] Furthermore, *in vitro* analysis identifies alterations in the expression of breast cancer-relevant genes, immune response genes, and hematologic system development genes in response to *CRY2* knockdown.[107,108] Some of these genes are predicted to have significant effects on several disease processes, including cancer.[107] Taken together, these findings suggest that CRY2 may exert significant effects on cancer susceptibility.

Genome-wide association study identified an intronic SNP in *CRY2* to be significantly associated with fasting glucose levels in nondiabetic adults.[109] Subsequent studies reported this locus to be correlated with T2D, as well as fasting glucose in healthy children and adolescents.[110,111]

Three SNPs in *CRY2* are linked to winter depression, including one of the SNPs that have been reported to be associated with the risk of breast cancer and non-Hodgkin lymphoma.[112] Molecular analysis revealed that the levels of *CRY2* mRNA are decreased in depressed bipolar patients. While a night of total sleep deprivation results in significant upregulation of *CRY2* transcript in control subjects, it fails to do so in depressed bipolar patients. Both the genetic and molecular studies suggest that dysregulation of *CRY2* expression may be involved in vulnerability to depression.

10. REV-ERBα

REV-ERBα is primarily implicated in BP. A haplotype comprised of two SNPs located in intron 1 and 5′-UTR of *REV-ERBα*, respectively, is significantly associated with BP.[113] Furthermore, a SNP in the intronic region of *REV-ERBα*, rs2314339 C/T, is associated with long-term efficacy of lithium carbonate therapy in BP.[114] The frequency of the T allele is significantly increased in nonresponders, and patients carrying the T allele are 3.5 × more likely to show no improvement or even worsening of the illness. Consistently, another SNP located in the promoter region of *REV-ERBα* correlates with good treatment response and changes in REV-ERBα expression in response to lithium treatment.[115] These findings support a role for REV-ERBα in the therapeutic mechanism of lithium.

11. CK1ε

A SNP in the 3′-UTR of *CK1ε* is significantly associated with self-reported response to D-amphetamine.[116] Consistently, quantitative trait loci (QTL) analysis in mice selectively bred for high versus low sensitivity to methamphetamine identified a QTL in the *Ck1ε* gene that may cause the difference in response to methamphetamine.[117] Expression differences of *Ck1ε* is also observed in mouse lines displaying high versus low sensitivity to methamphetamine. Collectively, human and animal studies suggest that CK1ε contributes to variability in stimulant response.

An intronic SNP of *CK1ε* is linked to BP and prostate cancer.[23,43] Furthermore, another variant in *CK1ε* correlates with testosterone to dihydrotestosterone ratio in the serum, which may contribute to the pathology of prostate cancer.[69]

12. CK1δ

Exon sequencing of circadian genes for individuals that belong to a moderate-sized family with FASP led to the identification of a second mutation that causes FASP. The mutation is a threonine-to-alanine alteration at amino acid 44 of CK1δ (CK1δ-T44A), and this threonine is conserved in other mammalian CK1s and *Drosophila* CK1 (dDBT).[14] *In vitro* kinase assay demonstrates that this mutation results in decreased phosphorylation of both exogenous substrates (phosvitin and alpha-casein) and circadian substrates (PER1–3). To examine the effects of this mutation on circadian rhythms *in vivo*, BAC transgenic mice carrying either the wild type (h*CK1δ-WT*) or the mutant (h*CK1δ-T44A*) h*CK1δ* were generated. The behavioral period under free-running condition is significantly shorter in the mutant transgenic mice compared to wild type, consistent with the phase-advanced phenotype of human subjects carrying this mutation. Neither $CK1\delta^{+/-}$ nor h*CK1δ-WT* transgenic mice exhibit altered period, suggesting that the period is not affected by wild-type *CK1δ* gene dosage. Thus, the shorter period observed in h*CK1δ-T44A* transgenic mice is likely due to the T44A mutation and not altered *CK1δ* gene dosage. Interestingly, expression of h*CK1δ-T44A* in *Drosophila* circadian neurons results in longer period compared to expression of h*CK1δ-WT*. This may reflect differences in the regulatory mechanism of the mammalian clock versus invertebrate clock.

The aforementioned h*PER2-S662G* and h*CK1δ-T44A* mutations indicate that phosphorylation of PER2 by CK1 is critical for circadian timing in humans. To characterize the functional relevance of the interaction between PER2 and CK1 *in vivo*, h*PER2* transgenic mice were crossed with both h*CK1δ-WT* transgenic and *CK1δ* knockout mice.[77] As described earlier in this chapter, h*PER2-S662G* transgenic mice exhibit a short period of ~22 h, whereas neither h*CK1δ-WT* transgenic nor $CK1\delta^{+/-}$ exhibits altered circadian period. However, in mice carrying both h*PER2-S662G* and h*CK1δ-WT* transgenes, the period is shorter than h*PER2-S662G* single transgenic animals by over 1 h. Consistently, expressing h*PER2-S662G* on the $CK1\delta^{+/-}$ background slightly lengthens the period compared to expressing h*PER2-S662G* on a wild-type background. On the other hand, h*PER2-S662D* transgenic mice, which show long period on wild-type background, exhibit even longer period in $CK1\delta^{+/-}$ background and a shorter period in h*CK1δ-WT* background. Therefore, decreasing *CK1δ* dosage lengthens period for both h*PER2-S662G* and h*PER2-S662D* transgenic mice. Similarly, increasing *CK1δ* dosage shortens the endogenous period of both S662 mutants but not wild type.

Taken together, these results lead to the proposal of the following model regarding how CK1 acts on PER2 to regulate circadian period: CK1 phosphorylates the serine residues downstream of S662 on PER2 after S662 is phosphorylated by a priming kinase. Phosphorylation in this region of PER2 increases *PER2* mRNA and thus protein, while CK1 likely phosphorylates some other site(s) that results in degradation of PER2. The net effect of these two opposing processes determines the level of PER2 and in turn sets circadian period. In wild-type background, the balance of these opposing effects can be maintained, thus decreasing or increasing *CK1δ* gene dosage does not change the period. In the presence of S662G mutation, the S662 residue can no longer be phosphorylated by the priming kinase, leading to hypophosphorylation of the downstream residues by CK1. Therefore, the net effect of CK1 on mutant PER2 results in reduced PER2 levels and shorter period. Decreasing *CK1δ* gene dosage partially suppresses the period shortening effect by reducing phosphorylation-mediated PER2 degradation, whereas increasing *CK1δ* gene dosage further shortens period by enhancing phosphorylation-mediated PER2 degradation.

13. CUL1

A SNP in intron 3 of *CUL1* is significantly associated with rheumatoid arthritis (RA).[118] In lymphocytic cell lines, this SNP affects transcriptional efficiency of *CUL1* promoter activity. CUL1 is highly expressed in lymphoid tissues, and suppression of CUL1 inhibits IL-8 induction, which plays an important role in migration of inflammatory cells into the affected area as seen in RA. Therefore, this SNP in intron 3 of *CUL1* could be affecting susceptibility to RA by modulating expression levels of *CUL1*. In another independent study, this SNP, along with two other, constitutes a haplotype that is significantly associated with RA and response to methotrexate treatment, a commonly prescribed drug for RA patients.[119]

14. β-TrCP

β-TrCP mutations have been implicated in cancer. Missense somatic mutations in *β-TrCP* were identified in gastric cancers.[120] In tissues carrying these mutations, β-catenin levels are increased with aberrant subcellular distribution, which may contribute to the development of gastric cancer. Further evidence came from an association study demonstrating that a 9-bp deletion polymorphism in the 3′-UTR of *β-TrCP* correlates with reduced risk of hepatocellular carcinoma (HCC).[121] Molecular analysis revealed that HCC tumor tissues with the deletion display reduced levels of β-TrCP compared to those that do not carry the deletion. Because β-TrCP is believed to be oncogenic, reduced β-TrCP levels associated with the deletion variant could explain the reduced risk of HCC. Additionally, duplication of *β-TrCP* gene is associated with split hand-split foot malformation.[122]

15. DEC1

DEC1 was identified as one of the genes located in a commonly deleted chromosomal region in a wide panel of esophageal squamous cell carcinoma.[123] *DEC1* transcript levels were significantly reduced in the majority of esophageal cancer cell lines, while introducing *DEC1* cDNA into cancer cells that lack *DEC1* expression significantly suppresses cell growth. Consistently, a polymorphism in the promoter region of *DEC1* (-249T/C) is significantly associated with the risk of squamous cell carcinoma of the head and neck (SCCHN), and human subjects homozygous for -249 *C* show significantly reduced susceptibility to SCCHN.[124] *In silico* analysis predicts that the -249 T to C change leads to a gain of a transcription factor-binding site. Indeed, further functional analysis demonstrated that the T–C change results in increased transcriptional activity at *DEC1* promoter and enhanced protein-DNA binding. In summary, these results suggest that DEC1 functions as a tumor suppressor and genetic variations in *DEC1* could alter susceptibility to cancer.

16. DEC2

The first human mutation identified to cause a sleep homeostasis phenotype is in *DEC2*.[125] Individuals carrying a proline to arginine mutation at amino acid position 384 (P384R) of DEC2 have approximately 2-h shorter sleep time per 24-h day compared to family members who do not carry the mutation (Fig. 3.2). Studies in cell culture demonstrated that the P384R mutation results in attenuated DEC2 repressor activity of CLOCK/BMAL1-driven transcription.

To validate that the P384R mutation is indeed causing the short-sleep phenotype and not merely associated with the phenotype, BAC transgenic

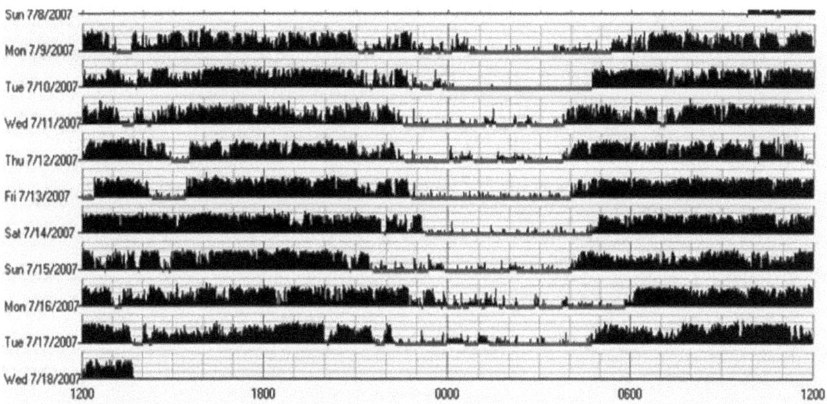

Figure 3.2 Activity recording of a *DEC2-P384R* mutation carrier. Filled bars indicate periods of activity by wrist actigraphy. Extended periods of activity can be observed. *Adapted from Ref. 125.* (See Color Insert.)

mice were generated to carry wild-type h*DEC2* (h*DEC2-WT)* or h*DEC2-P384R*. h*DEC2-P384R* mice do not exhibit altered free-running period, but the duration of the activity period (alpha) is 1.2-h longer relative to *hDEC2-WT* transgenic, wild-type littermates, and *Dec2* knockout mice. This recapitulates the shorter sleep duration (i.e., inactive period) phenotype observed in humans. Moreover, when h*DEC2-P384R* is expressed in a *Dec2* knockout background, alpha is further lengthened to ~2.5 h longer than controls.

To examine the effects of the *DEC2-P384R* mutation on sleep, EEG and electromyography were performed on mutant transgenic mice and littermate controls. h*DEC2-P384R* mice were awake for a significantly longer period of time during the light phase compared to wild type, accompanied by significant reduction of both NREM and REM sleep. Analysis of sleep architecture demonstrated decreased wake duration and an increase in the number of wake episodes in h*DEC2-P384R* mice relative to wild type. In addition, these animals exhibited significantly more NREM episodes during the light phase, but each episode is shorter in duration. These results indicate that sleep (in particular NREM sleep) is more fragmented in h*DEC2-P384R* mice than that of wild type. To better understand the role of DEC2 in sleep regulation, h*DEC2-P384R* mice and wild-type littermates were subjected to acute sleep deprivation. h*DEC2-P384R* mice showed significantly less rebound in both NREM and REM sleep, and a slower recovery of acute sleep loss. h*DEC2-P384R* mice also exhibited lower NREM delta power density change after sleep deprivation compared to wild type, which indicates that the depth of the rebound of NREM sleep is affected in h*DEC2-P384R* animals. Consistent with the mammalian data, expressing m*Dec2-P384R* in the sleep/rest center of *Drosophila* brain leads to significantly less sleep-like behavior with decreased sleep bout duration and increased sleep bout number versus flies expressing m*Dec2-WT*. In summary, these results demonstrate DEC2 as an important player in the regulation of sleep homeostasis.

17. TIMELESS

TIMELESS is associated with depression and sleep disturbances.[126] Four SNPs in or near the *TIMELESS* gene are linked to depression with fatigue in females, while two of these SNPs (rs7486220 A/G and rs1082214 C/T) are also linked to depression with early morning awakening in males. Notably, rs7486220 A and rs1082214 C correlate with depression with fatigue in females, whereas rs7486220 G and rs1082214 T correlate with depression with early morning awakening in males. In a separate set of individuals that do not have depression, rs1082214 C is correlated with higher levels of seasonal changes in mood in females, while rs1082214 T is correlated with early morning awakening and fatigue in males. Collectively, these data implicate a connection between TIMELESS and gender-dependent depression and sleep regulation.

18. CONCLUDING REMARKS

Studies of human clock-gene variants reveal that besides circadian timing, clock genes may also be involved in a number of other biological processes (Table 3.1). Most of the clock-gene polymorphisms are associated with sleep regulation, cancer development, metabolic traits, and mood disorders, implying that these processes may have particularly close connections with the circadian clock, and thus are more sensitive to alterations of the clock caused by genetic variations. In addition, *CLOCK*, *PER1–3*, and *CK1ε* polymorphisms are linked to addiction, suggesting a role for the clock in reward circuitry of the brain. *BMAL1* and *NPAS2* polymorphisms are related to fertility and seasonal variations, supporting the long-held view that circadian clock participates in seasonal adaptability. Furthermore, studies using mice deficient for clock genes verified the involvement of clock genes

Table 3.1 Genetic associations between the clock genes and human phenotypes

Gene	Human phenotype associated	Phenotype of mutant mouse model
BMAL1	Type 2 diabetes and hypertension	Hypoinsulinemia and diabetes
	Prostate cancer	
	Seasonal affective disorder	
	Number of pregnancies and miscarriages	Infertility
	Alcohol consumption	
CLOCK	Eveningness preference	Circadian phenotype
	Altered sleep	Reduced sleep
	Bipolar disorder, schizophrenia, attention deficit hyperactivity disorder, and fluvoxamine response in major depressive disorder patients	Mania-like behavior
	Metabolic traits, hypertension, and nonalcoholic fatty liver disease	Obesity and metabolic syndromes
	Prostate cancer, breast cancer, and survival rate of colorectal cancer	
NPAS2	Daily timing and sleepiness	Reduced sleep and enhanced light entrainment
	Seasonal affective disorder and depression	
	Autistic disorder	
	Chronic fatigue syndrome	
	Number of miscarriages	
	Non-Hodgkin's lymphoma, prostate cancer, and breast cancer	
PER1	Extreme morning preference	Circadian phenotype
	Alcohol consumption under psychosocial stress	Increased alcohol consumption in response to social defeat
	Prostate cancer	
	Autistic disorder	
PER2	Familial advanced sleep phase	Phase advance during light/dark cycle and shorter free-running period
	Morningness preference	
	Duration of sleep (nurses on day shift)	
	Depression	
	Reduced activity in medial frontal cortex of adolescents	Hypersensitized response to cocaine and strong cocaine-induced place preference
	Metabolic traits	
	Alcohol consumption	Increased alcohol consumption
	Seasonal affective disorder	
	Prostate cancer and breast cancer	

Continued

Table 3.1 Genetic associations between the clock genes and human phenotypes—cont'd

Gene	Human phenotype associated	Phenotype of mutant mouse model
PER3	Morningness/eveningness preference, timing of sleep phase, and sensitivity to blue-enriched light	
	Sleep homeostasis phenotypes	Altered sleep patterns
	Type 2 diabetes and body mass index	Increased adipose tissue and decreased muscle tissue
	Prostate cancer	
	Heroin dependence	
	Postpartum onset of bipolar disorder	
CRY1	Prostate cancer and breast cancer	
	Major depressive disorder	
CRY2	Prostate cancer, breast cancer, and non-Hodgkin's lymphoma	
	Fasting glucose levels	
	Winter depression	
REV-ERBα	Bipolar disorder and efficacy of lithium treatment in bipolar disorder	
CK1ε	Self-reported response to D-amphetamine	A quantitative trait locus in *Ck1ε* may cause differences in response to methamphetamine
	Bipolar disorder	
	Prostate cancer	
CK1δ	Familial advanced sleep phase	Shorter free-running period
CUL1	Rheumatoid arthritis and response to methotrexate treatment	
B-TrCP	Gastric cancers and hepatocellular carcinoma	
	Split hand-split foot malformation	
DEC1	Squamous cell carcinoma of the head and neck	
DEC2	Shorter sleep duration	Reduced sleep amount
TIMELESS	Depression and sleep disturbances	

in a number of the processes implicated by human genetic studies, including sleep, metabolic syndromes, mood disorders, addiction, and fertility.

It is worthwhile at this point to detail approaches to Mendelian genetics versus association studies as the distinction between the two is not well understood by many people who are interested in the current topic. Mendelian genetics deals with identification of genetic variants of strong effect and are sufficient to *cause* a phenotype. For example, genetic studies of rodents with spontaneous mutations (e.g., the *Ck1ε* mutant hamster) and forward genetic screens in model organisms where mutagenesis is performed and animals are screened for a phenotype (e.g., short/long period or arrhythmia) are focused on identifying the genes and mutations which *cause* the phenotype. Similar studies have been successful in humans where identification of FASP allowed cloning of causative genes and mutations.

In complex genetics, genetic variants are sought where there is a statistical association of the variant with a phenotype. The variant itself is only *associated* with an increased risk of the phenotype. Thus, having the variant does not mean that the carrier will have the phenotype. Neither does it mean that one without the variant cannot have the phenotype. Such a finding does not simply imply that variant is itself causative of the increased risk. Rather, it suggests that the associated variant and/or a genetic variant in the vicinity of the associated variant leads to increased risk. Consequently, we must be very careful when interpreting these data, as many such findings (positive associations) have (or will turn out to be) false positives. In some cases, a variant in a gene will be associated with the phenotype because that gene is truly linked to the biology underlying the phenotype of interest. In other cases, a genetic variant may truly be associated with the phenotype but only because it is in linkage disequilibrium with a variant in another gene. Mutations in some clock genes have been generated and result in behavioral and/or physiological phenotypes in animal models (such as the mouse studies described earlier in this chapter). These studies support the argument that the recognized clock gene associations in humans with similar or related phenotypes occur as a result of genetic variants in the respective clock genes. To validate whether these genetic variations found by the association studies lead to phenotypic changes will require generation and characterization of appropriate animal models carrying equivalent polymorphisms. Unlike Mendelian traits with very prominent phenotypes as in the cases of FASP, however, many of the behavioral and physiological phenotypes observed in association studies are relatively subtle and may exhibit complex allelic interactions, imposing great complications and challenges on studies in animal models. Nevertheless, as we learn more regarding the molecular underpinnings of the biological processes involved and as our phenotyping techniques improve, using animal models to investigate the mechanistic alterations caused by these human clock gene variants will become more effective and fruitful.

In the past few decades, much effort has been devoted into understanding *what* constitutes the circadian clock and *how* the clock functions. Thus, we currently have a handful of "clock genes" and a relatively clear picture of the molecular mechanisms regarding how these genes act together to set the phase and amplitude of the clock. One of the next big challenges in the field is answering the "*why*" question, that is, *why* is the clock built this way, or an even more fundamental question, *why* do we need a clock. At the individual level, investigating the broad consequences of alterations in clock genes will help us understand the function of the clock and the role it plays in our overall well-being. At a population level, studying the distribution of clock genotypes and associated phenotypes across the world will facilitate unveiling the interactions between the molecular clock and environment. Insights gained from these studies shall provide answers to some of the most fundamental questions in human circadian biology. Only with such understanding can we maximize the health benefits and therapeutic values of the circadian clock.

REFERENCES

1. Bell-Pedersen D, Cassone VM, Earnest DJ, et al. Circadian rhythms from multiple oscillators: lessons from diverse organisms. *Nat Rev Genet.* 2005;6(7):544–556.
2. Takahashi JS, Hong HK, Ko CH, McDearmon EL. The genetics of mammalian circadian order and disorder: implications for physiology and disease. *Nat Rev Genet.* 2008;9(10):764–775.
3. Ko CH, Takahashi JS. Molecular components of the mammalian circadian clock. *Hum Mol Genet.* 2006;15(Spec No 2):R271–R277.
4. Reick M, Garcia JA, Dudley C, McKnight SL. NPAS2: an analog of clock operative in the mammalian forebrain. *Science.* 2001;293(5529):506–509.
5. Honma S, Kawamoto T, Takagi Y, et al. Dec1 and Dec2 are regulators of the mammalian molecular clock. *Nature.* 2002;419(6909):841–844.
6. Noshiro M, Furukawa M, Honma S, et al. Tissue-specific disruption of rhythmic expression of Dec1 and Dec2 in clock mutant mice. *J Biol Rhythms.* 2005;20 (5):404–418.
7. Yamaguchi S, Mitsui S, Yan L, Yagita K, Miyake S, Okamura H. Role of DBP in the circadian oscillatory mechanism. *Mol Cell Biol.* 2000;20(13):4773–4781.
8. Mitsui S, Yamaguchi S, Matsuo T, Ishida Y, Okamura H. Antagonistic role of E4BP4 and PAR proteins in the circadian oscillatory mechanism. *Genes Dev.* 2001;15 (8):995–1006.
9. Yamajuku D, Shibata Y, Kitazawa M, et al. Cellular DBP and E4BP4 proteins are critical for determining the period length of the circadian oscillator. *FEBS Lett.* 2011;585 (14):2217–2222.

10. Gotter AL. A Timeless debate: resolving TIM's noncircadian roles with possible clock function. *Neuroreport*. 2006;17(12):1229–1233.
11. Akashi M, Tsuchiya Y, Yoshino T, Nishida E. Control of intracellular dynamics of mammalian period proteins by casein kinase I epsilon (CKIepsilon) and CKIdelta in cultured cells. *Mol Cell Biol*. 2002;22(6):1693–1703.
12. Lowrey PL, Shimomura K, Antoch MP, et al. Positional syntenic cloning and functional characterization of the mammalian circadian mutation tau. *Science*. 2000;288 (5465):483–492.
13. Meng QJ, Logunova L, Maywood ES, et al. Setting clock speed in mammals: the CK1 epsilon tau mutation in mice accelerates circadian pacemakers by selectively destabilizing PERIOD proteins. *Neuron*. 2008;58(1):78–88.
14. Xu Y, Padiath QS, Shapiro RE, et al. Functional consequences of a CKIdelta mutation causing familial advanced sleep phase syndrome. *Nature*. 2005;434(7033):640–644.
15. Eide EJ, Woolf MF, Kang H, et al. Control of mammalian circadian rhythm by CK1epsilon-regulated proteasome-mediated PER2 degradation. *Mol Cell Biol*. 2005;25:2795–2807.
16. Shirogane T, Jin J, Ang XL, Harper JW. $SCF^{\beta TRCP}$ controls clock-dependent transcription via casein kinase 1-dependent degradation of the mammalian Period-1 (Per1) protein. *J Biol Chem*. 2005;280:26863–26872.
17. Reischl S, Vanselow K, Westermark PO, et al. Beta-TrCP1-mediated degradation of PERIOD2 is essential for circadian dynamics. *J Biol Rhythms*. 2007;22(5):375–386.
18. Busino L, Bassermann F, Maiolica A, et al. SCFFbxl3 controls the oscillation of the circadian clock by directing the degradation of cryptochrome proteins. *Science*. 2007;316(5826):900–904.
19. Godinho SI, Maywood ES, Shaw L, et al. The after-hours mutant reveals a role for Fbxl3 in determining mammalian circadian period. *Science*. 2007;316(5826):897–900.
20. Siepka SM, Yoo SH, Park J, et al. Circadian mutant overtime reveals F-box protein FBXL3 regulation of cryptochrome and period gene expression. *Cell*. 2007;129 (5):1011–1023.
21. Woon PY, Kaisaki PJ, Braganca J, et al. Aryl hydrocarbon receptor nuclear translocator-like (BMAL1) is associated with susceptibility to hypertension and type 2 diabetes. *Proc Natl Acad Sci USA*. 2007;104(36):14412–14417.
22. Marcheva B, Ramsey KM, Buhr ED, et al. Disruption of the clock components CLOCK and BMAL1 leads to hypoinsulinaemia and diabetes. *Nature*. 2010;466 (7306):627–631.
23. Zhu Y, Stevens RG, Hoffman AE, et al. Testing the circadian gene hypothesis in prostate cancer: a population-based case-control study. *Cancer Res*. 2009;69 (24):9315–9322.
24. Partonen T, Treutlein J, Alpman A, et al. Three circadian clock genes Per2, Arntl, and Npas2 contribute to winter depression. *Ann Med*. 2007;39(3):229–238.
25. Kovanen L, Saarikoski ST, Aromaa A, Lonnqvist J, Partonen T. ARNTL (BMAL1) and NPAS2 gene variants contribute to fertility and seasonality. *PLoS One*. 2010;5 (4):e10007.
26. Alvarez JD, Hansen A, Ord T, et al. The circadian clock protein BMAL1 is necessary for fertility and proper testosterone production in mice. *J Biol Rhythms*. 2008;23 (1):26–36.
27. Boden MJ, Varcoe TJ, Voultsios A, Kennaway DJ. Reproductive biology of female Bmal1 null mice. *Reproduction*. 2010;139(6):1077–1090.
28. Kovanen L, Saarikoski ST, Haukka J, et al. Circadian clock gene polymorphisms in alcohol use disorders and alcohol consumption. *Alcohol Alcohol*. 2010;45(4):303–311.
29. Katzenberg D, Young T, Finn L, et al. A CLOCK polymorphism associated with human diurnal preference. *Sleep*. 1998;21(6):569–576.

30. Mishima K, Tozawa T, Satoh K, Saitoh H, Mishima Y. The 3111T/C polymorphism of hClock is associated with evening preference and delayed sleep timing in a Japanese population sample. *Am J Med Genet B Neuropsychiatr Genet.* 2005;133B(1):101–104.
31. Benedetti F, Radaelli D, Bernasconi A, et al. Clock genes beyond the clock: CLOCK genotype biases neural correlates of moral valence decision in depressed patients. *Genes Brain Behav.* 2008;7(1):20–25.
32. Garaulet M, Sanchez-Moreno C, Smith CE, Lee YC, Nicolas F, Ordovas JM. Ghrelin, sleep reduction and evening preference: relationships to CLOCK 3111 T/C SNP and weight loss. *PLoS One.* 2011;6(2):e17435.
33. Allebrandt KV, Roenneberg T. The search for circadian clock components in humans: new perspectives for association studies. *Braz J Med Biol Res.* 2008;41(8):716–721.
34. von Schantz M. Phenotypic effects of genetic variability in human clock genes on circadian and sleep parameters. *J Genet.* 2008;87(5):513–519.
35. Garaulet M, Corbalan MD, Madrid JA, et al. CLOCK gene is implicated in weight reduction in obese patients participating in a dietary programme based on the Mediterranean diet. *Int J Obes (Lond).* 2010;34(3):516–523.
36. Serretti A, Benedetti F, Mandelli L, et al. Genetic dissection of psychopathological symptoms: insomnia in mood disorders and CLOCK gene polymorphism. *Am J Med Genet B Neuropsychiatr Genet.* 2003;121(1):35–38.
37. Serretti A, Cusin C, Benedetti F, et al. Insomnia improvement during antidepressant treatment and CLOCK gene polymorphism. *Am J Med Genet B Neuropsychiatr Genet.* 2005;137(1):36–39.
38. Lattuada E, Cavallaro R, Benedetti F, Cocchi F, Lorenzi C, Smeraldi E. Genetic dissection of drug effects in clinical practice: CLOCK gene and clozapine-induced diurnal sleepiness. *Neurosci Lett.* 2004;367(2):152–155.
39. Allebrandt KV, Teder-Laving M, Akyol M, et al. CLOCK gene variants associate with sleep duration in two independent populations. *Biol Psychiatry.* 2010;67(11):1040–1047.
40. Naylor E, Bergmann BM, Krauski K, et al. The circadian clock mutation alters sleep homeostasis in the mouse. *J Neurosci.* 2000;20(21):8138–8143.
41. Lee KY, Song JY, Kim SH, et al. Association between CLOCK 3111T/C and preferred circadian phase in Korean patients with bipolar disorder. *Prog Neuropsychopharmacol Biol Psychiatry.* 2010;34(7):1196–1201.
42. Benedetti F, Serretti A, Colombo C, et al. Influence of CLOCK gene polymorphism on circadian mood fluctuation and illness recurrence in bipolar depression. *Am J Med Genet B Neuropsychiatr Genet.* 2003;123(1):23–26.
43. Shi J, Wittke-Thompson JK, Badner JA, et al. Clock genes may influence bipolar disorder susceptibility and dysfunctional circadian rhythm. *Am J Med Genet B Neuropsychiatr Genet.* 2008;147B(7):1047–1055.
44. Soria V, Martinez-Amoros E, Escaramis G, et al. Differential association of circadian genes with mood disorders: CRY1 and NPAS2 are associated with unipolar major depression and CLOCK and VIP with bipolar disorder. *Neuropsychopharmacology.* 2010;35(6):1279–1289.
45. Takao T, Tachikawa H, Kawanishi Y, Mizukami K, Asada T. CLOCK gene T3111C polymorphism is associated with Japanese schizophrenics: a preliminary study. *Eur Neuropsychopharmacol.* 2007;17(4):273–276.
46. Kissling C, Retz W, Wiemann S, et al. A polymorphism at the 3′-untranslated region of the CLOCK gene is associated with adult attention-deficit hyperactivity disorder. *Am J Med Genet B Neuropsychiatr Genet.* 2008;147(3):333–338.
47. Xu X, Breen G, Chen CK, Huang YS, Wu YY, Asherson P. Association study between a polymorphism at the 3′-untranslated region of CLOCK gene and attention deficit hyperactivity disorder. *Behav Brain Funct.* 2010;6:48.

48. Kishi T, Kitajima T, Ikeda M, et al. CLOCK may predict the response to fluvoxamine treatment in Japanese major depressive disorder patients. *Neuromolecular Med.* 2009;11 (2):53–57.
49. McClung CA, Sidiropoulou K, Vitaterna M, et al. Regulation of dopaminergic transmission and cocaine reward by the Clock gene. *Proc Natl Acad Sci USA.* 2005;102 (26):9377–9381.
50. Roybal K, Theobold D, Graham A, et al. Mania-like behavior induced by disruption of CLOCK. *Proc Natl Acad Sci USA.* 2007;104(15):6406–6411.
51. Monteleone P, Tortorella A, Docimo L, et al. Investigation of 3111T/C polymorphism of the CLOCK gene in obese individuals with or without binge eating disorder: association with higher body mass index. *Neurosci Lett.* 2008;435(1):30–33.
52. Sookoian S, Gemma C, Gianotti TF, Burgueno A, Castano G, Pirola CJ. Genetic variants of Clock transcription factor are associated with individual susceptibility to obesity. *Am J Clin Nutr.* 2008;87(6):1606–1615.
53. Tortorella A, Monteleone P, Martiadis V, Perris F, Maj M. The 3111T/C polymorphism of the CLOCK gene confers a predisposition to a lifetime lower body weight in patients with anorexia nervosa and bulimia nervosa: a preliminary study. *Am J Med Genet B Neuropsychiatr Genet.* 2007;144B(8):992–995.
54. Tsuzaki K, Kotani K, Sano Y, Fujiwara S, Takahashi K, Sakane N. The association of the Clock 3111 T/C SNP with lipids and lipoproteins including small dense low-density lipoprotein: results from the Mima study. *BMC Med Genet.* 2010;11:150.
55. Garaulet M, Lee YC, Shen J, et al. CLOCK genetic variation and metabolic syndrome risk: modulation by monounsaturated fatty acids. *Am J Clin Nutr.* 2009;90(6):1466–1475.
56. Sookoian S, Castano G, Gemma C, Gianotti TF, Pirola CJ. Common genetic variations in CLOCK transcription factor are associated with nonalcoholic fatty liver disease. *World J Gastroenterol.* 2007;13(31):4242–4248.
57. Turek FW, Joshu C, Kohsaka A, et al. Obesity and metabolic syndrome in circadian Clock mutant mice. *Science.* 2005;308(5724):1043–1045.
58. Hoffman AE, Yi CH, Zheng T, et al. CLOCK in breast tumorigenesis: genetic, epigenetic, and transcriptional profiling analyses. *Cancer Res.* 2010;70(4):1459–1468.
59. Dai H, Zhang L, Cao M, et al. The role of polymorphisms in circadian pathway genes in breast tumorigenesis. *Breast Cancer Res Treat.* 2011;127(2):531–540.
60. Zhou F, He X, Liu H, et al. Functional polymorphisms of circadian positive feedback regulation genes and clinical outcome of Chinese patients with resected colorectal cancer. *Cancer.* 2012;118(4):937–946.
61. Gamble KL, Motsinger-Reif AA, Hida A, et al. Shift work in nurses: contribution of phenotypes and genotypes to adaptation. *PLoS One.* 2011;6(4):e18395.
62. Dudley CA, Erbel-Sieler C, Estill SJ, et al. Altered patterns of sleep and behavioral adaptability in NPAS2-deficient mice. *Science.* 2003;301(5631):379–383.
63. Johansson C, Willeit M, Smedh C, et al. Circadian clock-related polymorphisms in seasonal affective disorder and their relevance to diurnal preference. *Neuropsychopharmacology.* 2003;28(4):734–739.
64. Nicholas B, Rudrasingham V, Nash S, Kirov G, Owen MJ, Wimpory DC. Association of Per1 and Npas2 with autistic disorder: support for the clock genes/social timing hypothesis. *Mol Psychiatry.* 2007;12(6):581–592.
65. Smith AK, Fang H, Whistler T, Unger ER, Rajeevan MS. Convergent genomic studies identify association of GRIK2 and NPAS2 with chronic fatigue syndrome. *Neuropsychobiology.* 2011;64(4):183–194.
66. Zhu Y, Leaderer D, Guss C, et al. Ala394Thr polymorphism in the clock gene NPAS2: a circadian modifier for the risk of non-Hodgkin's lymphoma. *Int J Cancer.* 2007;120 (2):432–435.

67. Chu LW, Zhu Y, Yu K, et al. Variants in circadian genes and prostate cancer risk: a population-based study in China. *Prostate Cancer Prostatic Dis.* 2008;11(4):342–348.
68. Zhu Y, Stevens RG, Leaderer D, et al. Non-synonymous polymorphisms in the circadian gene NPAS2 and breast cancer risk. *Breast Cancer Res Treat.* 2008;107(3):421–425.
69. Chu LW, Zhu Y, Yu K, et al. Correlation between circadian gene variants and serum levels of sex steroids and insulin-like growth factor-I. *Cancer Epidemiol Biomarkers Prev.* 2008;17(11):3268–3273.
70. Carpen JD, von Schantz M, Smits M, Skene DJ, Archer SN. A silent polymorphism in the PER1 gene associates with extreme diurnal preference in humans. *J Hum Genet.* 2006;51(12):1122–1125.
71. Dong L, Bilbao A, Laucht M, et al. Effects of the circadian rhythm gene period 1 (per1) on psychosocial stress-induced alcohol drinking. *Am J Psychiatry.* 2011;168(10):1090–1098.
72. Medicine AAoS . *International Classification of Sleep Disorders, 2nd ed.: Diagnostic and Coding Manual.* Westchester, IL: American Academy of Sleep Medicine; 2005.
73. Jones CR, Campbell SS, Zone SE, et al. Familial advanced sleep-phase syndrome: a short-period circadian rhythm variant in humans. *Nat Med.* 1999;5(9):1062–1065.
74. Campbell SS, Gillin JC, Kripke DF, Erikson P, Clopton P. Gender differences in the circadian temperature rhythms of healthy elderly subjects: relationships to sleep quality. *Sleep.* 1989;12(6):529–536.
75. Haimov I, Lavie P. Circadian characteristics of sleep propensity function in healthy elderly: a comparison with young adults. *Sleep.* 1997;20(4):294–300.
76. Toh KL, Jones CR, He Y, et al. An hPer2 phosphorylation site mutation in familial advanced sleep phase syndrome. *Science.* 2001;291(5506):1040–1043.
77. Xu Y, Toh KL, Jones CR, Shin J-Y, Fu Y-H, Ptacek LJ. Modeling of a human circadian mutation yields insights into clock regulation by PER2. *Cell.* 2007;128:59–70.
78. Carpen JD, Archer SN, Skene DJ, Smits M, von Schantz M. A single-nucleotide polymorphism in the 5′-untranslated region of the hPER2 gene is associated with diurnal preference. *J Sleep Res.* 2005;14(3):293–297.
79. Lee HJ, Kim L, Kang SG, et al. PER2 variation is associated with diurnal preference in a Korean young population. *Behav Genet.* 2011;41(2):273–277.
80. Lavebratt C, Sjoholm LK, Partonen T, Schalling M, Forsell Y. PER2 variantion is associated with depression vulnerability. *Am J Med Genet B Neuropsychiatr Genet.* 2010;153B(2):570–581.
81. Forbes EE, Dahl RE, Almeida JR, et al. PER2 rs2304672 polymorphism moderates circadian-relevant reward circuitry activity in adolescents. *Biol Psychiatry.* 2012;71(5):451–457.
82. Abarca C, Albrecht U, Spanagel R. Cocaine sensitization and reward are under the influence of circadian genes and rhythm. *Proc Natl Acad Sci USA.* 2002;99(13):9026–9030.
83. Garaulet M, Corbalan-Tutau MD, Madrid JA, et al. PERIOD2 variants are associated with abdominal obesity, psycho-behavioral factors, and attrition in the dietary treatment of obesity. *J Am Diet Assoc.* 2010;110(6):917–921.
84. Garcia-Rios A, Perez-Martinez P, Delgado-Lista J, et al. A Period 2 genetic variant interacts with plasma SFA to modify plasma lipid concentrations in adults with metabolic syndrome. *J Nutr.* 2012;142(7):1213–1218.
85. Spanagel R, Pendyala G, Abarca C, et al. The clock gene *Per2* influences the glutamatergic system and modulates alcohol consumption. *Nat Med.* 2005;11:35–42.

86. Comasco E, Nordquist N, Gokturk C, et al. The clock gene PER2 and sleep problems: association with alcohol consumption among Swedish adolescents. *Ups J Med Sci.* 2010;115(1):41–48.
87. Ebisawa T, Uchiyama M, Kajimura N, et al. Association of structural polymorphisms in the human period3 gene with delayed sleep phase syndrome. *EMBO Rep.* 2001;2 (4):342–346.
88. Archer SN, Robilliard DL, Skene DJ, et al. A length polymorphism in the circadian clock gene Per3 is linked to delayed sleep phase syndrome and extreme diurnal preference. *Sleep.* 2003;26(4):413–415.
89. Pereira DS, Tufik S, Louzada FM, et al. Association of the length polymorphism in the human Per3 gene with the delayed sleep-phase syndrome: does latitude have an influence upon it? *Sleep.* 2005;28(1):29–32.
90. Jones KH, Ellis J, von Schantz M, Skene DJ, Dijk DJ, Archer SN. Age-related change in the association between a polymorphism in the PER3 gene and preferred timing of sleep and waking activities. *J Sleep Res.* 2007;16(1):12–16.
91. Archer SN, Carpen JD, Gibson M, et al. Polymorphism in the PER3 promoter associates with diurnal preference and delayed sleep phase disorder. *Sleep.* 2010;33 (5):695–701.
92. Ellis J, von Schantz M, Jones KH, Archer SN. Association between specific diurnal preference questionnaire items and PER3 VNTR genotype. *Chronobiol Int.* 2009;26 (3):464–473.
93. Chellappa SL, Viola AU, Schmidt C, et al. Human melatonin and alerting response to blue-enriched light depend on a polymorphism in the clock gene PER3. *J Clin Endocrinol Metab.* 2012;97(3):E433–E437.
94. Viola AU, Archer SN, James LM, et al. PER3 polymorphism predicts sleep structure and waking performance. *Curr Biol.* 2007;17(7):613–618.
95. Groeger JA, Viola AU, Lo JC, von Schantz M, Archer SN, Dijk DJ. Early morning executive functioning during sleep deprivation is compromised by a PERIOD3 polymorphism. *Sleep.* 2008;31(8):1159–1167.
96. Vandewalle G, Archer SN, Wuillaume C, et al. Functional magnetic resonance imaging-assessed brain responses during an executive task depend on interaction of sleep homeostasis, circadian phase, and PER3 genotype. *J Neurosci.* 2009;29 (25):7948–7956.
97. Viola AU, James LM, Archer SN, Dijk DJ. PER3 polymorphism and cardiac autonomic control: effects of sleep debt and circadian phase. *Am J Physiol Heart Circ Physiol.* 2008;295(5):H2156–H2163.
98. Lazar AS, Slak A, Lo JC, et al. Sleep, diurnal preference, health, and psychological well-being: a prospective single-allelic-variation study. *Chronobiol Int.* 2012;29(2):131–146.
99. Brower KJ, Wojnar M, Sliwerska E, Armitage R, Burmeister M. PER3 polymorphism and insomnia severity in alcohol dependence. *Sleep.* 2012;35(4):571–577.
100. Hasan S, van der Veen DR, Winsky-Sommerer R, Dijk DJ, Archer SN. Altered sleep and behavioral activity phenotypes in PER3-deficient mice. *Am J Physiol Regul Integr Comp Physiol.* 2011;301(6):R1821–R1830.
101. Below JE, Gamazon ER, Morrison JV, et al. Genome-wide association and meta-analysis in populations from Starr County, Texas, and Mexico City identify type 2 diabetes susceptibility loci and enrichment for expression quantitative trait loci in top signals. *Diabetologia.* 2011;54(8):2047–2055.
102. Costa MJ, So AY, Kaasik K, et al. Circadian rhythm gene period 3 is an inhibitor of the adipocyte cell fate. *J Biol Chem.* 2011;286(11):9063–9070.
103. Guess J, Burch JB, Ogoussan K, et al. Circadian disruption, Per3, and human cytokine secretion. *Integr Cancer Ther.* 2009;8(4):329–336.

104. Zou Y, Liao G, Liu Y, et al. Association of the 54-nucleotide repeat polymorphism of hPer3 with heroin dependence in Han Chinese population. *Genes Brain Behav*. 2008;7 (1):26–30.
105. Dallaspezia S, Lorenzi C, Pirovano A, Colombo C, Smeraldi E, Benedetti F. Circadian clock gene Per3 variants influence the postpartum onset of bipolar disorder. *Eur Psychiatry*. 2011;26(3):138–140.
106. Lin DW, FitzGerald LM, Fu R, et al. Genetic variants in the LEPR, CRY1, RNASEL, IL4, and ARVCF genes are prognostic markers of prostate cancer-specific mortality. *Cancer Epidemiol Biomarkers Prev*. 2011;20(9):1928–1936.
107. Hoffman AE, Zheng T, Stevens RG, et al. Clock-cancer connection in non-Hodgkin's lymphoma: a genetic association study and pathway analysis of the circadian gene cryptochrome 2. *Cancer Res*. 2009;69(8):3605–3613.
108. Hoffman AE, Zheng T, Yi CH, et al. The core circadian gene Cryptochrome 2 influences breast cancer risk, possibly by mediating hormone signaling. *Cancer Prev Res (Phila)*. 2010;3(4):539–548.
109. Dupuis J, Langenberg C, Prokopenko I, et al. New genetic loci implicated in fasting glucose homeostasis and their impact on type 2 diabetes risk. *Nat Genet*. 2010;42 (2):105–116.
110. Hu C, Zhang R, Wang C, et al. Variants from GIPR, TCF7L2, DGKB, MADD, CRY2, GLIS3, PROX1, SLC30A8 and IGF1 are associated with glucose metabolism in the Chinese. *PLoS One*. 2010;5(11):e15542.
111. Barker A, Sharp SJ, Timpson NJ, et al. Association of genetic Loci with glucose levels in childhood and adolescence: a meta-analysis of over 6,000 children. *Diabetes*. 2011;60 (6):1805–1812.
112. Lavebratt C, Sjoholm LK, Soronen P, et al. CRY2 is associated with depression. *PLoS One*. 2010;5(2):e9407.
113. Severino G, Manchia M, Contu P, et al. Association study in a Sardinian sample between bipolar disorder and the nuclear receptor REV-ERBalpha gene, a critical component of the circadian clock system. *Bipolar Disord*. 2009;11(2):215–220.
114. Campos-de-Sousa S, Guindalini C, Tondo L, et al. Nuclear receptor rev-erb-α circadian gene variants and lithium carbonate prophylaxis in bipolar affective disorder. *J Biol Rhythms*. 2010;25(2):132–137.
115. McCarthy MJ, Nievergelt CM, Shekhtman T, Kripke DF, Welsh DK, Kelsoe JR. Functional genetic variation in the Rev-Erbalpha pathway and lithium response in the treatment of bipolar disorder. *Genes Brain Behav*. 2011;10(8):852–861.
116. Veenstra-VanderWeele J, Qaadir A, Palmer AA, Cook Jr EH, de Wit H. Association between the casein kinase 1 epsilon gene region and subjective response to D-amphetamine. *Neuropsychopharmacology*. 2006;31(5):1056–1063.
117. Palmer AA, Verbitsky M, Suresh R, et al. Gene expression differences in mice divergently selected for methamphetamine sensitivity. *Mamm Genome*. 2005;16(5):291–305.
118. Kawaida R, Yamada R, Kobayashi K, et al. CUL1, a component of E3 ubiquitin ligase, alters lymphocyte signal transduction with possible effect on rheumatoid arthritis. *Genes Immun*. 2005;6(3):194–202.
119. Negi S, Kumar A, Thelma BK, Juyal RC. Association of Cullin1 haplotype variants with rheumatoid arthritis and response to methotrexate. *Pharmacogenet Genomics*. 2011;21(9):590–593.
120. Kim CJ, Song JH, Cho YG, et al. Somatic mutations of the beta-TrCP gene in gastric cancer. *Apmis*. 2007;115(2):127–133.
121. Chen S, He Y, Ding J, et al. An insertion/deletion polymorphism in the 3′ untranslated region of beta-transducin repeat-containing protein (betaTrCP) is associated with susceptibility for hepatocellular carcinoma in Chinese. *Biochem Biophys Res Commun*. 2010;391(1):552–556.

122. de Mollerat XJ, Gurrieri F, Morgan CT, et al. A genomic rearrangement resulting in a tandem duplication is associated with split hand-split foot malformation 3 (SHFM3) at 10q24. *Hum Mol Genet.* 2003;12(16):1959–1971.
123. Nishiwaki T, Daigo Y, Kawasoe T, Nakamura Y. Isolation and mutational analysis of a novel human cDNA, DEC1 (deleted in esophageal cancer 1), derived from the tumor suppressor locus in 9q32. *Genes Chromosomes Cancer.* 2000;27(2):169–176.
124. Huang YJ, Niu J, Wei S, et al. A novel functional DEC1 promoter polymorphism -249T > C reduces risk of squamous cell carcinoma of the head and neck. *Carcinogenesis.* 2010;31(12):2082–2090.
125. He Y, Jones CR, Fujiki N, et al. The transcriptional repressor DEC2 regulates sleep length in mammals. *Science.* 2009;325(5942):866–870.
126. Utge SJ, Soronen P, Loukola A, et al. Systematic analysis of circadian genes in a population-based sample reveals association of TIMELESS with depression and sleep disturbance. *PLoS One.* 2010;5(2):e9259.

CHAPTER FOUR

Peripheral Circadian Oscillators: Time and Food

Ruud Buijs*, Roberto Salgado†, Elizabeth Sabath*, Carolina Escobar‡
*Departamento de Biología Celular y Fisiología, Instituto de Investigaciones Biomédicas, Universidad Nacional Autónoma de México, Distrito Federal, Mexico
†Departamento de Biología Celular, Facultad de Ciencias, Universidad Autónoma de San Luis Potosí, San Luis Potosí, Mexico
‡Departamento de Anatomía, Facultad de Medicina, Universidad Nacional Autónoma de México, Distrito Federal, Mexico

Contents

Abstract

The suprachiasmatic nucleus (SCN) provides timing to the brain and to the whole organism. Its rhythmic signal to mainly hypothalamic structures results in a synchronized hormonal and autonomic output to the body that coordinates behavior and physiology. As a result of this, the expression of clock genes in all organs has a rhythm that is dictated by the SCN. Together with these clock genes, a number of cellular processes follow a similar rhythm, whereby it has been proposed that these events are driven at least, in part, by clock genes.

Together, this forms a multiple oscillating system that interacts and under normal conditions is synchronized by the SCN. The autonomic and hormonal outputs from the SCN are examples of messages that are clearly targeted; the behaviors driven by the SCN are examples of messages that may have more diffuse targets. For example, food intake

Progress in Molecular Biology and Translational Science, Volume 119
ISSN 1877-1173
http://dx.doi.org/10.1016/B978-0-12-396971-2.00004-X

and locomotor activity, which are normally driven by the SCN, have the capacity to drive the rhythm of clock genes in cells of the liver. The influence of food has been shown by offering food outside the normal activity–food intake period. If such a condition persists, desynchronization follows between centrally and peripherally dictated rhythms because the SCN keeps transmitting temporal signals according to the day–night cycle. These circumstances promote pathologies such as the metabolic syndrome, which is characterized by the progressive onset of hypertension, insulin resistance, and diabetes. As clock genes are proposed to drive the rhythms of metabolic genes, it is very attractive to give the clock genes a central place in this desynchronization and pathology picture. Therefore, in this chapter, we pay special attention to the question of how the SCN is able to transmit its message to the cells of the body and focus on the liver, because of its essential role in metabolism. Here, we review recent evidence that shows how desynchronization may lead to the uncoupling of cellular processes within the liver cells. The basis for this cellular dissociation, we argue, is the fact that the network of brain–body interaction is desynchronized, leading also to an uncoupling of normally coupled systems within the cell.

1. INTRODUCTION

In mammals, the circadian system is coordinated by the suprachiasmatic nucleus (SCN). The SCN is synchronized to daily light–dark cycles by direct photic inputs received from the retina through the retinohypothalamic tract. The endogenous timekeeping mechanism in SCN cells is modeled as a network of interlocked transcription–translation feedback loops that oscillate with a 24-h periodicity. In this mammalian clock machinery, the transcription factors CLOCK and BMAL1 heterodimerize to drive the transcription of genes containing E-box enhancer elements, and among these are the Period (Per) and Cryptochrome (Cry) genes. PER and CRY proteins, in turn, multimerize and inhibit the action of CLOCK–BMAL1, resulting in a rhythmic oscillation of its own transcription and of many downstream targets.[1,2]

This molecular circadian machinery is present not only in the SCN but also in peripheral organs. While in the SCN this clock machinery can produce self-sustained oscillations for indefinite cycles, in peripheral organs when tested *in vitro*, in general, the evidence points to a progressive loss of rhythm in the absence of the SCN.[3]

Many behavioral and physiological aspects, that is, feeding/fasting, rest/activity, neuroendocrine secretion, and autonomic control, are used by the SCN to induce circadian expression of clock genes in peripheral organs;

nevertheless, feeding has been shown to be one of the most important stimuli[4,5] to induce a rhythm. Therefore, it is logical to assume that clock genes in the periphery somehow are associated with energy metabolism in the cells of the different tissues. *In vitro* studies have mainly evidenced that reciprocal interaction loops between clock genes and metabolic genes could underlie this food-clock connection.[6]

Thus, a system has been suggested with highly intertwined relationships whereby rhythmic behavioral (neuro)endocrine and autonomic functions support each other and, in turn, are supported by rhythmic cellular and molecular processes. All these processes are coordinated and driven by the SCN, resulting not only in tightly associated sleep–wake and fasting–feeding cycles but also in rhythmic patterns of core temperature, heart rate, and hormonal cycles, resulting in a myriad of functions that are influenced by the SCN. Thus, we propose that circadian disruption may be an important cause of severe health impairment, as consequence of a faulty relationship between the oscillatory processes in the cells of the body and with the rhythm of the SCN.

2. THE SCN AS MASTER CLOCK

The SCN has the capacity to maintain a 24-h rhythm in electrical activity, which results in a rhythmic release of its neurotransmitters from the nerve terminals.[7] The rhythm in neuronal activity is maintained *in vitro* and is proposed to be dependent on a molecular machinery of clock genes that, by mean of positive and negative feedback loops, give a possibility for the SCN to have this 24-h rhythm in electrical activity. Although it is assumed that individual neurons may have the capacity to generate an endogenous rhythm,[8] it has become likely that only when SCN neurons form a network, these neurons have the possibility to be rhythmic for long-time periods.[9] Consequently, it is not only the molecular machinery that gives the SCN neurons their capacity to have an endogenous rhythm of 24 h, but in addition, these neurons need to function within a network. That a neuronal network is essential for the SCN rhythmic properties can also be concluded from studies in which knocking out a vasoactive intestinal polypeptide (VIP) receptor, which is one of the most abundant neuropeptides within the SCN, results in the incapacity to maintain rhythmic behavioral patterns.[10] As these VIP receptors are highly expressed in the SCN and the electrical activity in the SCN of the knockout animals is strongly disturbed,[11] it can be concluded that intra-SCN neuronal communication is essential to sustain the SCN rhythm. On this basis, we propose that similar network connections are also important in the functioning of the whole circadian system, that is, the SCN needs signals back from the body and the brain in order to function optimally.

Without a doubt, the rhythmic message of the SCN to mainly hypothalamic structures drives hormonal and autonomic output in a 24-h pattern, resulting in a rhythm in behavior that is synchronized with, for example, the adequate hormone or glucose levels together with adequate temperature and cardiovascular parameters to ensure an optimal physiology.[12–14] In addition, these rhythmic patterns are not fixed but are highly adaptive and take into account the homeostatic situation of the animal. For example, the response of corticosterone to psychological stress is the highest during the beginning of the sleep period,[14–16] while the corticosterone response to hypoglycemia is the highest in the beginning of the active period.[17] Such interactions lead to a rhythmic organization of body functions, which is dependent on the integrity of the SCN and its output pathways.

3. THE PERIPHERAL OSCILLATORS AND THEIR RELATIONSHIP WITH CLOCK GENES

The phase of the rhythmic expression of clock genes in the periphery is opposite or phase delayed to the rhythm of the same clock genes expressed in the SCN.[15,16] In spite of their demonstration in all mammalian tissues, already more than 10 years ago, the link between the rhythmic clock gene expression to functional rhythms in these tissues is still weak. Evidence that supports the relevance of peripheral clock genes in physiology is provided in studies that show the involvement of clock genes in cell division processes,[18,19] cardiovascular function,[20,21] and adrenal function.[22] Possible connections of clock genes with metabolic processes have been explored especially in the liver, one of the most active organs of the body associated with metabolism. For example, peroxisome proliferator-activated receptor α (PPARα), involved in lipid and lipoprotein metabolism, binds directly to the Bmal1 promoter,[23] indicating a possible mechanism via which metabolism may influence clock mechanisms in the periphery. Especially in the liver, clock genes interact with metabolic genes and regulate their transcription, endowing cells with rhythmic molecular mechanisms that allow adapting to the daily cycles in metabolic demand.[6,24–26] In particular, genes that regulate gluconeogenesis and fatty acid oxidation are suggested to interact with clock genes for their rhythmic expression. This has been demonstrated for the PPARs, like Pparγ, Pparα, Pparβ, Pgc-α, and silent mating-type information regulator homolog1 (Sirt1), which is a member of the Sirtuin family. Decreased expression of both Sirt1 and Pparγ is associated with development of the metabolic syndrome.[27–29] Likewise, changes in the metabolic condition, especially the redox state and energy availability, influence the expression of Bmal1 and Per2, resulting in a strong interaction between metabolic and circadian regulation in hepatocytes.[1,30] However, all this evidence has emerged only from *in vitro* studies, which may give a limited picture of what is really going on in the organ *in vivo*.

Circadian rhythms governing organ functions may be produced by other cycling processes than endogenous clock genes expression. This has been observed in specific liver Bmal1 knockout animals, where due to the absence of this clock gene many liver genes loose their rhythm, but the glucose rhythm in the circulation remains, indicating a rhythmic liver output.[31] A similar type of observation was made by Cailotto *et al.*[32,33] who demonstrated that, after sympathetic denervation, the liver lost its rhythmic glycogen storage and enzymatic expression, but that in spite of this, the clock genes did not lose their rhythmicity. These observations indicate that, when observed *in vivo*, the suggested link between clock genes and metabolic genes or metabolism is not as firm as suggested and that when the organ is embedded in the system of the whole body, there might be several alternative ways to drive the rhythmic physiology of the organ.

Many events, metabolic, endocrine, and homeostatic (e.g., temperature, corticosterone), may drive the rhythmic expression of the organs. For example, in an elegant parabiosis experiment between SCN lesioned and intact mice, Guo *et al.*[34] demonstrated that behavioral or blood-borne signals are sufficient to maintain circadian rhythms of clock gene expression in liver and kidney, but not in heart, spleen, or skeletal muscle. Therefore, in this chapter, we use the term peripheral oscillator to refer to rhythmic functions or processes within organs or tissues.

4. THE SCN AS THE DRIVING FORCE BEHIND THE OSCILLATORS

Many tissues show a different sensitivity or different output depending on the time the organ was removed. For example, Ungar and Halberg demonstrated already in 1962 that the adrenal, when removed from the rat in the evening, was more sensitive to ACTH than when it was removed in the morning.[35] Later, similar observations were reported on the liver and the heart.[36,37] These changing responses can be induced by the SCN by altering the sensitivity of the organs to circulating hormones[14]; in particular, the circadian sensitivity of the adrenal to ACTH is induced by the autonomic nervous system[38] by means of a polysynaptic pathway originating in the SCN.[39,40] A similar role for the SCN was further illustrated by Cailotto *et al.* who showed that light affected the expression of the liver metabolic enzymes GLUT2 and glucokinase time dependently (at ZT14 but not at ZT20 for GLUT2 and the reverse for glucokinase) in intact but not in denervated animals, demonstrating that the SCN via the autonomic nervous system may change directly the production of liver enzymes.[41] Consequently, temporal changes in liver metabolic enzymes are directly induced by SCN-mediated autonomic input. The same study revealed that light could only modify Per1 and Per2 expression at ZT14 and not at ZT20, while, for example, PEPCK was changed at both time points, indicating that light may induce changes in metabolic enzymes via other cellular mechanisms that are also influenced by the autonomic input to the liver. Also in support for autonomic influence of the clock gene expression in the liver are the experiments of Terazano[42] who demonstrated that electrical stimulation of the sympathetic output to the liver induces a phase shift of clock gene expression.

The SCN can also induce and control the rhythmicity of peripheral oscillators via humoral or metabolic signals. Corticosterone and melatonin, both driven by the SCN, have been proposed as important humoral signals for transmitting daily rhythms to the body.[43] In addition, the SCN, by determining the sleep/activity cycle, can drive feeding rhythms and metabolic rhythms that can serve as additional internal entraining signals.

The discovery of clock genes has been an enormous help to evaluate the possibilities of the SCN to influence the activity of peripheral tissues by means of metabolic signaling and hormones. At first, it was shown that fibroblasts can be induced to show a cyclic expression of clock genes by a serum shock, implying a metabolic stimulus.[44] Then, in view of the important role of glucocorticoids to transmit the daily signal of the SCN to the tissues of the body, their role in synchronizing peripheral clock genes was demonstrated.[43] Finally, temperature, another variable strongly under the influence of the SCN, is also able to sustain, however, not to induce *de novo*, clock gene expression.[45] Studies on isolated fibroblast suggest that temperature may also affect clock gene expression via cold-inducible mRNA.[46] All this evidence indicates that, under physiological conditions, the SCN may use multiple strategies to drive rhythmicity in peripheral oscillators.

By considering the possible mechanism that drives peripheral oscillators, it is important to keep in mind that organs can exchange metabolic information via hormones and the modulation of the autonomic nervous system, with or without the central nervous system as interface.[47–50] The relevance of the interaction and exchange of information between the different organs is that, at many different places in our body and brain, metabolic information can be gathered and integrated by influencing the functionality of this network. It will be clear that under normal conditions, all these different elements, hormones, metabolites, and autonomic nervous system, act in synchrony enforcing each other's action. Consequently, removing one of these factors will not change the oscillations as other stimuli remain in place. This cooperation between different factors makes it difficult to estimate the individual contribution of, for example, corticosterone as compared to sympathetic input. One would need to know and remove all other contributors (like food intake or temperature) to be able to make such approximation. Later, we will come back to this point by discussing the effect of changing only the food availability to another circadian time point on the expression of clock and metabolic genes.

5. SYNCHRONIZATION OF THE CIRCADIAN SYSTEM BY THE SCN

We have seen that we have to consider the peripheral oscillators and the SCN function as one system that is in perfect balance. Consequently, under normal conditions, the peripheral oscillators are strongly driven by the SCN and SCN-driven processes. The loss of one rhythmic input, for example, the rhythm in corticosterone, would not mean that all rhythmicity is lost because many other inputs to the organ would remain in place. In contrast to this seems the proposed organization at the cellular level where up till now attention has been focused on the role of the clock genes and their position has been emphasized as pivotal also for the rhythmicity at the organ level. For example, the observations of Kornmann *et al.*[51] who used mice with a conditionally active liver clock, in which REV-ERBalpha represses (among others) the transcription of the essential core clock gene Bmal1 and reported that most liver genes lost their rhythm and just a few genes remained rhythmic. The authors concluded, therefore, that the rhythm of most liver genes would depend on the rhythmicity of BMAL1 and thus on the functionality of hepatocyte clocks. In contrast to this statement was their observation that Per2 expression kept a strong rhythmicity.

Also challenging this observation is a recent study of the group of Hogenesch using a brain-specific rescue of *Clock* that revealed that this rescued the rhythm of many liver genes and consequently demonstrated a so-called system-driven rhythm[52] supporting the notion that indeed many different SCN-driven systems other than peripheral clock genes are able to drive peripheral oscillators.

The complexity and diversity of the processes driven by the SCN require a fine tuning and synchronization of these different rhythms. For this reason, one would suggest that the SCN is driving the rhythm of the peripheral oscillators and that peripheral oscillators have a way to talk back to the SCN in order to allow it to function as one homeostatic system.

6. TIMING BY FOOD?

As early as clock genes were described in peripheral tissues, it became clear that in addition to the SCN, they could be entrained by food-related signals. Shifting feeding schedules toward the day in mice changed circadian gene expression in liver, kidney, heart, and pancreas but not in the SCN.[4] This effect was confirmed in the liver of transgenic mice in which the Per1 gene promoter had been linked to a luciferase reporter. Daytime feeding rapidly entrained the liver, shifting its phase by 10 h toward the day within 3 days.[3] Short 2–4 h restricted feeding schedules in arrhythmic mice due to a bilateral SCN lesion also restored circadian rhythmicity in some clock genes in the liver.[5] Finally, daytime-restricted feeding in nocturnal rodents induced and inverted the phase of clock gene expression; thus, feeding rhythms appeared to be a dominant Zeitgeber for the entrainment of hepatic clock genes. Importantly, feeding time had no effect on the expression of clock genes in the SCN, suggesting that a daytime-feeding regime completely uncouples the phases of the clock genes in the liver from those in the biological clock.[4,5,53] It is important to realize that, in spite of the fact that the overall clock gene expression in the SCN might be unaltered, within the SCN changes in food regime may induce important regional changes that allow the circadian system to adapt to these circumstances; such changes may go undetected by overall analysis.[54,55]

The possible signaling cue elicited by feeding schedules to peripheral cells is suggested to have multiple origins. Several circulating metabolic and hormonal factors, originally controlled by the SCN, are also dependent on food intake, including glucose, free fatty acids, glucocorticoids, thyroid hormones, and others. Several exhibit shifted food-entrained rhythms when food is scheduled to the day.[56–58] However, one note of caution: all these data are collected in animals by which the food is restricted to 2–4 h/day. When food is restricted to the whole inactive period, several parameters, for example, corticosterone, glucose, temperature, and activity, either show phase small changes or do not show a rhythm.[59,60]

The above-mentioned parabiosis experiments[34] illustrated that under such conditions, indeed, liver and kidney clock genes may follow the circulating signals, but that these are not sufficient to synchronize clock genes in other tissues. Because the heart, spleen, and kidney were not entrained in this parabiosis model, it is also clear that the resetting of these tissues might require different combinations of signaling cues including neuronal signals from the SCN. This observation also suggests that circadian clocks in peripheral tissues require different combinations of temporal signals. Hereby one also should consider the role of glucocorticoids and maybe also melatonin.[61–63] Hereby it is still questionable whether their entraining effect is reached by their influence on Per1 (clock gene) expression or via other mechanisms.[64]

However, in spite of all the evidence for some role of circulating hormones, the circulating metabolite that evidently can be modified by food intake is glucose. Glucose is the main metabolite, providing energy to the cell; it exhibits a daily rhythm driven by the SCN[65] and may as such provide a time signal to the cells. Indeed, the oscillation of clock gene expression in cultured rat fibroblasts is induced or reset when these cells are exposed to a glucose bolus.[66] In the same study, it was shown that glucose as well as fructose, mannose, and lactose triggered robust circadian expression of Per2, Dbp, and Bmal1 that lasted for at least three cycles, while proteins did not. Moreover, in experimental animals, food intake after fasting was shown to immediately shift clock genes in the liver and not in the lung.[67] All these studies indicate that the liver is capable of responding immediately and with a wide range of oscillatory genes to food especially when animals have been fasting.[68] Still, it is also known that food is not the only driving force, as liver oscillatory genes maintain their rhythm under fasting conditions.[69,70] Consequently, a picture emerges in which food has the *capacity* to change especially clock gene expression in the liver, while on the other hand, food is not *essential* for sustaining rhythms of these genes. The rhythm imposed by the SCN (without the food intake) is sufficient to drive the rhythms of clock genes even though a robust rhythm is only observed when both SCN and food give the signal.[69] Up till now the precise cellular mechanisms that are responsible for the oscillations in both clock genes and metabolic genes are still not clear.

7. ENERGY OSCILLATION IN THE CELL

7.1. The NAD/SIRT1 clock gene link

Several molecules have been shown to be involved in the possible reciprocal relationship between clock genes and metabolic genes, and metabolic processes within the cell. Serious attention has been paid to the redox potential, which is directly linked to the energy state of the cells. Energy metabolism and consequently glucose modify cellular redox levels which are, for example, influenced by the NAD^+ to NADH ratio.[71] It is suggested that molecular oscillators in the liver and other peripheral organs may be directly influenced through elements of the redox potential; *in vitro*, the DNA binding of the CLOCK–BMAL1 or NPAS2–BMAL1 heterodimers is highly sensitive to the proportion of NAD^+ cofactors.[72] NAD^+ is a classic coenzyme that, in mammals, is produced by the conversion of nicotinamide and 5′-phosphoribosylpyrophosphate to nicotinamide mononucleotide (NMN) by the rate-limiting enzyme nicotinamide phosphoribosyltransferase (NAMPT).[73] Hereby the NAMPT-mediated NAD^+ biosynthesis is suggested to play a critical role in a number of biological processes through the NAD^+-dependent deacetylase of SIRT1.[30] These associations of one of the most important cycles involved in the redox potential with the clock genes[24,74,75] are now widely seen as the logical connection between the interaction of the rhythms induced by the SCN and the rhythms induced by food (Fig. 4.1). This agrees with the observation that liver cytoplasm and mitochondria exhibit a reduced redox which is immediately reverted to an oxidized state after feeding.[76]

7.2. Metabolism, AMPK, and clock genes

Adenosine triphosphate (ATP) is considered as the energy currency of the cells; the hydrolysis of it to ADP (adenosine diphosphate) and/or AMP (adenosine monophosphate) can be coupled to transport across membranes, synthesis reactions, and other processes that require energy. Thus, the ratio between AMP and ATP is one of the most reliable indicators of the energetic state of the cell. In eukaryotes, the AMP-activated protein kinase (AMPK) is a conserved enzyme which exquisitely senses small variations in these molecules and, in turn, phosphorylates a broad range of downstream targets resulting in the overall effect of increasing ATP-producing pathways while decreasing ATP-consuming ones.[77–79] The system which senses the

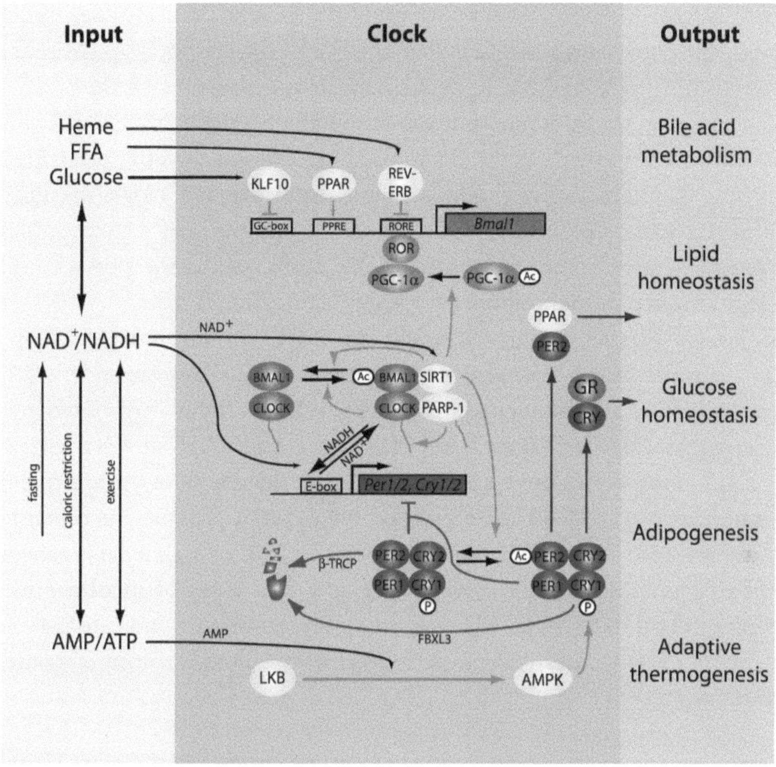

Figure 4.1 The relationship between clock genes and the metabolism/energy requirements of the cell. Many energy- or metabolism-associated molecules have the capacity to influence the clock gene cycle; on the other hand, also the output is associated with the energy cycle of the cell (see text and [26] for details). Figure from [26] with permission. (See Color Insert.)

energetic AMP/ATP status of the cell is under circadian control. The catalytic subunit alpha1 of the AMP/ATP sensor enzyme AMPK exhibits a robust circadian rhythm of nuclear localization in mouse liver, peaking synchronously with the beta2 subunit expression. The phosphorylation of AMPK substrates Raptor-Ser792 and ACC1-Ser79 also has a diurnal rhythm.[25] One of the first studies that showed the reciprocal connection of AMPK with the clock system demonstrated *in vitro* that AMPK phosphorylates Casein kinase I epsilon (an important regulator of the period length), resulting in increased activity of this enzyme which, in turn, phosphorylates and induces the degradation of Per2. *In vivo*, injection of the AMPK-activating drug metformin leads to mPer2 degradation in peripheral tissues and a phase advance in the circadian expression pattern of clock genes in wild-type mice but not in AMPK alpha 2 knockout mice.[80] It has been demonstrated also that AMPK phosphorylates and destabilizes the clock component cryptochrome 1. *In vivo*, stimulation of AMPK destabilized cryptochromes in the liver and altered circadian rhythms, and mice in which the AMPK pathway was genetically disrupted showed alterations in peripheral clocks.[25] The following are other sets of interactions of AMPK with the molecular clock machinery. AMPK is a master controller of PGC-1a, a transcriptional coactivator that orchestrates a constellation of transcription factors, such as the PPARs among others, to induce mitochondrial gene expression. Additionally, AMPK seems to influence SIRT1 activity through an AMPK-induced modulation of NAD^+ metabolites. Pharmacological or physiological activation of AMPK is followed by a robust increase in NAD^+ within hours, promoted by an increase in fatty acid oxidation rates. This increase in NAD^+ levels is sustained by the induction of Nampt expression, the gene that encodes the enzyme that resynthesizes NAD^+ from its metabolic breakdown product, nicotinamide. This constitutes a two-way impact of AMPK on SIRT1 activity as it generates the SIRT1 activator NAD^+, while reducing the levels of nicotinamide, a physiological inhibitor of SIRT1 activity.[81] Recent reports indicate that indirectly, SIRT1 enhances AMPK activation in the liver, creating a positive-feedback loop.[82]

8. HEME AS ANOTHER METABOLIC CLOCK CONNECTION POSSIBILITY

Heme is better known for its major role as the prosthetic group of hemoglobin in erythrocytes; however, it has other multiple actions. Heme consists of an iron atom surrounded by a porphyrin ring structure. The first step in its biosynthesis is the condensation of one aminoacid (glycine) with one intermediate of Krebs cycle (succinyl CoA) to form 5-aminolevulinate (ALA) and CO_2. The enzyme responsible for this rate-limiting step is the ALA synthase (ALAS).[83] A connection with the molecular clock is suggested by the fact that the ALAS1 is a PGC1alpha target gene.[84] A second connection is that heme is the endogenous ligand for Rev-erbalpha, the transcription factor which is part of the accessory loop that regulates the transcription of Bmal1 gene.[85] NPAS2, a mammalian transcription factor part of the molecular clock system which dimerizes with BMAL1, binds heme as a prosthetic group and at least *in vitro*, the heme status controls its DNA binding.[86] Interestingly, it seems that heme is connected to the clock from its beginnings to its end. The enzyme responsible for heme degradation is the heme oxygenase; this enzyme catalyzes the conversion of heme into biliverdin, iron, and carbon monoxide. Upon exposure to carbon monoxide, inactive BMAL1 homodimers are formed at the expense of NPAS2–BMAL1 heterodimers, indicating that the heterodimerization of NPAS2 is regulated by CO through the heme-based sensor of its PAS domains.[86]

Finally, heme is considered a link between metabolism and the clock because its production and destiny can be influenced by the following factors. First, it is known that the activity of ALAS1 enzyme is inversely regulated by glucose.[87] Second, insulin blunts hepatocyte ALAS1 induction, by disrupting the interaction of FOXO1 (forkhead box O1) and PGC-1α.

All this together suggests a relevant role of the energy state or the redox potential as an important signal associated with the fasting/feeding cycle driving not only cellular oscillations but also the rhythm in clock gene expression. However, as we see in Section 9, in spite of this wealth of data illustrating the influence of the feeding pattern on clock gene expression especially in the liver, and the evidence for this link by *in vitro* studies, in conditions where an animal is living under desynchronized conditions, the suggested connection between clock genes and metabolic genes is lost.

9. THE LINK BETWEEN CLOCK GENES AND METABOLIC GENES

Circadian disruption induced by night and shift work leads in the long term to overweight, increased abdominal fat deposition and development of indicators of metabolic syndrome.[88,89] This and the observation that the clock gene expression can be driven by food[4] have motivated many scientists to search for the relationship between clock genes and metabolic processes within the cell. Using a rat model of shift work, we have examined to what extent food intake during the sleep phase could be a factor contributing to the symptoms of the metabolic syndrome that is associated with shift work.[90] Using a model of forced activity in slowly rotating drums during the rest phase, we demonstrated that this induces a circadian misalignment based on a spontaneous temporal shift of food intake, metabolic parameters, and a loss in blood glucose rhythm, while hormonal rhythms, predominantly dependent on the SCN, remained unchanged.[60] This could be concluded as shift-working animals that only had food access in their active phase did not develop this circadian misalignment.[59] In contrast, rats exposed to forced activity during the active phase do not show signs of circadian disruption. These observations suggested that the activity during the rest phase and especially the resulting inverted feeding patterns causes a loss of synchrony between the SCN and the liver which is, as we have seen, importantly driven by metabolic signals. Therefore, we investigated metabolic parameters, clock gene expression rhythms, and metabolic gene expression rhythms in this shift-work rat model. Also as we have seen above, NAMPT plays a critical role in a number of biological processes through the NAD^+-dependent deacetylase SIRT1.[73,91] Therefore, we focused in our shift-work model on the relationship between clock gene expression and the NAMPT–NAD–SIRT1 cascade, including the changes in Pparα, Pparγ, and Pgc1α. We hypothesized that an altered activity pattern combined with a shifted feeding pattern would lead to the loss of synchrony between clock genes and metabolic genes in the liver and especially affect this cascade.

Circadian misalignment was induced by placing rats in a slow-rotating wheel for 8 h during their rest phase; this was compared with rats placed in a slow-rotating wheel during their active phase and with undisturbed controls.[92] Another set of rats was exclusively exposed to restricted food intake in the rest period or in the active period. In agreement with our expectation, we observed that clock genes reversed their rhythm and became synchronized by food. Surprisingly, in spite of the data suggesting the direct production of NAMPT via CLOCK/BMAL1,[30,74] we observed in animals that work and eat during the inactive period that NAD^+ and *Nampt* did not follow the inversion of the rhythm in the core clock genes: they lost their rhythm together with the metabolic genes *Pgc-1α*, *Pparα*, and *Pparγ*. Similarly, animals that only eat during the rest phase also lost their rhythm in NAD^+ and *Nampt*, but *Pgc-1α*, *Pparα*, and *Pparγ* (Fig. 4.2) did not lose their rhythm but had a flattened rhythm following the food (Fig. 4.2). Such changes in animals eating and/or working during their active phase indicate that the change in food intake is the main stimulus that induced a desynchronization within the liver. In addition, the expression of *Nampt* in the liver in animals eating in their sleep phase is decreased, which agrees with the decrease of *Nampt* seen in *Clock* mutant mice[30] and might consequently also be related to the disturbed food intake of these *Clock* mutants. This study also showed that inhibition of *Nampt* is also associated with a decrease in *Per2* levels possibly via SIRT1 (Fig. 4.1). As Per2 binds to and affects the activity of many nuclear receptors such as PPARα, this may also explain why there is a chance in the rhythm in the PPARs. The coenzyme NAD^+ together with its rate-limiting biosynthetic enzyme NAMPT stimulates the production of SIRT1 which is a protein that coordinates metabolic programs like gluconeogenesis, glycolysis, and lipid metabolism[29] and

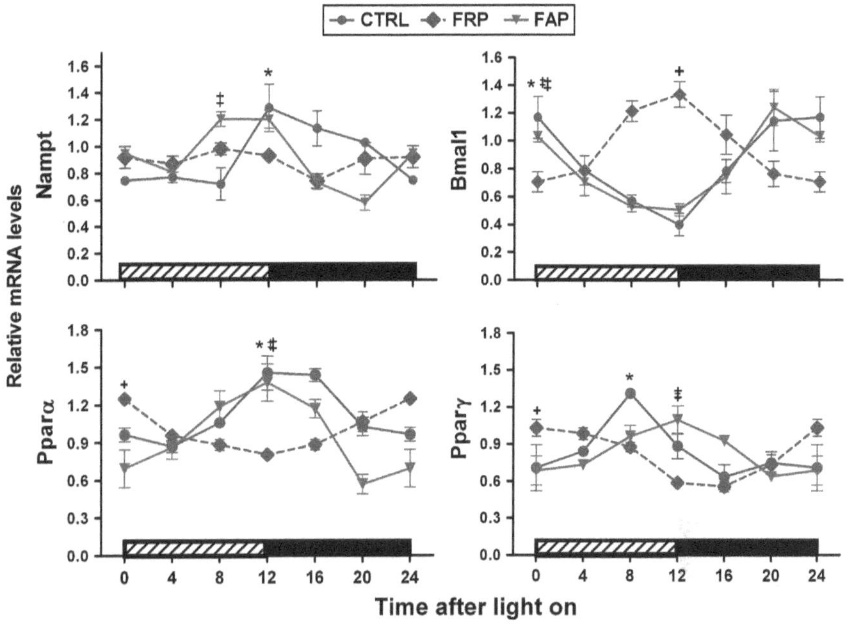

Figure 4.2 In a model with food during the rest period, the relationship between clock and metabolic genes changes. Animals with food *ad libitum* (CTRL) or during the active period (FAA) have clear rhythms in NAD, Pparα, and Pparγ that are in opposite to the rhythm of Bmal1. However, when food is restricted to the rest period (FRP), this relationship is lost for NAD or flattened for PPARα and PPARγ. See text and Salgado *et al.* (2013) for further details. (See Color Insert.)

has a proposed feedback role on clock genes by inhibiting *Clock* and *Bmal1*.[24] The observation of a simultaneous decrease in SIRT1, together with the decrease and loss of rhythm in *Per2* and *Nampt*, suggests an important disturbance of these three elements in the link between the core clock and metabolism, in circadian misaligned animals. The uncoupling of the rhythm of *Nampt* from that of *Clock* and *Bmal1* raises the question which other cellular messengers will form the link between food (i.e., glucose uptake), metabolic genes, and clock genes and demonstrate the necessity of using physiological models in order to understand and validate proposed interactions of the molecular networks in the cell. In this respect, it might be worth considering that peroxiredoxins, which are evolutionarily extremely conserved molecules associated with the oxidative state of the cell, show a rhythm in their oxidation level which, in turn, is associated with the metabolic cycle may influence the synthesis of NAMPT.[93]

The implications of these changes for liver metabolism are disastrous only 3 weeks in such working protocol induced an increase in triglyceride accumulation. These observation consonants with previous observations showing that rats without SIRT1 develop liver steatosis while mice that overexpress SIRT1 are protected against diet-induced liver steatosis.[29] The low levels of SIRT1 observed in rats eating in their inactive phase agree with their propensity to accumulate abdominal fat and with their significant overweight. Moreover, the decreased levels of *Pgc1α, Pparα, and Pparγ* that we observed in our model agree with the studies showing that SIRT1 stimulates the transcription of *Pgc1α, Pparα, and Pparγ*,[29,94,95] and therefore, we suggest that the consequence of low *SIRT1* and *Pgc-1α* together with the decrease of the PPAR's in ARP rats may lead to overweight and metabolic disturbances as observed, for example, in the *Sirt1* knockdown mice. Moreover in line with these observations, it has been shown that an increase of SIRT1 improves insulin sensitivity,[96] indicating that, indeed, the low levels of SIRT1 observed in the present study may reflect a state of lesser sensitivity for insulin. Therefore, the surprising outcome of our recent research was that, in spite of several studies that show that the nuclear receptor genes like *Pparα* and *Pparγ* are (in)direct targets of circadian clock genes,[97] this study illustrates that metabolic genes such as *Nampt*, *Sirt1* and *Pparα* may not always be linked with the core clock genes.

10. CONCLUSIONS

At present, the current literature suggests a tight coupling between food and the expression of clock genes and metabolic genes in organs like the liver. This review illustrates that indeed tight links exist between clock genes and metabolic genes. However, when the animal as a whole, and not organ tissues *in vitro*, is examined, many other network possibilities and elements play a role in determining how also at a cellular level the systems react to changes in food availability conditions. This is especially evident under so-called desynchronization conditions whereby food and activity are imposed at times when the animal normally rests. Such *in vivo* conditions show that when the network is in place and the SCN is able to influence also the cellular events in the liver but when the network because of the conditions is desynchronized, the coupling between clock genes and metabolic genes is lost showing that *in vivo* other relationships may prevail than *in vitro*. This suggests that these relationships between clock genes are not hardwired but can be malleable by the conditions. It is interesting that especially food has a powerful influence on the network, such that when it is taken outside the period dictated by the SCN, it has the potency to desynchronize the complete network.

ACKNOWLEDGMENTS

The authors gratefully acknowledge the support of PAPIIT Grant IN 209711 and Conacyt M 79797.

REFERENCES

1. Marcheva B, Ramsey KM, Buhr ED, et al. Disruption of the clock components CLOCK and BMAL1 leads to hypoinsulinaemia and diabetes. *Nature*. 2010;466:627–631.
2. Maywood ES, O'neill JS, Chesham JE, Hastings MH. The circadian clockwork of the suprachiasmatic nuclei: analysis of a cellular oscillator that drives endocrine rhythms. *Endocrinology*. 2007;148:5624–5634.
3. Stokkan KA, Yamazaki S, Tei H, Sakaki Y, Menaker M. Entrainment of the circadian clock in the liver by feeding. *Science*. 2001;291:490–493.
4. Damiola F, Le Minh N, Preitner N, Kornmann B, Fleury-Olela F, Schibler U. Restricted feeding uncouples circadian oscillators in peripheral tissues from the central pacemaker in the suprachiasmatic nucleus. *Genes Dev*. 2000;14:2950–2961.
5. Hara R, Wan K, Wakamatsu H, et al. Restricted feeding entrains liver clock without participation of the suprachiasmatic nucleus. *Genes Cells*. 2001;6:269–278.
6. Asher G, Schibler U. Crosstalk between components of circadian and metabolic cycles in mammals. *Cell Metab*. 2011;13:125–137.
7. Gillette MU, Reppert SM. The hypothalamic suprachiasmatic nuclei: circadian patterns of vasopressin secretion and neuronal activity *in vitro*. *Brain Res Bull*. 1987;19:135–139.
8. Welsh DK, Logothetis DE, Meister M, Reppert SM. Individual neurons dissociated from rat suprachiasmatic nucleus express independently phased circadian firing rhythms. *Neuron*. 1995;14:697–706.
9. Webb AB, Angelo N, Huettner JE, Herzog ED. Intrinsic, nondeterministic circadian rhythm generation in identified mammalian neurons. *Proc Natl Acad Sci USA*. 2009;106:16493–16498.
10. Harmar AJ, Marston HM, Shen S, et al. The VPAC(2) receptor is essential for circadian function in the mouse suprachiasmatic nuclei. *Cell*. 2002;109:497–508.
11. Brown T, Colwell CS, Waschek J, Piggins HD. Disrupted neuronal activity rhythms in the suprachiasmatic nuclei of vasoactive intestinal polypeptide-deficient mice. *J Neurophysiol*. 2006;97(3):2553–2558.
12. Buijs RM, Van Eden CG, Goncharuk VD, Kalsbeek A. The biological clock tunes the organs of the body: timing by hormones and the autonomic nervous system. *J Endocrinol*. 2003;177:17–26.
13. Hastings MH, Reddy AB, Maywood ES. A clockwork web: circadian timing in brain and periphery, in health and disease. *Nat Rev Neurosci*. 2003;4:649–661.
14. Buijs RM, Kalsbeek A. Hypothalamic integration of central and peripheral clocks. *Nat Rev Neurosci*. 2001;2:521–526.
15. Oishi K, Sakamoto K, Okada T, Nagase T, Ishida N. Antiphase circadian expression between BMAL1 and period homologue mRNA in the suprachiasmatic nucleus and peripheral tissues of rats. *Biochem Biophys Res Commun*. 1998;253:199–203.

16. Zylka MJ, Shearman LP, Weaver DR, Reppert SM. Three period homologs in mammals: differential light responses in the suprachiasmatic circadian clock and oscillating transcripts outside of brain. *Neuron.* 1998;20:1103–1110.
17. Kalsbeek A, Ruiter M, La Fleur SE, Van Heijningen C, Buijs RM. The diurnal modulation of hormonal responses in the rat varies with different stimuli. *J Neuroendocrinol.* 2003;15:1144–1155.
18. Fu L, Patel MS, Bradley A, Wagner EF, Karsenty G. The molecular clock mediates leptin-regulated bone formation. *Cell.* 2005;122:803–815.
19. Matsuo T, Yamaguchi S, Mitsui S, Emi A, Shimoda F, Okamura H. Control mechanism of the circadian clock for timing of cell division in vivo. *Science.* 2003;302:255–259.
20. Rudic RD, Curtis AM, Cheng Y, FitzGerald G. Peripheral clocks and the regulation of cardiovascular and metabolic function. *Methods Enzymol.* 2005;393:524–539.
21. Paschos GK, FitzGerald GA. Circadian clocks and vascular function. *Circ Res.* 2010;106:833–841.
22. Doi M, Takahashi Y, Komatsu R, et al. Salt-sensitive hypertension in circadian clock-deficient Cry-null mice involves dysregulated adrenal Hsd3b6. *Nat Med.* 2010;16:67–74.
23. Canaple L, Rambaud J, Dkhissi-Benyahya O, et al. Reciprocal regulation of brain and muscle Arnt-like protein 1 and peroxisome proliferator-activated receptor alpha defines a novel positive feedback loop in the rodent liver circadian clock. *Mol Endocrinol.* 2006;20:1715–1727.
24. Nakahata Y, Sahar S, Astarita G, Kaluzova M, Sassone-Corsi P. Circadian control of the NAD+ salvage pathway by CLOCK-SIRT1. *Science.* 2009;324:654–657.
25. Lamia KA, Sachdeva UM, Ditacchio L, et al. AMPK regulates the circadian clock by cryptochrome phosphorylation and degradation. *Science.* 2009;326:437–440.
26. Albrecht U. Timing to perfection: the biology of central and peripheral circadian clocks. *Neuron.* 2012;74:246–260.
27. Semple RK, Chatterjee VK, O'Rahilly S. PPARgamma and human metabolic disease. *J Clin Invest.* 2006;116:581–589.
28. Liang F, Kume S, Koya D. SIRT1 and insulin resistance. *Nat Rev Endocrinol.* 2009;5:367–373.
29. Purushotham A, Schug TT, Xu Q, Surapureddi S, Guo X, Li X. Hepatocyte-specific deletion of SIRT1 alters fatty acid metabolism and results in hepatic steatosis and inflammation. *Cell Metab.* 2009;9:327–338.
30. Ramsey KM, Yoshino J, Brace CS, et al. Circadian clock feedback cycle through NAMPT-mediated NAD+ biosynthesis. *Science.* 2009;324:651–654.
31. Lamia KA, Storch KF, Weitz CJ. Physiological significance of a peripheral tissue circadian clock. *Proc Natl Acad Sci USA.* 2008;105:15172–15177.
32. Cailotto C, La Fleur SE, Van Heijningen C, et al. The suprachiasmatic nucleus controls the daily variation of plasma glucose via the autonomic output to the liver: are the clock genes involved? *Eur J Neurosci.* 2005;22:2531–2540.
33. Cailotto C, Van Heijningen C, Van der Vliet J, et al. Daily rhythms in metabolic liver enzymes and plasma glucose require a balance in the autonomic output to the liver. *Endocrinology.* 2007;149:1914–1925.
34. Guo H, Brewer JM, Champhekar A, Harris RB, Bittman EL. Differential control of peripheral circadian rhythms by suprachiasmatic-dependent neural signals. *Proc Natl Acad Sci USA.* 2005;102:3111–3116.
35. Ungar F, Halberg F. Circadian rhythm in the in vitro response of mouse adrenal to adrenocorticotropic hormone. *Science.* 1962;137:1058–1060.
36. Lemmer B, Witte K. Circadian rhythm of the in vitro stimulation of adenylate cyclase in rat heart tissue. *Eur J Pharmacol.* 1989;159:311–315.
37. Litynska A, Wojczyk B. Daily changes of protein synthesis and secretion rates in rat liver slices. *Int J Biochem.* 1986;18:569–571.

38. Jasper MS, Engeland WC. Splanchnic neural activity modulates ultradian and circadian rhythms in adrenocortical secretion in awake rats. *Neuroendocrinology*. 1994;59:97–109.
39. Buijs RM, Wortel J, Van Heerikhuize JJ, et al. Anatomical and functional demonstration of a multisynaptic suprachiasmatic nucleus adrenal (cortex) pathway. *Eur J Neurosci*. 1999;11:1535–1544.
40. Ishida A, Mutoh T, Ueyama T, et al. Light activates the adrenal gland: timing of gene expression and glucocorticoid release. *Cell Metab*. 2005;2:297–307.
41. Cailotto C, Lei J, van der Vliet J, et al. Effects of nocturnal light on (clock) gene expression in peripheral organs: a role for the autonomic innervation of the liver. *PLoS One*. 2009;4:e5650.
42. Terazono H, Mutoh T, Yamaguchi S, et al. Adrenergic regulation of clock gene expression in mouse liver. *Proc Natl Acad Sci USA*. 2003;100:6795–6800.
43. Balsalobre A, Brown SA, Marcacci L, et al. Resetting of circadian time in peripheral tissues by glucocorticoid signaling. *Science*. 2000;289:2344–2347.
44. Balsalobre A, Damiola F, Schibler U. A serum shock induces circadian gene expression in mammalian tissue culture cells. *Cell*. 1998;93:929–937.
45. Brown S, Zumbrunn G, Fleury-Olela F, Preitner N, Schibler U. Rhythms of mammalian body temperature can sustain peripheral circadian clocks. *Curr Biol*. 2002;12:1574.
46. Morf J, Rey G, Schneider K, et al. Cold-inducible RNA-binding protein modulates circadian gene expression posttranscriptionally. *Science*. 2012;338:379–383.
47. Cettour-Rose P, Samec S, Russell AP, et al. Redistribution of glucose from skeletal muscle to adipose tissue during catch-up fat: a link between catch-up growth and later metabolic syndrome. *Diabetes*. 2005;54:751–756.
48. Uno K, Katagiri H, Yamada T, et al. Neuronal pathway from the liver modulates energy expenditure and systemic insulin sensitivity. *Science*. 2006;312:1656–1659.
49. Samec S, Seydoux J, Dulloo AG. Interorgan signaling between adipose tissue metabolism and skeletal muscle uncoupling protein homologs: is there a role for circulating free fatty acids? *Diabetes*. 1998;47:1693–1698.
50. Veniant MM, Hale C, Helmering J, et al. FGF21 promotes metabolic homeostasis via white adipose and leptin in mice. *PLoS One*. 2012;7:e40164.
51. Kornmann B, Schaad O, Bujard H, Takahashi JS, Schibler U. System-driven and oscillator-dependent circadian transcription in mice with a conditionally active liver clock. *PLoS Biol*. 2007;5:e34.
52. Hughes ME, Hong HK, Chong JL, et al. Brain-specific rescue of Clock reveals system-driven transcriptional rhythms in peripheral tissue. *PLoS Genet*. 2012;8:e1002835.
53. Wu T, Jin Y, Ni Y, Zhang D, Kato H, Fu Z. Effects of light cues on re-entrainment of the food-dominated peripheral clocks in mammals. *Gene*. 2008;419:27–34.
54. Kalsbeek A, Barassin S, Van Heerikhuize JJ, Van der Vliet J, Buijs RM. Restricted daytime feeding attenuates reentrainment of the circadian melatonin rhythm after an 8-h phase advance of the light-dark cycle. *J Biol Rhythms*. 2000;15:57–66.
55. Kalsbeek A, Van Heerikhuize JJ, Wortel J, Buijs RM. Restricted daytime feeding modifies suprachiasmatic nucleus vasopressin release in rats. *J Biol Rhythms*. 1998;13:18–29.
56. Aceves C, Escobar C, Rojas-Huidobro R, et al. Liver 5'-deiodinase activity is modified in rats under restricted feeding schedules: evidence for post-translational regulation. *J Endocrinol*. 2003;179:91–96.
57. Escobar C, Diaz-Munoz M, Encinas F, Aguilar-Roblero R. Persistence of metabolic rhythmicity during fasting and its entrainment by restricted feeding schedules in rats. *Am J Physiol*. 1998;274:R1309–R1316.
58. Krieger DT. Food and water restriction shifts corticosterone, temperature, activity and brain amine periodicity. *Endocrinology*. 1974;95:1195–1201.

59. Salgado-Delgado R, Angeles-Castellanos M, Saderi N, Buijs RM, Escobar C. Food intake during the normal activity phase prevents obesity and circadian desynchrony in a rat model of night work. *Endocrinology*. 2010;151:1019–1029.
60. Salgado-Delgado R, Angeles-Castellanos M, Buijs MR, Escobar C. Internal desynchronization in a model of night-work by forced activity in rats. *Neuroscience*. 2008;154:922–931.
61. Le Minh N, Damiola F, Tronche F, Schutz G, Schibler U. Glucocorticoid hormones inhibit food-induced phase-shifting of peripheral circadian oscillators. *EMBO J*. 2001;20:7128–7136.
62. Kiessling S, Eichele G, Oster H. Adrenal glucocorticoids have a key role in circadian resynchronization in a mouse model of jet lag. *J Clin Invest*. 2010;120:2600–2609.
63. Angeles-Castellanos M, Amaya JM, Salgado-Delgado R, Buijs RM, Escobar C. Scheduled food hastens re-entrainment more than melatonin does after a 6-h phase advance of the light-dark cycle in rats. *J Biol Rhythms*. 2011;26:324–334.
64. Yamamoto T, Nakahata Y, Tanaka M, et al. Acute physical stress elevates mouse period1 mRNA expression in mouse peripheral tissues via a glucocorticoid-responsive element. *J Biol Chem*. 2005;280:42036–42043.
65. La Fleur SE, Kalsbeek A, Wortel J, Fekkes ML, Buijs RM. A daily rhythm in glucose tolerance: a role for the suprachiasmatic nucleus. *Diabetes*. 2001;50:1237–1243.
66. Hirota T, Okano T, Kokame K, Shirotani-Ikejima H, Miyata T, Fukada Y. Glucose down-regulates Per1 and Per2 mRNA levels and induces circadian gene expression in cultured Rat-1 fibroblasts. *J Biol Chem*. 2002;277:44244–44251.
67. Oike H, Nagai K, Fukushima T, Ishida N, Kobori M. Feeding cues and injected nutrients induce acute expression of multiple clock genes in the mouse liver. *PLoS One*. 2011;6:e23709.
68. Davidson AJ, Poole AS, Yamazaki S, Menaker M. Is the food-entrainable circadian oscillator in the digestive system? *Genes Brain Behav*. 2003;2:32–39.
69. Vollmers C, Gill S, Ditacchio L, Pulivarthy SR, Le HD, Panda S. Time of feeding and the intrinsic circadian clock drive rhythms in hepatic gene expression. *Proc Natl Acad Sci USA*. 2009;106:21453–21458.
70. Horikawa K, Minami Y, Iijima M, Akiyama M, Shibata S. Rapid damping of food-entrained circadian rhythm of clock gene expression in clock-defective peripheral tissues under fasting conditions. *Neuroscience*. 2005;134:335–343.
71. Satoh Y, Kawai H, Kudo N, Kawashima Y, Mitsumoto A. Time-restricted feeding entrains daily rhythms of energy metabolism in mice. *Am J Physiol Regul Integr Comp Physiol*. 2006;290:R1276–R1283.
72. Rutter J, Reick M, Wu LC, McKnight SL. Regulation of clock and NPAS2 DNA binding by the redox state of NAD cofactors. *Science*. 2001;293:510–514.
73. Revollo JR, Korner A, Mills KF, et al. Nampt/PBEF/Visfatin regulates insulin secretion in beta cells as a systemic NAD biosynthetic enzyme. *Cell Metab*. 2007;6:363–375.
74. Asher G, Gatfield D, Stratmann M, et al. SIRT1 regulates circadian clock gene expression through PER2 deacetylation. *Cell*. 2008;134:317–328.
75. Belden WJ, Dunlap JC. SIRT1 is a circadian deacetylase for core clock components. *Cell*. 2008;134:212–214.
76. Diaz-Munoz M, Vazquez-Martinez O, Aguilar-Roblero R, Escobar C. Anticipatory changes in liver metabolism and entrainment of insulin, glucagon, and corticosterone in food-restricted rats. *Am J Physiol Regul Integr Comp Physiol*. 2000;279:R2048–R2056.
77. Cardaci S, Filomeni G, Ciriolo MR. Redox implications of AMPK-mediated signal transduction beyond energetic clues. *J Cell Sci*. 2012;125:2115–2125.
78. Carling D, Thornton C, Woods A, Sanders MJ. AMP-activated protein kinase: new regulation, new roles? *Biochem J*. 2012;445:11–27.

79. Hardie DG, Ross FA, Hawley SA. AMPK: a nutrient and energy sensor that maintains energy homeostasis. *Nat Rev Mol Cell Biol.* 2012;13:251–262.
80. Um JH, Pendergast JS, Springer DA, et al. AMPK regulates circadian rhythms in a tissue- and isoform-specific manner. *PLoS One.* 2011;6:e18450.
81. Canto C, Auwerx J. Targeting sirtuin 1 to improve metabolism: all you need is NAD(+)? *Pharmacol Rev.* 2012;64:166–187.
82. Lan F, Cacicedo JM, Ruderman N, Ido Y. SIRT1 modulation of the acetylation status, cytosolic localization, and activity of LKB1. Possible role in AMP-activated protein kinase activation. *J Biol Chem.* 2008;283:27628–27635.
83. Khan AA, Quigley JG. Control of intracellular heme levels: heme transporters and heme oxygenases. *Biochim Biophys Acta.* 1813;2011:668–682.
84. Estall JL, Ruas JL, Choi CS, et al. PGC-1alpha negatively regulates hepatic FGF21 expression by modulating the heme/Rev-Erb(alpha) axis. *Proc Natl Acad Sci USA.* 2009;106:22510–22515.
85. Raghuram S, Stayrook KR, Huang P, et al. Identification of heme as the ligand for the orphan nuclear receptors REV-ERBalpha and REV-ERBbeta. *Nat Struct Mol Biol.* 2007;14:1207–1213.
86. Dioum EM, Rutter J, Tuckerman JR, Gonzalez G, Gilles-Gonzalez MA, McKnight SL. NPAS2: a gas-responsive transcription factor. *Science.* 2002;298:2385–2387.
87. Bonkowsky HL, Sinclair PR, Sinclair JF. Hepatic heme metabolism and its control. *Yale J Biol Med.* 1979;52:13–37.
88. Tenkanen L, Sjoblom T, Harma M. Joint effect of shift work and adverse life-style factors on the risk of coronary heart disease. *Scand J Work Environ Health.* 1998;24:351–357.
89. Karlsson BH, Knutsson AK, Lindahl BO, Alfredsson LS. Metabolic disturbances in male workers with rotating three-shift work. Results of the WOLF study. *Int Arch Occup Environ Health.* 2003;76:424–430.
90. Ha M, Park J. Shiftwork and metabolic risk factors of cardiovascular disease. *J Occup Health.* 2005;47:89–95.
91. Revollo JR, Grimm AA, Imai S. The NAD biosynthesis pathway mediated by nicotinamide phosphoribosyltransferase regulates Sir2 activity in mammalian cells. *J Biol Chem.* 2004;279:50754–50763.
92. Salgado-Delgado RC, Saderi N, Basualdo MC, Guerrero-Vargas NN, Escobar C, Buijs RM. Shift work or food intake during the rest phase promotes metabolic disruption and desynchrony of liver genes in male rats. *PLoS One.* 2013;8:e60052.
93. Edgar RS, Green EW, Zhao Y, et al. Peroxiredoxins are conserved markers of circadian rhythms. *Nature.* 2012;485:459–464.
94. Picard F, Kurtev M, Chung N, et al. Sirt1 promotes fat mobilization in white adipocytes by repressing PPAR-gamma. *Nature.* 2004;429:771–776.
95. Nemoto S, Fergusson MM, Finkel T. SIRT1 functionally interacts with the metabolic regulator and transcriptional coactivator PGC-1{alpha}. *J Biol Chem.* 2005;280:16456–16460.
96. Sun C, Zhang F, Ge X, et al. SIRT1 improves insulin sensitivity under insulin-resistant conditions by repressing PTP1B. *Cell Metab.* 2007;6:307–319.
97. Grimaldi B, Bellet MM, Katada S, et al. PER2 controls lipid metabolism by direct regulation of PPARgamma. *Cell Metab.* 2010;12:509–520.

CHAPTER FIVE

Circadian Clocks, Food Intake, and Metabolism

Etienne Challet
Neurobiology of Rhythms, Institute of Cellular and Integrative Neurosciences, CNRS UPR3212 Associated with University of Strasbourg, Strasbourg, France

Contents

Abstract

Circadian rhythmicity that has been shaped by evolution over millions of years generates an internal timing controlling the sleep–wake and metabolism cycles. The daily variations between sleep/fasting/catabolism and wakefulness/feeding/anabolism are coordinated by a master hypothalamic clock, mainly reset by ambient light. Secondary clocks, including liver and adipose tissue, are normally synchronized by the master clock, but they are also sensitive to feeding time, especially when meals take place during the usual resting period. Cellular metabolism and circadian clocks are tightly interconnected at the molecular levels. Although the suprachiasmatic clock is not shifted by mealtime under light–dark conditions, nutritional cues can feedback onto it and modulate its function under hypo- and hypercaloric (high-fat) conditions. Food-related reward cues

Progress in Molecular Biology and Translational Science, Volume 119
ISSN 1877-1173
http://dx.doi.org/10.1016/B978-0-12-396971-2.00005-1

are other modulators of the master clock. Circadian disturbances (e.g., desynchronization induced by shift work or chronic jet lag) are frequently associated with metabolic dysfunctions (chronobesity) and vice versa. Pharmacological tools and natural synchronizers (i.e., light and mealtime) can be useful as chronotherapeutic treatments to limit the occurrence of metabolic risk factors.

1. INTRODUCTION

Energy metabolism, food intake, and circadian clocks are tightly interconnected. By providing energy substrates to the organism, feeding is essential for maintaining energy homeostasis. Most often, this does not occur randomly at any time of the astronomical day, but takes place periodically during a certain temporal niche (e.g., daytime or nighttime), depending on whether the species is diurnal or nocturnal, respectively. The daily period of feeding and food foraging also coincides with the period of wakefulness, exercise, high metabolic activity, and anabolism. Conversely, the daily period of fasting corresponds to sleep, low metabolic activity, and catabolism. At a cellular level, glucose availability is maintained with a quite narrow margin of variations throughout 24 h, despite the daily rhythm of food ingestion reported above. The two main sources of energy stores include carbohydrate (i.e., glycogen synthesized in the liver and muscle) and lipid (i.e., triacylglycerols synthesized in the white adipose tissue). During the 12-h period of activity/feeding, glucose supply comes mostly from dietary carbohydrate supply, as well as from glycogen for short-term needs (e.g., exercise). By contrast, during the 12-h period of sleep and fasting (glycogenolysis and lipolysis), energy used to cover basal energy expenditure comes from energy substrates stored in anticipation during the previous period of feeding (concomitant with glycogenesis and lipogenesis).

The 24-h temporal segregation of physiology and behavior is controlled by the circadian system. This timing system is actually comprised of a network of endogenous circadian clocks that generate, via their local or distributed outputs, an internal rhythmicity close to 24 h. At the top of the circadian system is a master clock located in the suprachiasmatic nuclei (SCN) of the hypothalamus. Most circadian rhythms in behavior (e.g., sleep–wake cycle) and physiology (e.g., hormonal rhythms, like pineal melatonin, or adrenal glucocorticoids) are controlled by this hypothalamic structure.[1–3] Besides, almost all peripheral tissues, such as liver, muscle, and skin, express circadian oscillations of the molecular actors of the clockwork, called clock genes.[4,5] The mammalian clock machinery generates autoregulatory transcriptional loops, leading to rhythmic expression of clock genes and clock-controlled genes (i.e., downstream targets of the clockwork). In the core of these oscillatory mechanisms are transcription factors, BMAL1 (Brain and Muscle Aryl hydrocarbon receptor nuclear translocator-Like protein 1) and CLOCK (Circadian Locomotor Output Cycles Kaput), or its analog NPAS2 (Neuronal PAS domain protein 2), that dimerize together to activate the transcription of other clock genes, including three Period (*Per1–3*) and two *Cryptochrome* (*Cry1–2*) genes via E-box sequences in their promoter, thus defining a main positive loop.[4,6] The PER and CRY proteins then form complexes that are translocated in the nucleus where they inhibit their own CLOCK (NPAS2)/BMAL1-induced transactivation, defining a main negative loop. There are also reinforcing loops comprising other transcription factors, REV-ERB (*Reverse Viral Erythroblastosis oncogene products*) $\alpha-\beta$, and ROR (*Retinoic acid-related Orphan Receptors*) $\alpha-\beta-\gamma$ that modulate the transcription of *Bmal1*, *Npas2*, and *Clock* via retinoic acid-related orphan receptor response elements. Furthermore, PER/CRY repressor complexes are inactivated via ubiquitination and proteasome degradation by F-box proteins 3 and 21 for the CRYs,[7,8] and β-transducin repeat containing proteins 1 and 2 for the PERs.[9] Besides its role as a transcriptional activator, CLOCK is also a histone acetyltransferase that drives the cyclic acetylation of various targets, including BMAL1.[10] The internal coordination of circadian rhythmicity is structured as a multistep network, in which the master suprachiasmatic clock is a conductor that provides temporal signals to the secondary clocks/oscillators in the brain and peripheral organs via nervous and endocrine messages.[3,11,12] Light perceived by the retina, that contains its own clock, is widely recognized as the most potent synchronizer of the master clock. Nevertheless, several cues associated with feeding (and fasting) also impact circadian functioning at different steps of the circadian system.

The first purpose of this chapter is to provide an overview of the complex physiological interactions between feeding–fasting cycles and the various clocks/oscillators, including feedback effects of nutritional cues on the circadian clocks that control feeding rhythmicity (Fig. 5.1). Another issue that will be covered is the reciprocal disturbances between circadian rhythmicity and metabolic pathologies of energy metabolism. Finally, emerging chronotherapeutic approaches in the field of dieting and prevention of metabolic risks will be briefly introduced.

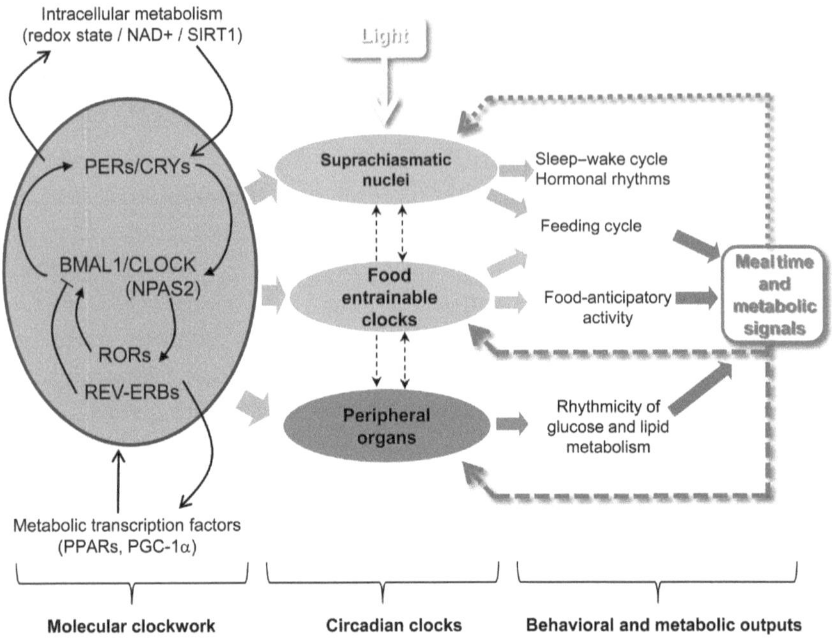

Figure 5.1 The multilevel interactions between circadian clocks and metabolism in mammals. The molecular clock is present in virtually all tissues. The master clock in the suprachiasmatic nuclei and other secondary clock in the brain adjust the phase of behavioral rhythms (sleep–wake and feeding cycles, food-anticipatory activity), while peripheral clocks/oscillators participate in the rhythms of metabolic processes (e.g., glucose tolerance, insulin sensitivity, fatty acid oxidation, fat storage). In turn, mealtime cues and metabolic signals feedback on circadian clocks to modulate their oscillations. (See Color Insert.)

2. PHYSIOLOGY

2.1. Daily rhythm of food intake

Homeostasis of food intake and energy metabolism is regulated by the brain, via a network of cerebral nuclei located in the basal hypothalamus (e.g., arcuate and ventromedial nuclei) and brainstem (e.g., nucleus of the solitary tract and parabrachial nucleus). These structures receive and detect blood-borne metabolic signals from peripheral tissues, such as metabolites (glucose, nonesterified fatty acids) and metabolic hormones (stomach ghrelin, fat leptin and pancreatic insulin, and intestinal gluco-incretins).[13] Although feeding-sensitive variations in clock gene expression have been detected in these neural structures,[14–16] their functional implication in the daily rhythm of ingestive behaviors remains to be established.

Bilateral lesions of the SCN as well as knife-cuts around them abolish circadian rhythmicity of the feeding/fasting cycle, indicating that the suprachiasmatic clock controls the circadian rhythm of feeding.[17,18] As aforementioned, because suprachiasmatic lesions also produce arrhythmicity of sleep–wake cycle, the concomitant loss of feeding rhythm may partly be an indirect consequence of behavioral arrhythmicity (in that view, destructured sleep timing would be the direct cause of the destructured pattern of food intake). There are experimental arguments outlined below, suggesting that a feeding-entrainable system outside the SCN may also participate in the daily rhythm of food foraging/intake (Fig. 5.1).

Timing of food availability has a major impact on overt rhythmicity. When food access is limited to a few hours every day at the same time (temporal restricted feeding), animals display food-anticipatory activity, that is, a bout of arousal accompanied with food-appetitive behaviors and physical activity prior to the expected food presentation.[19] Such a rhythmic behavior anticipating food presentation on a daily basis is manifest not only in food-restricted adult animals but also in pups nursed daily by the mother shortly once a day.[20,21] Other physiological parameters such as body temperature and corticosterone release also rise before food presentation, in phase with anticipatory behavior.[19,22] In rodents arrhythmic after suprachiasmatic lesions, temporal restricted feeding provides timing cues to the rest of the circadian system, thus restoring behavioral rhythmicity via daily food-anticipatory activity, hormonal rhythmicity, and/or sympathetic activation.[22,23] There are food-entrainable clocks throughout the brain that likely define a multi-oscillatory network coupling several neural oscillators most sensitive to feeding cues. Albeit the anatomic brain substrate that initiates food-anticipatory behavioral activity has been difficult to ascertain, possibly due to its distributed nature, experimental data favor the participation of some structures in the metabolic hypothalamus, the brainstem, and cerebellum.[19,24–26]

Alterations in the diurnal pattern of feeding have been detected in mice with functionally impaired clock genes. The daily pattern of food intake under a light–dark cycle is markedly attenuated in *Clock* mutant mice and in mice with adipocyte-specific deletion of *Bmal1*, food intake during daytime being found greater than that in wild-type mice.[27,28] Other mutations of clock genes, such *Rev-erbα*, do not impair daily pattern of feeding.[29] In mammals, a major modulator of the ultradian meal pattern during the feeding period is the size of the preceding meal. Indeed, larger meals lead to long intervals until the initiation of the next meal.[30] It is not known yet with certainty whether these ultradian patterns of intermeal lags are disturbed in clock mutant or mice knockout for clock genes.

Furthermore, in addition to the circadian and homeostatic control, food intake can be influenced by the ambient lighting conditions, defining direct effects of light or "masking." In nocturnal mammals, light exposure at night acutely reduces food intake during the active period (i.e., negative masking of feeding), while dark exposure during daytime enhances food intake during the usual sleep/fasting period (i.e., positive masking of feeding).[31]

2.2. Daily variations in energy metabolism

Mammals maintain a relatively high metabolic rate with narrow daily variations, albeit they do not feed continuously. Nonetheless, it is possible to measure with indirect calorimetry daily oscillations in energy expenditure (via oxygen consumption) and in respiratory exchange ratio, also called respiratory quotient (RQ; i.e., the ratio of carbon dioxide produced and oxygen consumed), which is an indicator of metabolized fuels. Lesions of the suprachiasmatic clock suppress circadian rhythmicity of energy expenditure as well as RQ.[32] This observation has been interpreted as meaning that the master clock controls the daily variations in energy metabolism. Alternatively, this arrhythmicity may result from the arrhythmic sleep–wake cycle that would prevent the detection of circadian variations.

When mammals are food deprived for several days under a light–dark cycle, daily rhythms of energy metabolites persist, thus demonstrating that the metabolic rhythmicity does not rely solely on daily feeding–fasting cycles.[33] Temporal restricted feeding can shift the daily rhythms of energy metabolites and RQ as well.[33,34]

Per2-null mice have been found to be leaner due to increased energy expenditure, while 24-h RQ values do not differ from wild-type littermates.[35] *Clock* mutant mice are less active during nighttime compared to wild-type mice and, accordingly, display a reduction in nocturnal energy expenditure.[27] In mice with adipocyte-specific deletion of *Bmal1*, the diurnal rhythm of energy expenditure is dampened, due to reduced expenditure at night without modification of the level of nocturnal activity.[28] The daily variations of the RQ show large interindividual differences in *Cry1−/−; Cry2−/−* mice, leading to flattened 24-h average.[36] A mutation of *Rev-erbα* leads to an altered daily rhythm of *in vivo* carbohydrate/lipid utilization, as highlighted by larger and lower RQ values during night and day, respectively. Thus, lack of REV-ERBα causes not only an increased utilization of fatty acids during both resting (daytime) and acute fasting but also an enhanced nocturnal glucose utilization (used for *de novo* lipogenesis from dietary carbohydrates), as well as diet-induced obesity. Thus, these findings indicate that REV-ERBα is crucial for the daily variations of fuel utilization.[29] Notwithstanding, the molecular mechanisms underlying this temporal partitioning of fuel utilization remain to be further elucidated.

Glucose, the main source of energy for cells, comes from the liver or from dietary carbohydrate via the intestine and circulates easily in the bloodstream. In most tissues, cellular uptake of glucose via facilitated diffusion is controlled by insulin, except for the liver and the brain. Another major fuel source is fat. As hydrophobic molecules, lipids cannot circulate readily in the aqueous blood. As a matter of fact, lipids are transported as particles associating them with hydrophilic molecules, called apolipoproteins. The expression of apolipoproteins is activated and repressed by the circadian factors RORα and REV-ERBβ, respectively.[37–39] The lipid particles in the plasma include chylomicrons, very low-density lipoproteins (VLDL), low-density lipoproteins, and high-density lipoproteins.

Lipoprotein lipase (LPL) is an enzyme that catalyses the hydrolysis of the triacylglycerols in chylomicrons and VLDL, releasing nonesterified fatty acids and facilitating their cellular uptake.[40] *Lpl* is strongly expressed in tissues that store lipids (i.e., white adipose tissue) or use them as fuel (i.e., muscles and brown fat). Of note, LPL activity displays marked daily variations according to organs, but with opposite timing depending on their role: LPL activity in epididymal fat of nocturnal rats is highest at night for lipogenesis, while it rises progressively from morning to early night in skeletal muscle for fat oxidation.[41] LPL expression and activity are modulated by a number of factors, such as PPARs, insulin, glucose, and fatty acids.[42] Furthermore, CLOCK alone or together with BMAL1 can also transactivate *Lpl* expression.[29,40]

Heme, being notably an iron-containing compound embedded to the hemoglobin, is a crucial component of the intermediary metabolism. Moreover, circadian rhythmicity can be modulated by heme bioavailability, illustrating a systemic link between metabolic and circadian pathways. A decrease in heme biosynthesis shortens the circadian period of *in vivo* body temperature rhythm.[43] Also, heme and inhibitors of heme oxygenase dose-dependently dampen circadian oscillations rhythms of SCN explants from PER2:LUC mice, while pharmacological inhibition of heme synthesis lengthens the circadian period of SCN PER2:LUC rhythms.[44] At a molecular level, heme was first shown to bind to NPAS2, thereby modulating DNA binding of NPAS2/BMAL1 in response to the presence of carbon monoxide.[45] Further work involved PER2 in heme effects on the molecular clock, but this is still a subject of debate.[46,47] Heme can also serve as a physiological ligand for REV-ERBα and REV-ERBβ to modulate their transcriptional efficiency.[48,49] REV-ERBs are also responsive to carbon monoxide and nitric oxide.[50,51] Other examples of intracellular interactions between the metabolic and circadian systems will be mentioned below.

2.3. Cross talk between molecular clocks and intracellular metabolic pathways

Peroxiredoxins are ubiquitous antioxidant enzymes that detoxify reactive oxygen species, such as hydrogen peroxide. Reduction–oxidation (redox) cycles of peroxiredoxins define 24-h metabolic cycles that can work in the absence of the transcriptional/translational circadian clockwork, for instance in red blood cells.[52] A comparative analysis reveals that these 24-h redox cycles are conserved in all living organisms studied so far, including bacteria.[53] In mammalian nucleated cells, peroxiredoxin oscillations are influenced by the transcriptional/translational circadian clockwork, as shown by altered phase relationships in fibroblasts from *Cry1−/−; Cry2−/−* mice.[52]

Redox reactions are involved in multiple biological processes, including the molecular clockwork itself. For instance, the DNA-binding activity of CLOCK/BMAL1 and NPAS2/BMAL1 heterodimers *in silico* is enhanced by the reduced form (NDAH, from NAD^+) of nicotinamide adenine dinucleotide (NAD).[54] Intracellular levels of NAD^+ show circadian oscillations in fibroblasts from wild-type mice, but they are arrhythmic in fibroblasts sampled from *Clock* mutant or *Cry1−/−;Cry2−/−* mice.[55] Synthesis of NAD^+ is largely controlled by the enzyme nicotinamide phosphoribosyltransferase (NAMPT) whose circadian expression is regulated by CLOCK/BMAL1 heterodimers. In turn, NAMPT modulates the molecular clockwork of peripheral clocks (fibroblasts, hepatocytes), thus defining a new feedback loop.[55,56] Another example of interactions between clock mechanisms and redox reactions is given by the fact that transcription repression mediated by heme-bound REV-ERBs is sensitive to redox states.[50]

5′-Adenosine monophosphate-activated protein kinase (AMPK) is an enzyme that plays a key role in the cellular regulation of fatty acid and glucose metabolism aiming at keeping energy homeostasis via phosphorylation of a number of metabolic enzymes. AMPK is viewed as a cellular sensor of energy status because, beside AMP, it is also activated by many physiological stimuli, such as stress, food deprivation, acute exercise, or hormones (e.g., leptin). Some of AMPK targets are clock components. CRY1 is phosphorylated and destabilized by AMPK.[57] When phosphorylated by AMPK, casein kinase Iε degrades the clock protein PER2, thereby impacting circadian oscillating. Fibroblasts treated with metformin, an activator of AMPK, display a shortened circadian period. Furthermore, *in vivo* injections of metformin produce phase advances of clock gene oscillations in peripheral tissues.[58]

The NAD^+-dependent SIRT1 (Sirtuin 1) histone deacetylase is a redox sensor that has been involved in a multitude of processes related to cellular metabolism, stress, and senescence.[59] In particular, SIRT1 modulates the activity of the metabolic transcription factor PGC-1α (peroxisome proliferator-activated receptor (i.e., PPARγ) coactivator 1α),[60,61] also identified as a modifier of the transcription of clock genes, such as *Bmal1* and *Rev-erbα*, in part via coactivation of RORs.[62] Moreover, SIRT1 binds circadianly to CLOCK/BMAL1 heterodimers, thus adding another functional link between cellular energy state and the molecular clockwork.[63,64]

Strikingly, nuclear receptors recognized as circadian factors, that is, REV-ERBs and RORs, have also been reported having regulatory roles in metabolic function. A pioneer work discovered that REV-ERBα promotes and RORα inhibits *in vitro* adipocyte differentiation.[65,66] BMAL1, whose transcription is regulated by both REV-ERbs and RORs, was also considered to play a prominent role in adipogenesis.[67] More recently, PER2 has been shown to promote adipogenesis via PPARγ (see below).[35] A point that remains to be clarified is whether the effects reported above result from abnormal metabolic function due to noncircadian roles of these circadian factors, or from altered circadian control of metabolic pathways. Furthermore, it is noteworthy that metabolic factors activated by fatty acids, such as PPAR, are tightly linked bidirectionally to the molecular clockwork. Actually, *Rev-erbα* expression in the adipocytes and hepatocytes is induced respectively by PPARγ and PPARα.[65,68] In addition, PPARα in the liver activates the transcription of *Bmal1*, while daily expression of *Pparα* involves BMAL1.[69] Although not exhaustive, this brief overview of multiple and interconnected regulatory loops highlights the now established functional and molecular cross talk between the circadian and metabolic systems (Fig. 5.1).

2.4. Peripheral organs and most brain regions: Clocks entrainable by mealtime

Time when food is eaten has potent phase-resetting effects on the clockwork of all peripheral tissues studied so far, including liver, white adipose tissue, gastrointestinal tract, heart, lung, and kidney[70–72] (Fig. 5.2). The characteristics of the feeding–fasting cycles are critical because both food volume and interval of food deprivation matter to reset the liver oscillations.[73] However, the nature of the feeding-associated signals capable of resetting peripheral oscillators is not yet fully identified. Hormones, such as glucocorticoids and metabolites, like glucose, as well as nutrient sensors, like AMPK, are good candidates.[57,74,75] In the liver, transient upregulation of *Per2* and *Dec1* transcription is observed in the first hour after feeding.[76] Moreover, refeeding-induced insulin secretion leads not only to an upregulation of *Per2* expression but also to downregulated *Rev-erv*α mRNA hepatic levels.[77]

Timing of circadian oscillations is markedly modified by restricted feeding in many, but not all, cerebral regions out of the SCN. For example, daytime feeding modifies the phase of molecular oscillations in the

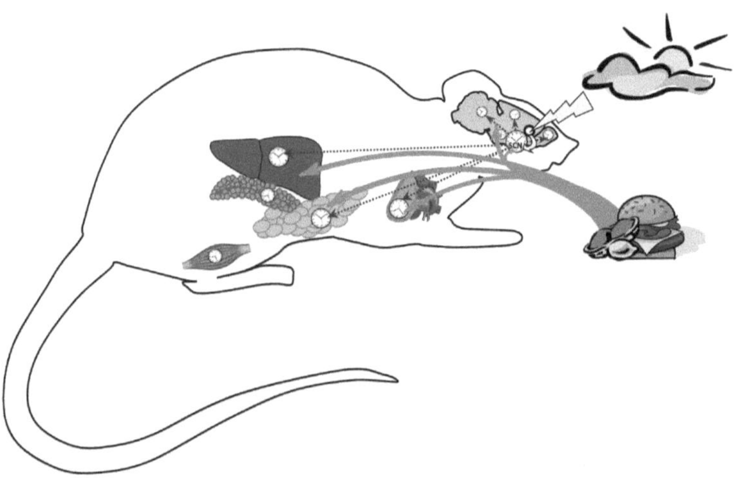

Figure 5.2 The suprachiasmatic nuclei (SCN) contain the master clock that controls sleep–wake cycle and hormonal rhythms. The SCN are the conductor of the many secondary clocks/oscillators in the brain and peripheral organs, in part via temporal messages transmitted by nervous pathways (dotted arrows). Light and feeding time act as synchronizers (filled arrows) at different levels of the multi-oscillatory circadian network. (See Color Insert.)

arcuate and dorsomedial hypothalamic nuclei,[15,78] central amygdala,[25,79] and cerebellum.[80,81] By contrast, circadian oscillations in few brain structures, such as the SCN and the hippocampus, are less sensitive to timed mealtime.[14] The cerebral oscillators or clocks that are sensitive to mealtime define a feeding-entrainable network, some of them being likely involved in the mechanisms of behavioral meal anticipation, as evoked above.

2.5. SCN: The master light-entrainable clock is sensitive to metabolic and reward cues

The mammalian clock machinery described in Section 1 is present in cells of the master suprachiasmatic clock. Compared to the rapid dampening of circadian oscillations in peripheral organs, the very robust self-sustained rhythmicity of suprachiasmatic cells, even when isolated *in vitro*, involves likely a strong intercellular coupling.[82]

Activity of mitochondrial cytochrome oxidase in suprachiasmatic cells is higher during daytime, while lactate dehydrogenase activity increases at night.[83,84] Because the daily variations of cytochrome oxidase activity are no longer detectable in constant darkness, this suggest that the photic inputs play a regulatory role in the rhythmicity of this metabolic process.[83]

The capacity of the suprachiasmatic cells to generate endogenous rhythmicity was initially demonstrated *ex vivo* and *in vitro* with a metabolic readout, namely, the circadian rhythm of 2-deoxyglucose uptake.[85] Of interest, cultured fibroblasts cannot generate such metabolic rhythmicity, despite synchronized clock gene oscillations. However, sustained oscillations of 2-deoxyglucose uptake can be triggered in fibroblasts when they are cocultured without physical contact with immortalized suprachiasmatic cells.[86]

Another powerful demonstration of self-sustained rhythmicity generated by the SCN came from the circadian rhythm of neuronal firing rate in suprachiasmatic slices kept *in vitro*.[87] Electrical properties of neurons are controlled by various regulatory mechanisms, such as conductances of voltage-gated ion channels. Inhibition of oxidative phosphorylation or glycolysis blocks the Na/K pump to depolarize resting potential and increase spontaneous firing in suprachiasmatic cells, thus indicating a metabolic modulation of the Na/K pump.[88] However, the connection between the molecular clock and the rhythmic electrical activity within the master clock remained elusive. It turns out that redox rhythmicity, by itself driven by the molecular clockwork as aforementioned, has a direct influence on the excitability of suprachiasmatic neurons via a modulation of K^+ conductance.[89]

Among stimuli capable of resetting the central clock, light is the most potent (Fig. 5.2). Light cues are first perceived by the retina that contains a circadian clock.[90] Photosensitive ganglion cells containing the photopigment melanopsin send fibers that project to the SCN, either directly via the retinohypothalamic tract or indirectly via the intergeniculate leaflet of the thalamus.[91,92] The way the SCN clock is synchronized to light is characterized by a photosensitive daily period (mainly at night) during which light cues can shift the clock, while the temporal window around midday defines a period during which light has no phase-resetting effect.[91] A newly identified modulator of photic resetting in the SCN is the metabolic transcription factor, PPARβ/δ.[93]

Under light–dark cycle, light can indirectly affect rhythmicity in peripheral organs through signals coming from the SCN via sympathetic projections. This is the case for plasma glucose that, in addition to its circadian control, can be increased by light exposure or stress.[94,95] Furthermore, as already evoked for food intake, the apparent daily sleep–wake cycle can also be modulated by direct, clock-independent responses to light, called "masking" in the circadian field.[96,97]

Daily rhythmicity of release of hormones in the bloodstream is the rule, rather than the exception. In nocturnal rats, plasma levels of both insulin and leptin increase during the early activity period (night).[98–100] Because suprachiasmatic lesions in rodents abolish this rhythmicity (i.e., hormonal rhythms become flattened, usually around the mean level), this reveals a control by the suprachiasmatic clock.[98,100] Studies in functional neuroanatomy have shown that the suprachiasmatic control of endocrine rhythmicity is largely mediated by the sympathetic innervation via hypothalamic relays (i.e., paraventricular and dorsomedial nuclei, and subparaventricular region) receiving vasopressinergic, glutamatergic, and GABAergic inputs from the master clock.[3]

In humans, hormonal rhythmicity is known to depend not only on circadian clocks but also on food intake, sleep, and light. To limit these so-called confounding effects from a strict circadian point of view, human subjects can be maintained for almost 2 days awake to avoid sleep-induced effects, fed hourly isocaloric meals to avoid synchronizing effects of daily mealtime, and in constant dim light to prevent direct or synchronizing effects of light (this experimental situation is the so-called condition of constant routine). Using this paradigm, plasma leptin and insulin show a peak around the minimal body temperature and close to the usual time of awakening, respectively.[101] Similarly, plasma fatty acids are also found to be under circadian control, with higher levels during subjective daytime.[102]

Contrary to the high sensitivity of peripheral clocks to the feeding cues, the suprachiasmatic clock of food-restricted animals under a light–dark cycle appears to be buffered against any synchronizing effect of mealtime, as shown with the lack of phase shift of either the circadian rhythm of firing rate or clock gene expression in the SCN.[70,103,104] This does not mean, however, that feeding-associated signals do not reach at all the suprachiasmatic cells. When the photic synchronizer is absent, that is in constant dark or constant light, timed meals can, although not systematically, entrain the suprachiasmatic clock.[19,22] Diurnal parenteral nutrition and *in vivo* glucose infusion produce shifts of clock gene oscillations in the master clock.[105]

Calorie restriction and starvation both lead to a major mobilization of energy stores and affect the timing of the sleep–wake cycle. When challenged with calorie restriction, nocturnal animals become active during their usual sleep period (i.e., they become partially diurnal), independently of the time of feeding.[106,107] Conversely, calorie-restricted diurnal rodents change their behavioral timing of activity to nighttime.[108] These modifications in behavioral timing in case of negative energy balance are due in part to the fact that metabolic cues associated with calorie restriction affect the suprachiasmatic clock machinery and its synchronization to light.[107,108] A ketogenic diet is another example of negative energy balance leading to body mass loss, lipid mobilization, and phase-advanced sleep–wake cycle.[109] Nocturnal mice, that have to work for getting food with increasing levels of workload over days, become also partially diurnal. Interestingly, the switch from nocturnal to diurnal pattern of activity coincides with a gradual shift toward a negative energy balance.[110] Whatever the cause, chronic hypocaloric conditions may ultimately change the cellular metabolic state of suprachiasmatic cells, therefore, altering the mechanisms of circadian oscillations. Alternatively or in combination, circulating metabolites (glucose, nonesterified fatty acids) and metabolic hormones may modulate photic resetting according to the metabolic status.

A daily palatable snack in addition to regular food (chow pellets) provided *ad libitum* is able to entrain behavioral rhythms of rats and mice in constant darkness conditions.[111,112] In mice, ingestion of the attractive and palatable snack activates both the reward and arousal systems in the brain, suggesting that the modulatory effects on the master clock involve somehow dopaminergic and orexinergic pathways.[112] The timing of suprachiasmatic clock and its synchronization to light can be modified by both the negative metabolic drive associated with hypocaloric feeding and the positive hedonic drive associated with palatable pellets. However, the direction of the modulation of light resetting is opposite: the amplitude of the circadian responses to light of the master clock are increased and reduced by hypocaloric and food-related reward cues, respectively. As discussed elsewhere, the orexinergic neurons in the lateral hypothalamus can integrate both kinds of feeding-related cues and may provide a main modulatory afferent pathway to the SCN.[26]

Together, these findings highlight the fact that the multi-oscillatory circadian network is involved in the daily variation of metabolism at all levels of the circadian system. Moreover, mealtime and other nutritional cues can act on the timing of various clocks (Fig. 5.1). This may explain why alterations in circadian timing have deleterious consequences on metabolic health, as detailed in the next part.

3. PATHOLOGY

3.1. Circadian disturbances are associated with metabolic dysfunctions

In most instances, circadian disturbances result from an altered endogenous clockwork or from altered exogenous timing cues. Mutations or KO of clock genes have been frequently linked with disturbances of metabolism. For instance, *Clock* mutant mice show increased adiposity, possibly due to the hypoactivity and hyperphagia reported above.[27] These mutant mice also display reduced gluconeogenesis, but enhanced insulin sensitivity.[113] Mice synthesizing nocturnal melatonin and having disrupted expression of *Clock* in the liver and skeletal muscles show lower glucose tolerance and signs of impaired glycolysis and gluconeogenesis.[114] *Cry1−/−;Cry2−/−* mice suffer from hypertension.[115] *Per2(Brdm1)* mutant mice have impaired glucose homeostasis, characterized with hypoglycemia and hyperinsulinemia relative to wild-type mice,[116] while *Per2−/−* mice have reduced lipid stores.[35] Knockout of *Rev-erbα* leads to increased adiposity and chronic hyperglycemia, despite the same daily energy intake of chow diet and daily level of motor activity as those in wild-type littermates.[29] A cardiomyocyte-specific *Clock* mutation in mice leads to increased fatty acid oxidation and decreased cardiac efficiency.[117] Mice bearing a liver-specific deletion of *Bmal1* show a mild hypoglycemia in the afternoon and increased glucose tolerance.[118] Mice with pancreas-specific deletion of *Bmal1* mutant mice are intolerant to glucose and have diminished insulin secretion[119] (Fig. 5.3).

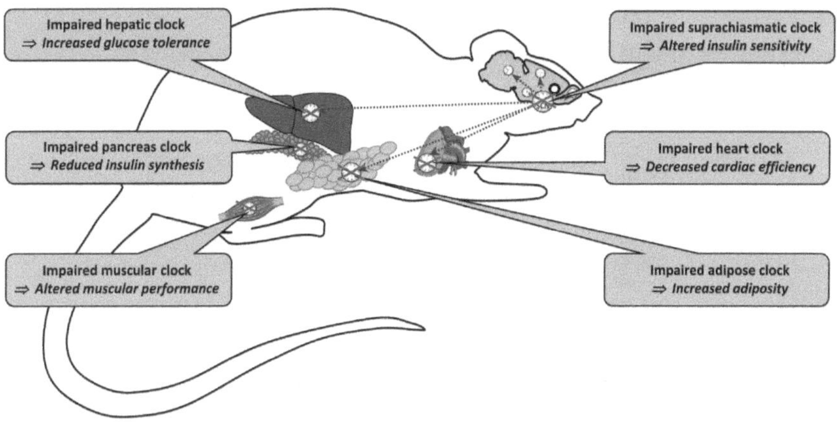

Figure 5.3 Examples of metabolic consequences resulting from impairments of specific circadian clocks in rodents. SCN, suprachiasmatic nuclei. (See Color Insert.)

In humans, polymorphisms in some clock genes have been correlated with metabolic risks. For instance, two *Bmal1* haplotypes are linked with type 2 diabetes and hypertension. Polymorphisms in *Clock* or *Rev-erbα* are significantly associated with high-fat mass.[120,121]

Light being the most potent synchronizer for the master clock, it is not surprising that changes in lighting conditions markedly affect circadian timing, and possibly energy metabolism. The photoperiod (i.e., the relative duration of the light period per 24 h) is perceived by the retina and integrated by the suprachiasmatic clock. Natural fluctuations in food intake, body mass, and adiposity occur recurrently according to seasons in so-photoperiodic species, whose physiology is specifically regulated on a seasonal basis.[122] In these cases, photoperiodic information is translated into neuroendocrine changes via the nocturnal secretion of melatonin by the pineal gland.[123] In rats, whose reproduction is not photoperiod-dependent, free access to sucrose and imposed high-fat feeding do not have the same obesogenic effects depending on the photoperiod. Exposure to short photoperiod (shorter day and longer night) compared to long photoperiod (longer day and shorter night) stimulates spontaneous ingestion of sucrose and promotes fat accretion, suggesting that the photoperiodic environment can modulate metabolic responses.[124] Another important parameter of lighting conditions is the light intensity during the periods of wake and sleep. For example, in spite of nondifferent levels of caloric intake and daily motor activity, mice exposed to dim light at night consume relatively more food during the light phase than mice exposed to regular light–dark cycle.[125] Also, mice exposed to constant light show larger gain in body gain and changes in daily insulin sensitivity, as compared to individuals kept under a light–dark cycle.[126]

Disruption in circadian rhythmicity can be triggered by light exposure at unusual times of the daily cycle. In humans, transmeridian fast travels (across more than two or three time zones) have become very common following the considerable development of air transport. Physiologically, they cause a transient loss of circadian synchronization, internal regulations being initially out of phase with respect to the new light–dark cycle. Then, day after day, the suprachiasmatic clock will resynchronize to the new cycle and impose an appropriate adjustment of peripheral clocks and oscillators to the local time. The transient period of resynchronisation, relatively proportional to the number of time zones crossed (but dependent on the east–west direction of travel), is accompanied by sleep quality problems, digestive disorders, and several metabolic and hormonal alterations.[127] In particular, carbohydrate oxidation was increased in human subjects exposed to 3 days of jet lag, while protein oxidation was decreased.[128]

Shift work and rotating work schedules trigger chronobiological conflicts between the endogenous clockwork and the ambient light environment as well as mealtime, leading to situations of altered internal temporal organization (e.g., between the master clock and peripheral oscillators or between different peripheral oscillators) and occurrences of desynchronization (misalignment between internal timing and local time). The deleterious effects of chronic desynchronization on metabolic health have been identified in animals. Rats undergoing desynchronization caused by a long-term biweekly change of the light–dark cycle are overweight or have impaired insulin secretion compared to animals maintained in a fixed light–dark cycle.[129,130] Moreover, rats forced to exercise during their usual sleep period show a reversed rhythm of triacylglycerols and increased gain in body mass.[131] Obesity and increased body mass index are commonly observed in large-scale epidemiological studies on night workers and workers with rotating schedules.[132–136] The obesogenic properties of repeated light–dark shifts in animals or chronic shift work in humans leads to the concept that we called "chronobesity."[137]

Mice exposed to light–dark cycles that are too short (i.e., 20 h) to allow daily photic resetting of their master clock show cardiovascular disease, larger gain in body mass, large insulin/glucose ratio, the latter indicating insulin resistance.[138,139] In animal studies, ultimate desynchrony (i.e., arrhythmicity) can be induced by complete suprachiasmatic destruction. Of note, mice rendered arrhythmic by suprachiasmatic lesions display hepatic insulin resistance.[140] The consequences of circadian misalignment between the endogenous clockwork and the sleep–wake and feeding–fasting cycles have been studied in humans, using a protocol of forced desynchrony. For that purpose, healthy subjects were exposed to seven recurring 28-h sleep–wake cycles in dim light; this paradigm triggered metabolic alterations in postprandial glucose responses evoking a prediabetic state.[141]

Mealtime being an efficient time giver for peripheral clocks, unusual times of meals in individuals exposed to a light–dark cycle will induce a state of internal desynchronization, the master clock being synchronized by light, while peripheral timing being phase adjusted by feeding times. Thus, it is consistent to find marked chronobiological effects when, regardless of the cause, food intake occurs at unusual hours compared to the normal cycle of sleep–wake cycle. In nocturnal rodents, the metabolic impact of eating chow pellets only during daytime is weak, as in most case, food-restricted rodents do not change their body mass. Different conclusions can be drawn in obese rodents (see below). In them, the spontaneous intake in late afternoon (end of resting period) seems to have the most detrimental effects on energy balance. In humans, the critical period is rather at the beginning of the resting period (early night). Large intake of calories for dinner is associated with increased body mass index.[142] Another study performed in the same girls between childhood and adolescence found that larger energy intake in evening/night meal of children was positively correlated with body mass index few years later.[143] It should be also noted that patients with night eating syndrome have a higher risk of developing obesity.[144] Whether the discrepancies in consequences of meal timing between rodents and humans rely on interspecies or nocturnal–diurnal differences remain to be established. Finally, in shift workers, a high intake at lunch has been identified as a particularly deleterious factor (i.e., it increases the risk of developing a cardiometabolic syndrome). Of note, besides the more fractionated pattern of energy intake, shift workers ingest usually more (+10%) saturated lipids than regular day workers.[134]

Sleep restriction in rats kept in a regular light–dark cycle alters glucose homeostasis (i.e., hyperglycemia and impaired glucose tolerance), while leading to body mass loss without significant change in energy intake.[145] In mice, repeated sleep deprivation during early daytime leads to some metabolic disruption, such as impaired gluconeogenesis.[146] In humans, sleep curtailment is increasing worldwide. Chronic partial sleep deprivation has been shown to have adverse effects on glucose metabolism, such as impairment of glucose tolerance and insulin sensitivity, both being major risk factors for type 2 diabetes, and sleep deprivation leads to an increase in hunger feeling.[147] Shorter periods of sleep will also increase the daily period available for eating.

Together, these data reveal that unusual timing of light exposure and/or meals in healthy individuals are major contributors of circadian misalignment, perturbing clock rhythmicity, and sleep homeostasis, whose alterations increase the likelihood of metabolic risk factors.

3.2. Metabolic pathologies are frequently associated with circadian disturbances

Now will be described some of the circadian abnormalities observed in genetic and experimental models of obesity and diabetes.

The Zucker rat is an animal model of genetic obesity, caused by a mutation (*fa*) in the gene encoding the receptor of leptin, an anorexigenic hormone synthesized by adipose tissue and acting notably on the metabolic hypothalamus. The *fa* mutation leads to hyperphagia and excessive adiposity. In addition, the Zucker rat maintained under a light–dark cycle displays alterations in daily timing, characterized by phase advances of feeding, locomotor activity and body temperature rhythms.[148,149] In these leptin-resistant rats, the amplitude of clock gene expression is dampened in the liver, but neither in the white adipose tissue nor in the SCN.[150] Mice carrying the *ob* mutation become obese because their adipocytes cannot synthesize leptin. These mice have an increased ultradian activity at the expense of the circadian pattern and an increased daytime activity, while the endogenous period is not affected.[151] The daily pattern of feeding is modified in *ob/ob* mice, with increased intake in the second half of the light period, and greater energy intake in the early and late parts of the dark period.[152] The amplitude of clock gene oscillations in *ob/ob* mice is decreased in the liver and adipose tissue, but not in the SCN. These circadian abnormalities are observed before any detectable metabolic dysfunctions, ruling out a causal role of obesity in the appearance of the circadian perturbations.[153]

Experimental studies in mice have shown that excess energy intake of a high-fat diet is associated with several circadian abnormalities. The period of the endogenous clock suprachiasmatic is elongated relative to that of control mice fed with a standard chow diet.[137,154] Furthermore, the daily period of feeding behavior is lengthened, due to increased food intake during the late part of usual period of rest (daytime in mice).[154] In addition, daily variations of metabolic hormones in mice fed with high fat are attenuated, with higher and lower plasma levels of leptin and corticosterone, respectively.[137,154] The daily or day–night variations in clock gene expression of peripheral tissues in high-fat fed mice have been found to show either major[154,155] or minor changes.[29,156]

The mouse line carrying the mutation *db*, which invalidates the leptin receptor, is a classical model of obesity associated with severe diabetes mellitus and hypertension. The amplitude of activity–rest rhythm and blood pressure is dampened in *db/db* mice.[157,158] The characteristics of the clock gene oscillations in the liver are significantly altered compared to those observed in *db/+* control mice.[157] Experimental type 1 diabetes induced by streptozotocin, which destroys pancreatic β cells, is associated with several circadian disorders. In particular, the amplitude of oscillations of clock genes is reduced in the liver of diabetic mice.[159] Moreover, the phase-delaying effects of light are increased in streptozotocin-treated mice.[160] In both cases, acute treatment with insulin normalizes circadian alterations.

Obesity in humans is associated with a more flattened and fragmented rhythm of wrist temperature.[161] Moreover, the daily variations in glucose tolerance, which usually decrease throughout daytime in lean subjects, are reversed in obese subjects with or without type 2 diabetes.[162] At a molecular level, clock gene mRNAs in visceral adipose tissue have been correlated with adiposity, at least in women.[163,164] Using serial biopsies of white adipose tissue in the same individuals, no significant change is detected in the characteristics of clock gene oscillations in overweight/obese patients with or without type 2 diabetes, as compared to lean subjects.[165] This study thus contrasts with the majority of findings in animal studies reported above. Whether the differences are due to the type of white adipose tissue (e.g., subcutaneous vs. retroperitoneal), the severity of the metabolic diseases, or interspecific differences clearly warrants further investigations.

4. CHRONOTHERAPEUTICS

4.1. Pharmacology

Besides taking into account the pharmacokinetics of drugs according to the time of the day to improve their efficiency and reduce their side effects (chronopharmacology), targeting drugs that affect the circadian system (so-called chronobiotic drugs) is an emerging and active field of pharmacology. Below are mentioned a few examples of recent advances in that domain, reviewed elsewhere in detail.[166]

Agomelatine is an antidepressant drug with melatonergic (MT1/MT2) agonist and 5-HT(2C) receptor antagonist properties. Because both melatonin and serotonin are known regulators of the master clock and possibly secondary clocks, part of the antidepressant properties of agomelatine can be mediated by its resynchronizing effects on circadian rhythms. Furthermore, physiological doses of melatonin stimulate the activity of several antioxidant enzymes.[167] In the case of the metabolic syndrome, such melatonergic compounds may help to correct the altered sleep–wake cycle.[168]

With respect to the clock gene machinery, REV-ERB$\alpha-\beta$ are among the rare circadian factors with a known endogenous ligand (i.e., heme). Synthetic REV-ERBs ligands recently developed have significant effects on clock gene expression and less clear resetting effects, as they decrease nocturnal activity in mice, rather than shifting the sleep–wake cycle. Nevertheless, some of these promising compounds have been shown to improve the metabolic profile of obese mice.[169,170] Other hopeful targets to cure metabolic disorders are the closely related members of the ROR family.[171]

4.2. Food composition and feeding time

Even if most animal studies reported above concerned either chow or high-fat diet, the nature of diet-derived nutrients can play a role in the control of peripheral circadian timing. For instance, the daily rhythm of sterol regulatory element-binding protein-1 expression in the liver, where it regulates lipid metabolism, shows differential phase shifts according to various macronutriment regimens (i.e., standard vs. high-carbohydrate, -fat, or -protein).[172]

Several works highlight the fact that excessive food intake during the resting period is deleterious for metabolic health and, conversely, that avoiding it leads to beneficial effects in case of metabolic diseases. Zucker rats display a phase-advanced rhythm of food intake, which begins in the afternoon and not the evening (lights off), as in control rats. It is interesting to note that if food intake is limited exclusively to night (normal feeding period in rats), the overweight of Zucker rats is reduced by 23% compared to those that have a free access to food, despite similar energy intake for the two groups of rats.[149] In the case of *db/db* mice that are obese and severely diabetic, restricting food access to the dark period not only restores a robust rhythm of rest–activity but also ameliorates plasma glucose, insulin, and triacylglycerols.[157]

With respect to diet-induced obesity, the daily period during which high fat is consumed seems important for its obesogenic effects in rodents, as shown by the lack of major adverse metabolic consequences of high-fat feeding when it is restricted to the usual feeding period (nighttime for mice).[173,174] Furthermore, fatty acid composition matters in the altered eating pattern induced by high-fat feeding. In rats, this effect is specifically mediated by saturated fatty acids.[175]

Finally, in animal studies, the adverse effects of desynchronization can be alleviated by timed feeding. The increased gain in body mass of rats desynchronized by a biweekly shift of the light–dark cycle is not observed if food access is limited to the dark period (including during the shifted cycle).[130] Feeding restricted to the dark period in rats (i.e., corresponding to the usual period of food intake) limits body mass gain and desynchronization resulting from forced activity imposed during the resting period (daytime).[176]

4.3. Light and other (de)synchronizers

For applying reliable chronotherapy in humans, it is important to determine internal time for each subject. A recent study reveals that two blood samples taken at 12 h apart from each other are sufficient to estimate individual circadian timing.[177] Adequate timing of light exposure can promote phase adjustment of the master clock. In addition, timed light avoidance can be as useful by preventing photic resetting and allowing transiently endogenous free run. In view of the rather unique and exclusive synchronizing role of light for the suprachiasmatic clock, a means to prevent desynchronization implies strong and appropriate (timed), rather than weak and mismatched lighting information (e.g., light at night).[178] As discussed by these authors, light strategy should combine also appropriate timing of other putative (de)synchronizers (mealtime, exercise) in a global "Zeitgeber hygiene."

In humans also, timed carbohydrate-rich meals can act as a synchronizer of peripheral oscillators.[179] Apart from meal timing, dietary energy density during daytime may modulate overall energy intake. This observation led, for instance, to the recommendation that eating low-density foods in the morning and avoiding high-density foods at night might aid in reducing daily energy intake.[180]

5. CONCLUSION

In spite of the huge literature demonstrating the tight connections between circadian clocks and metabolism in animals, much work remains to be done to confirm in humans the conclusions drawn in nocturnal rodents. Nevertheless, epidemiological studies consistently report an increased prevalence of metabolic risk factors in shift workers and other desynchronizing conditions. Therefore, it is relevant from a clinical point of view to improve basic knowledge and develop models of shift work and other circadian disturbances in day-active animals. In view of the more and more recognized importance of (circadian) timing in pathophysiology, pharmacological and dietary interventions for limiting metabolic risks could take into account circadian rhythmicity to maintain and/or restore a temporal organization appropriately synchronized to local time.

ACKNOWLEDGMENTS

I would like to thank Pr. Martha Gillette for the invitation to write this chapter, and Pr. Paul Pévet and Pr. André Malan for guidance and support. I also thank for their help and discussion all my colleagues of the "Regulation of circadian clocks" team at the Institute of Cellular and Integrative Neurosciences, CNRS, and University of Strasbourg.

REFERENCES

1. Moore RY, Eichler VB. Loss of a circadian adrenal corticosterone rhythm following suprachiasmatic lesions in the rat. *Brain Res.* 1972;42:201–206.
2. Stephan FK, Zucker I. Circadian rhythms in drinking behavior and locomotor activity of rats are eliminated by hypothalamic lesions. *Proc Natl Acad Sci USA.* 1972;69:1583–1586.
3. Kalsbeek A, Scheer FA, Perreau-Lenz S, et al. Circadian disruption and SCN control of energy metabolism. *FEBS Lett.* 2011;585:1412–1426.
4. Dardente H, Cermakian N. Molecular circadian rhythms in central and peripheral clocks in mammals. *Chronobiol Int.* 2007;24:195–213.
5. Sandu C, Dumas M, Malan A, et al. Human skin keratinocytes, melanocytes, and fibroblasts contain distinct circadian clock machineries. *Cell Mol Life Sci.* 2012;69:3329–3339.
6. Ko CH, Takahashi JS. Molecular components of the mammalian circadian clock. *Hum Mol Genet.* 2006;15(Spec No 2):R271–R277.
7. Dardente H, Mendoza J, Fustin JM, Challet E, Hazlerigg DG. Implication of the F-Box Protein FBXL21 in circadian pacemaker function in mammals. *PLoS One.* 2008;3:e3530.
8. Godinho SI, Maywood ES, Shaw L, et al. The after-hours mutant reveals a role for Fbxl3 in determining mammalian circadian period. *Science.* 2007;316:897–900.
9. Eide EJ, Woolf MF, Kang H, et al. Control of mammalian circadian rhythm by CKIepsilon-regulated proteasome-mediated PER2 degradation. *Mol Cell Biol.* 2005;25:2795–2807.

10. Hirayama J, Sahar S, Grimaldi B, et al. CLOCK-mediated acetylation of BMAL1 controls circadian function. *Nature*. 2007;450:1086–1090.
11. Alonso-Vale MI, Andreotti S, Mukai PY, et al. Melatonin and the circadian entrainment of metabolic and hormonal activities in primary isolated adipocytes. *J Pineal Res*. 2008;45:422–429.
12. Pevet P, Challet E. Melatonin: both master clock output and internal time-giver in the circadian clocks network. *J Physiol Paris*. 2011;105:170–182.
13. Penicaud L, Leloup C, Lorsignol A, Alquier T, Guillod E. Brain glucose sensing mechanism and glucose homeostasis. *Curr Opin Clin Nutr Metab Care*. 2002;5:539–543.
14. Feillet CA, Mendoza J, Albrecht U, Pevet P, Challet E. Forebrain oscillators ticking with different clock hands. *Mol Cell Neurosci*. 2008;37:209–221.
15. Minana-Solis MC, Angeles-Castellanos M, Feillet C, Pevet P, Challet E, Escobar C. Differential effects of a restricted feeding schedule on clock-gene expression in the hypothalamus of the rat. *Chronobiol Int*. 2009;26:808–820.
16. Moriya T, Aida R, Kudo T, et al. The dorsomedial hypothalamic nucleus is not necessary for food-anticipatory circadian rhythms of behavior, temperature or clock gene expression in mice. *Eur J Neurosci*. 2009;29:1447–1460.
17. Nagai K, Nishio T, Nakagawa H, Nakamura S, Fukuda Y. Effect of bilateral lesions of the suprachiasmatic nuclei on the circadian rhythm of food-intake. *Brain Res*. 1978;142:384–389.
18. Nishio T, Shiosaka S, Nakagawa H, Sakumoto T, Satoh K. Circadian feeding rhythm after hypothalamic knife-cut isolating suprachiasmatic nucleus. *Physiol Behav*. 1979;23:763–769.
19. Mistlberger RE. Neurobiology of food anticipatory circadian rhythms. *Physiol Behav*. 2011;104:535–545.
20. Caba M, Gonzalez-Mariscal G. The rabbit pup, a natural model of nursing-anticipatory activity. *Eur J Neurosci*. 2009;30:1697–1706.
21. Trejo-Munoz L, Navarrete E, Montufar-Chaveznava R, Caldelas I. Determining the period, phase and anticipatory component of activity and temperature patterns in newborn rabbits that were maintained under a daily nursing schedule and fasting conditions. *Physiol Behav*. 2012;106:587–596.
22. Mendoza J. Circadian clocks: setting time by food. *J Neuroendocrinol*. 2007;19:127–137.
23. Feillet CA, Mendoza J, Pevet P, Challet E. Restricted feeding restores rhythmicity in the pineal gland of arrhythmic suprachiasmatic-lesioned rats. *Eur J Neurosci*. 2008;28:2451–2458.
24. Davidson AJ. Lesion studies targeting food-anticipatory activity. *Eur J Neurosci*. 2009;30:1658–1664.
25. Verwey M, Amir S. Food-entrainable circadian oscillators in the brain. *Eur J Neurosci*. 2009;30:1650–1657.
26. Challet E, Mendoza J. Metabolic and reward feeding synchronises the rhythmic brain. *Cell Tissue Res*. 2010;341:1–11.
27. Turek FW, Joshu C, Kohsaka A, et al. Obesity and metabolic syndrome in circadian Clock mutant mice. *Science*. 2005;308:1043–1045.
28. Paschos GK, Ibrahim S, Song WL, et al. Obesity in mice with adipocyte-specific deletion of clock component Arntl. *Nat Med*. 2012;18:1768–1777.
29. Delezie J, Dumont S, Dardente H, et al. The nuclear receptor REV-ERBalpha is required for the daily balance of carbohydrate and lipid metabolism. *FASEB J*. 2012;26:3321–3335.
30. Strubbe JH, Woods SC. The timing of meals. *Psychol Rev*. 2004;111:128–141.
31. Plata-Salaman CR, Oomura Y. Food intake dependence on acute changes in light schedule. *Physiol Behav*. 1987;41:135–140.

32. Nagai K, Nishio T, Nakagawa H. Bilateral lesions of suprachiasmatic nucleus eliminate circadian rhythms of oxygen consumption and the respiratory quotient in rats. *Experientia*. 1985;41:1136–1138.
33. Escobar C, Diaz-Munoz M, Encinas F, Aguilar-Roblero R. Persistence of metabolic rhythmicity during fasting and its entrainment by restricted feeding schedules in rats. *Am J Physiol Regul Integr Comp Physiol*. 1998;274:R1309–R1316.
34. Satoh Y, Kawai H, Kudo N, Kawashima Y, Mitsumoto A. Time-restricted feeding entrains daily rhythms of energy metabolism in mice. *Am J Physiol Regul Integr Comp Physiol*. 2006;290:R1276–R1283.
35. Grimaldi B, Bellet MM, Katada S, et al. PER2 controls lipid metabolism by direct regulation of PPARgamma. *Cell Metab*. 2010;12:509–520.
36. Vollmers C, Gill S, DiTacchio L, Pulivarthy SR, Le HD, Panda S. Time of feeding and the intrinsic circadian clock drive rhythms in hepatic gene expression. *Proc Natl Acad Sci USA*. 2009;106:21453–21458.
37. Vu-Dac N, Gervois P, Grotzinger T, et al. Transcriptional regulation of apolipoprotein A-I gene expression by the nuclear receptor RORalpha. *J Biol Chem*. 1997;272:22401–22404.
38. Raspe E, Duez H, Gervois P, et al. Transcriptional regulation of apolipoprotein C-III gene expression by the orphan nuclear receptor RORalpha. *J Biol Chem*. 2001;276:2865–2871.
39. Wang J, Li Y, Zhang M, et al. A zinc finger HIT domain-containing protein, ZNHIT-1, interacts with orphan nuclear hormone receptor Rev-erbbeta and removes Rev-erbbeta-induced inhibition of apoCIII transcription. *FEBS J*. 2007;274:5370–5381.
40. Gimble JM, Floyd ZE. Fat circadian biology. *J Appl Physiol*. 2009;107:1629–1637.
41. Tsutsumi K, Inoue Y, Kondo Y. The relationship between lipoprotein lipase activity and respiratory quotient of rats in circadian rhythms. *Biol Pharm Bull*. 2002;25:1360–1363.
42. Wang H, Eckel RH. Lipoprotein lipase: from gene to obesity. *Am J Physiol Endocrinol Metab*. 2009;297:E271–E288.
43. Iwadate R, Satoh Y, Watanabe Y, et al. Impairment of heme biosynthesis induces short circadian period in body temperature rhythms in mice. *Am J Physiol Regul Integr Comp Physiol*. 2012;303:R8–R18.
44. Guenthner CJ, Bickar D, Harrington ME. Heme reversibly damps PERIOD2 rhythms in mouse suprachiasmatic nucleus explants. *Neuroscience*. 2009;164:832–841.
45. Dioum EM, Rutter J, Tuckerman JR, Gonzalez G, Gilles-Gonzalez MA, McKnight SL. NPAS2: a gas-responsive transcription factor. *Science*. 2002;298:2385–2387.
46. Kaasik K, Lee CC. Reciprocal regulation of haem biosynthesis and the circadian clock in mammals. *Nature*. 2004;430:467–471.
47. Airola MV, Du J, Dawson JH, Crane BR. Heme binding to the Mammalian circadian clock protein period 2 is nonspecific. *Biochemistry*. 2010;49:4327–4338.
48. Raghuram S, Stayrook KR, Huang P, et al. Identification of heme as the ligand for the orphan nuclear receptors REV-ERBalpha and REV-ERBbeta. *Nat Struct Mol Biol*. 2007;14:1207–1213.
49. Yin L, Wu N, Curtin JC, et al. Rev-erbalpha, a heme sensor that coordinates metabolic and circadian pathways. *Science*. 2007;318:1786–1789.
50. Pardee KI, Xu X, Reinking J, et al. The structural basis of gas-responsive transcription by the human nuclear hormone receptor REV-ERBbeta. *PLoS Biol*. 2009;7:e43.
51. Gupta N, Ragsdale SW. Thiol-disulfide redox dependence of heme binding and heme ligand switching in nuclear hormone receptor rev-erb{beta}. *J Biol Chem*. 2011;286:4392–4403.

52. O'Neill JS, Reddy AB. Circadian clocks in human red blood cells. *Nature.* 2011;469:498–503.
53. Edgar RS, Green EW, Zhao Y, et al. Peroxiredoxins are conserved markers of circadian rhythms. *Nature.* 2012;485:459–464.
54. Rutter J, Reick M, Wu LC, McKnight SL. Regulation of clock and NPAS2 DNA binding by the redox state of NAD cofactors. *Science.* 2001;293:510–514.
55. Nakahata Y, Sahar S, Astarita G, Kaluzova M, Sassone-Corsi P. Circadian control of the NAD^+ salvage pathway by CLOCK-SIRT1. *Science.* 2009;324:654–657.
56. Ramsey KM, Yoshino J, Brace CS, et al. Circadian clock feedback cycle through NAMPT-mediated NAD^+ biosynthesis. *Science.* 2009;324:651–654.
57. Lamia KA, Sachdeva UM, DiTacchio L, et al. AMPK regulates the circadian clock by cryptochrome phosphorylation and degradation. *Science.* 2009;326:437–440.
58. Um JH, Yang S, Yamazaki S, et al. Activation of 5'-AMP-activated kinase with diabetes drug metformin induces casein kinase Iepsilon (CKIepsilon)-dependent degradation of clock protein mPer2. *J Biol Chem.* 2007;282:20794–20798.
59. Nogueiras R, Habegger KM, Chaudhary N, et al. Sirtuin 1 and sirtuin 3: physiological modulators of metabolism. *Physiol Rev.* 2012;92:1479–1514.
60. Nemoto S, Fergusson MM, Finkel T. SIRT1 functionally interacts with the metabolic regulator and transcriptional coactivator PGC-1alpha. *J Biol Chem.* 2005;280:16456–16460.
61. Rodgers JT, Lerin C, Haas W, Gygi SP, Spiegelman BM, Puigserver P. Nutrient control of glucose homeostasis through a complex of PGC-1alpha and SIRT1. *Nature.* 2005;434:113–118.
62. Liu C, Li S, Liu T, Borjigin J, Lin JD. Transcriptional coactivator PGC-1alpha integrates the mammalian clock and energy metabolism. *Nature.* 2007;447:477–481.
63. Asher G, Gatfield D, Stratmann M, et al. SIRT1 regulates circadian clock gene expression through PER2 deacetylation. *Cell.* 2008;134:317–328.
64. Nakahata Y, Kaluzova M, Grimaldi B, et al. The NAD^+-dependent deacetylase SIRT1 modulates CLOCK-mediated chromatin remodeling and circadian control. *Cell.* 2008;134:329–340.
65. Fontaine C, Dubois G, Duguay Y, et al. The orphan nuclear receptor Rev-Erbalpha is a peroxisome proliferator-activated receptor (PPAR) gamma target gene and promotes PPARgamma-induced adipocyte differentiation. *J Biol Chem.* 2003;278:37672–37680.
66. Duez H, Duhem C, Laitinen S, et al. Inhibition of adipocyte differentiation by RORalpha. *FEBS Lett.* 2009;583:2031–2036.
67. Shimba S, Ishii N, Ohta Y, et al. Brain and muscle Arnt-like protein-1 (BMAL1), a component of the molecular clock, regulates adipogenesis. *Proc Natl Acad Sci USA.* 2005;102:12071–12076.
68. Gervois P, Chopin-Delannoy S, Fadel A, et al. Fibrates increase human REV-ERBalpha expression in liver via a novel peroxisome proliferator-activated receptor response element. *Mol Endocrinol.* 1999;13:400–409.
69. Canaple L, Rambaud J, Dkhissi-Benyahya O, et al. Reciprocal regulation of brain and muscle Arnt-like protein 1 and peroxisome proliferator-activated receptor alpha defines a novel positive feedback loop in the rodent liver circadian clock. *Mol Endocrinol.* 2006;20:1715–1727.
70. Damiola F, Le Minh N, Preitner N, Kornmann B, Fleury-Olela F, Schibler U. Restricted feeding uncouples circadian oscillators in peripheral tissues from the central pacemaker in the suprachiasmatic nucleus. *Genes Dev.* 2000;14:2950–2961.
71. Hoogerwerf WA, Hellmich HL, Cornelissen G, et al. Clock gene expression in the murine gastrointestinal tract: endogenous rhythmicity and effects of a feeding regimen. *Gastroenterology.* 2007;133:1250–1260.

72. Zvonic S, Ptitsyn AA, Conrad SA, et al. Characterization of peripheral circadian clocks in adipose tissues. *Diabetes*. 2006;55:962–970.
73. Hirao A, Nagahama H, Tsuboi T, Hirao M, Tahara Y, Shibata S. Combination of starvation interval and food volume determines the phase of liver circadian rhythm in Per2::Luc knock-in mice under two meals per day feeding. *Am J Physiol Gastrointest Liver Physiol*. 2010;299:G1045–G1053.
74. Balsalobre A, Brown SA, Marcacci L, et al. Resetting of circadian time in peripheral tissues by glucocorticoid signaling. *Science*. 2000;289:2344–2347.
75. Hirota T, Okano T, Kokame K, Shirotani-Ikejima H, Miyata T, Fukada Y. Glucose down-regulates Per1 and Per2 mRNA levels and induces circadian gene expression in cultured Rat-1 fibroblasts. *J Biol Chem*. 2002;277:44244–44251.
76. Wu T, Ni Y, Kato H, Fu Z. Feeding-induced rapid resetting of the hepatic circadian clock is associated with acute induction of Per2 and Dec1 transcription in rats. *Chronobiol Int*. 2010;27:1–18.
77. Tahara Y, Otsuka M, Fuse Y, Hirao A, Shibata S. Refeeding after fasting elicits insulin-dependent regulation of Per2 and Rev-erbalpha with shifts in the liver clock. *J Biol Rhythms*. 2011;26:230–240.
78. Mieda M, Williams SC, Richardson JA, Tanaka K, Yanagisawa M. The dorsomedial hypothalamic nucleus as a putative food-entrainable circadian pacemaker. *Proc Natl Acad Sci USA*. 2006;103:12150–12155.
79. Waddington Lamont E, Harbour VL, Barry-Shaw J, et al. Restricted access to food, but not sucrose, saccharine, or salt, synchronizes the expression of Period2 protein in the limbic forebrain. *Neuroscience*. 2007;144:402–411.
80. Mendoza J, Pevet P, Felder-Schmittbuhl MP, Bailly Y, Challet E. The cerebellum harbors a circadian oscillator involved in food anticipation. *J Neurosci*. 2010;30:1894–1904.
81. Rath MF, Rohde K, Moller M. Circadian oscillations of molecular clock components in the cerebellar cortex of the rat. *Chronobiol Int*. 2012;29:1289–1299.
82. Hogenesch JB, Herzog ED. Intracellular and intercellular processes determine robustness of the circadian clock. *FEBS Lett*. 2011;585:1427–1434.
83. Ximenes da Silva A, Gendrot G, Serviere J, Lavialle M. Daily changes of cytochrome oxidase activity within the suprachiasmatic nucleus of the Syrian hamster. *Neurosci Lett*. 2000;286:139–143.
84. Isobe Y, Hida H, Nishino H. Circadian rhythm of metabolic oscillation in suprachiasmatic nucleus depends on the mitochondrial oxidation state, reflected by cytochrome C oxidase and lactate dehydrogenase. *J Neurosci Res*. 2011;89:929–935.
85. Schwartz WJ, Davidsen LC, Smith CB. In vivo metabolic activity of a putative circadian oscillator, the rat suprachiasmatic nucleus. *J Comp Neurol*. 1980;189:157–167.
86. Allen G, Rappe J, Earnest DJ, Cassone VM. Oscillating on borrowed time: diffusible signals from immortalized suprachiasmatic nucleus cells regulate circadian rhythmicity in cultured fibroblasts. *J Neurosci*. 2001;21:7937–7943.
87. Gillette MU. The suprachiasmatic nuclei: circadian phase-shifts induced at the time of hypothalamic slice preparation are preserved in vitro. *Brain Res*. 1986;379:176–181.
88. Wang YC, Yang JJ, Huang RC. Intracellular Na(+) and metabolic modulation of Na/K pump and excitability in the rat suprachiasmatic nucleus neurons. *J Neurophysiol*. 2012;108:2024–2032.
89. Wang TA, Yu YV, Govindaiah G, et al. Circadian rhythm of redox state regulates excitability in suprachiasmatic nucleus neurons. *Science*. 2012;337:839–842.
90. Paul KN, Saafir TB, Tosini G. The role of retinal photoreceptors in the regulation of circadian rhythms. *Rev Endocr Metab Disord*. 2009;10:271–278.
91. Meijer JH, Michel S, Vansteensel MJ. Processing of daily and seasonal light information in the mammalian circadian clock. *Gen Comp Endocrinol*. 2007;152:159–164.

92. Harrington ME. The ventral lateral geniculate nucleus and the intergeniculate leaflet: interrelated structures in the visual and circadian systems. *Neurosci Biobehav Rev.* 1997;21:705–727.
93. Challet E, Denis I, Rochet V, et al. The role of PPARbeta/delta in the regulation of glutamatergic signaling in the hamster suprachiasmatic nucleus. *Cell Mol Life Sci.* 2013;70:2003–2014.
94. Challet E, Malan A, Turek FW, Van Reeth O. Daily variations of blood glucose, acid-base state and PCO2 in rats: effect of light exposure. *Neurosci Lett.* 2004;355:131–135.
95. Rai D, Bhatia G, Sen T, Palit G. Comparative study of perturbations of peripheral markers in different stressors in rats. *Can J Physiol Pharmacol.* 2003;81:1139–1146.
96. Redlin U. Neural basis and biological function of masking by light in mammals: suppression of melatonin and locomotor activity. *Chronobiol Int.* 2001;18:737–758.
97. Tsai JW, Hannibal J, Hagiwara G, et al. Melanopsin as a sleep modulator: circadian gating of the direct effects of light on sleep and altered sleep homeostasis in Opn4(-/-) mice. *PLoS Biol.* 2009;7:e1000125.
98. Yamamoto H, Nagai K, Nakagawa H. Role of SCN in daily rhythms of plasma glucose, FFA, insulin and glucagon. *Chronobiol Int.* 1987;4:483–491.
99. Kalsbeek A, Strubbe JH. Circadian control of insulin secretion is independent of the temporal distribution of feeding. *Physiol Behav.* 1998;63:553–558.
100. Kalsbeek A, Fliers E, Romijn JA, et al. The suprachiasmatic nucleus generates the diurnal changes in plasma leptin levels. *Endocrinology.* 2001;142:2677–2685.
101. Shea SA, Hilton MF, Orlova C, Ayers RT, Mantzoros CS. Independent circadian and sleep/wake regulation of adipokines and glucose in humans. *J Clin Endocrinol Metab.* 2005;90:2537–2544.
102. Dallmann R, Viola AU, Tarokh L, Cajochen C, Brown SA. The human circadian metabolome. *Proc Natl Acad Sci USA.* 2012;109:2625–2629.
103. Inouye SI. Restricted daily feeding does not entrain circadian rhythms of the suprachiasmatic nucleus in the rat. *Brain Res.* 1982;232:194–199.
104. Stokkan KA, Yamazaki S, Tei H, Sakaki Y, Menaker M. Entrainment of the circadian clock in the liver by feeding. *Science.* 2001;291:490–493.
105. Iwanaga H, Yano M, Miki H, et al. Per2 gene expressions in the suprachiasmatic nucleus and liver differentially respond to nutrition factors in rats. *J Parenter Enteral Nutr.* 2005;29:157–161.
106. Mendoza J, Drevet K, Pevet P, Challet E. Daily meal timing is not necessary for resetting the main circadian clock by calorie restriction. *J Neuroendocrinol.* 2008;20:251–260.
107. Challet E. Interactions between light, mealtime and calorie restriction to control daily timing in mammals. *J Comp Physiol B.* 2010;180:631–644.
108. Mendoza J, Gourmelen S, Dumont S, Sage-Ciocca D, Pevet P, Challet E. Setting the main circadian clock of a diurnal mammal by hypocaloric feeding. *J Physiol (Lond).* 2012;590:3155–3168.
109. Oishi K, Uchida D, Ohkura N, et al. Ketogenic diet disrupts the circadian clock and increases hypofibrinolytic risk by inducing expression of plasminogen activator inhibitor-1. *Arterioscler Thromb Vasc Biol.* 2009;29:1571–1577.
110. Hut RA, Pilorz V, Boerema AS, Strijkstra AM, Daan S. Working for food shifts nocturnal mouse activity into the day. *PLoS One.* 2011;6:e17527.
111. Mendoza J, Angeles-Castellanos M, Escobar C. A daily palatable meal without food deprivation entrains the suprachiasmatic nucleus of rats. *Eur J Neurosci.* 2005;22:2855–2862.
112. Mendoza J, Clesse D, Pevet P, Challet E. Food-reward signalling in the suprachiasmatic clock. *J Neurochem.* 2010;112:1489–1499.

113. Rudic RD, McNamara P, Curtis AM, et al. BMAL1 and CLOCK, two essential components of the circadian clock, are involved in glucose homeostasis. *PLoS Biol.* 2004;2: e377.
114. Kennaway DJ, Owens JA, Voultsios A, Boden MJ, Varcoe TJ. Metabolic homeostasis in mice with disrupted Clock gene expression in peripheral tissues. *Am J Physiol Regul Integr Comp Physiol.* 2007;293:R1528–R1537.
115. Doi M, Takahashi Y, Komatsu R, et al. Salt-sensitive hypertension in circadian clock-deficient Cry-null mice involves dysregulated adrenal Hsd3b6. *Nat Med.* 2010;16:67–74.
116. Carvas JM, Vukolic A, Yepuri G, et al. Period2 gene mutant mice show compromised insulin-mediated endothelial nitric oxide release and altered glucose homeostasis. *Front Physiol.* 2012;3:337.
117. Bray MS, Shaw CA, Moore MW, et al. Disruption of the circadian clock within the cardiomyocyte influences myocardial contractile function, metabolism, and gene expression. *Am J Physiol Heart Circ Physiol.* 2008;294:H1036–H1047.
118. Lamia KA, Storch KF, Weitz CJ. Physiological significance of a peripheral tissue circadian clock. *Proc Natl Acad Sci USA.* 2008;105:15172–15177.
119. Marcheva B, Ramsey KM, Buhr ED, et al. Disruption of the clock components CLOCK and BMAL1 leads to hypoinsulinaemia and diabetes. *Nature.* 2010;466:627–631.
120. Garaulet M, Lee YC, Shen J, et al. CLOCK genetic variation and metabolic syndrome risk: modulation by monounsaturated fatty acids. *Am J Clin Nutr.* 2009;90:1466–1475.
121. Goumidi L, Grechez A, Dumont J, et al. Impact of REV-ERB alpha gene polymorphisms on obesity phenotypes in adult and adolescent samples. *Int J Obes (Lond).* 2013;37:666–672.
122. Pevet P, Jacob N, Lakhdar-Ghazal N, Vuillez P. How do the suprachiasmatic nuclei of the hypothalamus integrate photoperiodic information? *Biol Cell.* 1997;89:569–577.
123. Pevet P. Melatonin: from seasonal to circadian signal. *J Neuroendocrinol.* 2003;15:422–426.
124. Sinitskaya N, Schuster-Klein C, Guardiola-Lemaitre B, Gourmelen S, Pevet P, Challet E. Short day-length increases sucrose consumption and adiposity in rats fed a high-fat diet. *Psychoneuroendocrinology.* 2008;33:1269–1278.
125. Fonken LK, Workman JL, Walton JC, et al. Light at night increases body mass by shifting the time of food intake. *Proc Natl Acad Sci USA.* 2010;107:18664–18669.
126. Coomans CP, van den Berg SA, Houben T, et al. Detrimental effects of constant light exposure and high-fat diet on circadian energy metabolism and insulin sensitivity. *FASEB J.* 2013;27:1721–1732.
127. Haimov I, Arendt J. The prevention and treatment of jet lag. *Sleep Med Rev.* 1999;3:229–240.
128. Gonnissen HK, Rutters F, Mazuy C, Martens EA, Adam TC, Westerterp-Plantenga MS. Effect of a phase advance and phase delay of the 24-h cycle on energy metabolism, appetite, and related hormones. *Am J Clin Nutr.* 2012;96:689–697.
129. Tsai LL, Tsai YC, Hwang K, Huang YW, Tzeng JE. Repeated light-dark shifts speed up body weight gain in male F344 rats. *Am J Physiol Endocrinol Metab.* 2005;289: E212–E217.
130. Bartol-Munier I, Gourmelen S, Pevet P, Challet E. Combined effects of high-fat feeding and circadian desynchronization. *Int J Obes (Lond).* 2006;30:60–67.
131. Salgado-Delgado R, Angeles-Castellanos M, Buijs MR, Escobar C. Internal desynchronization in a model of night-work by forced activity in rats. *Neuroscience.* 2008;154:922–931.
132. van Amelsvoort LG, Schouten EG, Kok FJ. Duration of shiftwork related to body mass index and waist to hip ratio. *Int J Obes Relat Metab Disord.* 1999;23:973–978.

133. Karlsson B, Knutsson A, Lindahl B. Is there an association between shift work and having a metabolic syndrome? Results from a population based study of 27,485 people. *Occup Environ Med*. 2001;58:747–752.
134. Esquirol Y, Bongard V, Mabile L, Jonnier B, Soulat JM, Perret B. Shift work and metabolic syndrome: respective impacts of job strain, physical activity, and dietary rhythms. *Chronobiol Int*. 2009;26:544–559.
135. Di Lorenzo L, De Pergola G, Zocchetti C, et al. Effect of shift work on body mass index: results of a study performed in 319 glucose-tolerant men working in a Southern Italian industry. *Int J Obes Relat Metab Disord*. 2003;27:1353–1358.
136. Macagnan J, Pattussi MP, Canuto R, Henn RL, Fassa AG, Olinto MT. Impact of nightshift work on overweight and abdominal obesity among workers of a poultry processing plant in southern Brazil. *Chronobiol Int*. 2012;29:336–343.
137. Mendoza J, Pevet P, Challet E. High-fat feeding alters the clock synchronization to light. *J Physiol (Lond)*. 2008;586:5901–5910.
138. Martino TA, Tata N, Belsham DD, et al. Disturbed diurnal rhythm alters gene expression and exacerbates cardiovascular disease with rescue by resynchronization. *Hypertension*. 2007;49:1104–1113.
139. Karatsoreos IN, Bhagat S, Bloss EB, Morrison JH, McEwen BS. Disruption of circadian clocks has ramifications for metabolism, brain, and behavior. *Proc Natl Acad Sci USA*. 2011;108:1657–1662.
140. Coomans CP, van den Berg SA, Lucassen EA, et al. The suprachiasmatic nucleus controls circadian energy metabolism and hepatic insulin sensitivity. *Diabetes*. 2013;62:1102–1108.
141. Scheer FA, Hilton MF, Mantzoros CS, Shea SA. Adverse metabolic and cardiovascular consequences of circadian misalignment. *Proc Natl Acad Sci USA*. 2009;106:4453–4458.
142. Baron KG, Reid KJ, Kern AS, Zee PC. Role of sleep timing in caloric intake and BMI. *Obesity (Silver Spring)*. 2011;19:1374–1381.
143. Thompson OM, Ballew C, Resnicow K, et al. Dietary pattern as a predictor of change in BMI z-score among girls. *Int J Obes (Lond)*. 2006;30:176–182.
144. Colles SL, Dixon JB, O'Brien PE. Night eating syndrome and nocturnal snacking: association with obesity, binge eating and psychological distress. *Int J Obes (Lond)*. 2007;31:1722–1730.
145. Barf RP, Meerlo P, Scheurink AJ. Chronic sleep disturbance impairs glucose homeostasis in rats. *Int J Endocrinol*. 2010;2010:819414.
146. Barclay JL, Husse J, Bode B, et al. Circadian desynchrony promotes metabolic disruption in a mouse model of shiftwork. *PLoS One*. 2012;7:e37150.
147. Spiegel K, Tasali E, Leproult R, Van Cauter E. Effects of poor and short sleep on glucose metabolism and obesity risk. *Nat Rev Endocrinol*. 2009;5:253–261.
148. Fukagawa K, Sakata T, Yoshimatsu H, Fujimoto K, Uchimura K, Asano C. Advance shift of feeding circadian rhythm induced by obesity progression in Zucker rats. *Am J Physiol Regul Integr Comp Physiol*. 1992;263:R1169–R1175.
149. Mistlberger RE, Lukman H, Nadeau BG. Circadian rhythms in the Zucker obese rat: assessment and intervention. *Appetite*. 1998;30:255–267.
150. Motosugi Y, Ando H, Ushijima K, et al. Tissue-dependent alterations of the clock gene expression rhythms in leptin-resistant Zucker diabetic fatty rats. *Chronobiol Int*. 2011;28:968–972.
151. Sans-Fuentes MA, Diez-Noguera A, Cambras T. Light responses of the circadian system in leptin deficient mice. *Physiol Behav*. 2010;99:487–494.
152. Ho A, Chin A. Circadian feeding and drinking patterns of genetically obese mice fed solid chow diet. *Physiol Behav*. 1988;43:651–656.
153. Ando H, Kumazaki M, Motosugi Y, et al. Impairment of peripheral circadian clocks precedes metabolic abnormalities in ob/ob mice. *Endocrinology*. 2011;152:1347–1354.

154. Kohsaka A, Laposky AD, Ramsey KM, et al. High-fat diet disrupts behavioral and molecular circadian rhythms in mice. *Cell Metab.* 2007;6:414–421.
155. Hsieh MC, Yang SC, Tseng HL, Hwang LL, Chen CT, Shieh KR. Abnormal expressions of circadian-clock and circadian clock-controlled genes in the livers and kidneys of long-term, high-fat-diet-treated mice. *Int J Obes (Lond).* 2009;34:227–239.
156. Yanagihara H, Ando H, Hayashi Y, Obi Y, Fujimura A. High-fat feeding exerts minimal effects on rhythmic mRNA expression of clock genes in mouse peripheral tissues. *Chronobiol Int.* 2006;23:905–914.
157. Kudo T, Akiyama M, Kuriyama K, Sudo M, Moriya T, Shibata S. Night-time restricted feeding normalises clock genes and Pai-1 gene expression in the db/db mouse liver. *Diabetologia.* 2004;47:1425–1436.
158. Su W, Guo Z, Randall DC, Cassis L, Brown DR, Gong MC. Hypertension and disrupted blood pressure circadian rhythm in type 2 diabetic db/db mice. *Am J Physiol Heart Circ Physiol.* 2008;295:H1634–H1641.
159. Kuriyama K, Sasahara K, Kudo T, Shibata S. Daily injection of insulin attenuated impairment of liver circadian clock oscillation in the streptozotocin-treated diabetic mouse. *FEBS Lett.* 2004;572:206–210.
160. Challet E, van Reeth O, Turek FW. Altered circadian responses to light in streptozotocin-induced diabetic mice. *Am J Physiol Endocrinol Metab.* 1999;277: E232–E237.
161. Corbalan-Tutau MD, Madrid JA, Ordovas JM, Smith CE, Nicolas F, Garaulet M. Differences in daily rhythms of wrist temperature between obese and normal-weight women: associations with metabolic syndrome features. *Chronobiol Int.* 2011;28:425–433.
162. Van Cauter E, Polonsky KS, Scheen AJ. Roles of circadian rhythmicity and sleep in human glucose regulation. *Endocr Rev.* 1997;18:716–738.
163. Gomez-Abellan P, Hernandez-Morante JJ, Lujan JA, Madrid JA, Garaulet M. Clock genes are implicated in the human metabolic syndrome. *Int J Obes (Lond).* 2008;32:121–128.
164. Zanquetta MM, Correa-Giannella ML, Giannella-Neto D, et al. Expression of clock genes in human subcutaneous and visceral adipose tissues. *Chronobiol Int.* 2012;29:252–260.
165. Otway DT, Mantele S, Bretschneider S, et al. Rhythmic diurnal gene expression in human adipose tissue from individuals who are lean, overweight, and type 2 diabetic. *Diabetes.* 2011;60:1577–1581.
166. Chen Z, Yoo SH, Takahashi JS. Small molecule modifiers of circadian clocks. *Cell Mol Life Sci.* 2013; in press.
167. Hardeland R. Antioxidative protection by melatonin: multiplicity of mechanisms from radical detoxification to radical avoidance. *Endocrine.* 2005;27:119–130.
168. Cardinali DP, Cano P, Jimenez-Ortega V, Esquifino AI. Melatonin and the metabolic syndrome: physiopathologic and therapeutical implications. *Neuroendocrinology.* 2011;93:133–142.
169. Meng QJ, McMaster A, Beesley S, et al. Ligand modulation of REV-ERBalpha function resets the peripheral circadian clock in a phasic manner. *J Cell Sci.* 2008;121:3629–3635.
170. Solt LA, Wang Y, Banerjee S, et al. Regulation of circadian behaviour and metabolism by synthetic REV-ERB agonists. *Nature.* 2012;485:62–68.
171. Solt LA, Burris TP. Action of RORs and their ligands in (patho)physiology. *Trends Endocrinol Metab.* 2012;23:619–627.
172. Matsumoto E, Ishihara A, Tamai S, et al. Time of day and nutrients in feeding govern daily expression rhythms of the gene for sterol regulatory element-binding protein (SREBP)-1 in the mouse liver. *J Biol Chem.* 2010;285:33028–33036.

173. Hatori M, Vollmers C, Zarrinpar A, et al. Time-restricted feeding without reducing caloric intake prevents metabolic diseases in mice fed a high-fat diet. *Cell Metab.* 2012;15:848–860.
174. Sherman H, Genzer Y, Cohen R, Chapnik N, Madar Z, Froy O. Timed high-fat diet resets circadian metabolism and prevents obesity. *FASEB J.* 2012;26:3493–3502.
175. Hariri N, Thibault L. Dietary obesity caused by a specific circadian eating pattern. *Chronobiol Int.* 2011;28:216–228.
176. Salgado-Delgado R, Angeles-Castellanos M, Saderi N, Buijs RM, Escobar C. Food intake during the normal activity phase prevents obesity and circadian desynchrony in a rat model of night work. *Endocrinology.* 2010;151:1019–1029.
177. Kasukawa T, Sugimoto M, Hida A, et al. Human blood metabolite timetable indicates internal body time. *Proc Natl Acad Sci USA.* 2012;109:15036–15041.
178. Erren TC, Reiter RJ. Light hygiene: time to make preventive use of insights—old and new—into the nexus of the drug light, melatonin, clocks, chronodisruption and public health. *Med Hypotheses.* 2009;73:537–541.
179. Krauchi K, Cajochen C, Werth E, Wirz-Justice A. Alteration of internal circadian phase relationships after morning versus evening carbohydrate-rich meals in humans. *J Biol Rhythms.* 2002;17:364–376.
180. de Castro JM. When, how much and what foods are eaten are related to total daily food intake. *Br J Nutr.* 2009;102:1228–1237.

CHAPTER SIX

Control of Sleep and Wakefulness in Health and Disease

Jamie M. Zeitzer[*,†]
[*]Department of Psychiatry and Behavioral Sciences, Stanford University, Palo Alto, California, USA
[†]Mental Illness Research Education and Clinical Center, VA Palo Alto Health Care System, Palo Alto, California, USA

Contents

Abstract

Sleep and wake are actively promoted states of consciousness that are dependent on a network of state-modulating neurons arising from both the brain stem and hypothalamus. This network helps to coordinate the occurrence of a sleep state in billions of cortical neurons. In many neurological diseases, there is a specific disruption to one of the components of this network. Under conditions of such disruptions, we often gain an improved understanding of the underlying function of the specific component under nonpathological conditions. The loss or dysfunction of one of the hypothalamic or brain stem regions that are responsible for promotion of sleep or wake can lead to disruptions in sleep and wake states that are often subtle, but sometime quite pronounced and of significant medical importance. By understanding the neural substrate and its pathophysiology, one can more appropriately target therapies that might help the specific sleep disruption. This chapter reviews what is currently understood about the neurobiological underpinnings of sleep and wake regulation and how various pathologies evoke changes in these regulatory mechanisms.

Sleep and wake in mammals are created by a complex confluence of multiple neuromodulatory systems that push the brain into the various firing modes that constitute our experience of "sleep" and "wake," both of which are actively generated states. These systems include those found in the brain stem (serotonin, noradrenaline, acetylcholine, dopamine, glutamate),

Progress in Molecular Biology and Translational Science, Volume 119
ISSN 1877-1173
http://dx.doi.org/10.1016/B978-0-12-396971-2.00006-3

hypothalamus (hypocretin, histamine), and elsewhere in the brain (adenosine, acetylcholine, GABA). There are dozens of other neuroactive substances (peptides, immune molecules, hormones) that influence sleep, but significantly less is understood about their role. The functions of neuromodulators are likely redundant for the experience of sleep and wake, but they are also likely to have specific, nonoverlapping roles for specific physiologic events associated with each state. A better understanding of the physiology and neurochemistry of the systems underlying sleep and wake can lead to a greater understanding of the manner in which disease and its treatment can lead to reciprocal alterations in sleep and wake (Fig. 6.1).

1. THE BRAIN STEM

Seminal experiments done in the 1930s and 1940s[1,2] identified the brain stem as an area of critical importance for occurrence of normal sleep and wake. These early experiments "proved" the prevailing concept of sleep as being a default brain state and wake being actively stimulated by sensory input. Current scientific thought is that *both* sleep and wake are actively generated states. There are multiple discreet nuclei in the brain stem that contribute to the generation of sleep and wake. These nuclei innervate both cortical and subcortical structures and are primarily responsible for the synchronization or desynchronization of electroencephalographic (EEG) activity. EEG synchrony is coordinated by the activity of thalamocortical neurons that have two primary firing modes—single-spike firing and burst firing.[3] Single-spike firing occurs during wake and is permissive for independent (i.e., nonsynchronized) cortical EEG activity. During single-spike firing, the membrane potential of thalamocortical neurons is kept elevated by the excitatory influences of a variety of neuromodulators. The excitatory stimuli are primarily derived from nuclei in the basal forebrain (acetylcholine) and brain stem (noradrenaline, dopamine, serotonin, acetylcholine). During nonrapid eye movement sleep (NREMS), there is a reduction in the release of these excitatory transmitters, resulting in a decrease of thalamocortical neuron membrane potential that leads to the expression of T- and H-currents and a shift from single-spike firing to burst firing.[3] This regular burst firing allows the thalamocortical neurons to synchronize the firing of large numbers of cortical neurons, leading to the EEG synchronization that is indicative of NREMS. In rapid eye movement sleep (REMS), there is a return of acetylcholinergic tone, which is sufficient to raise the

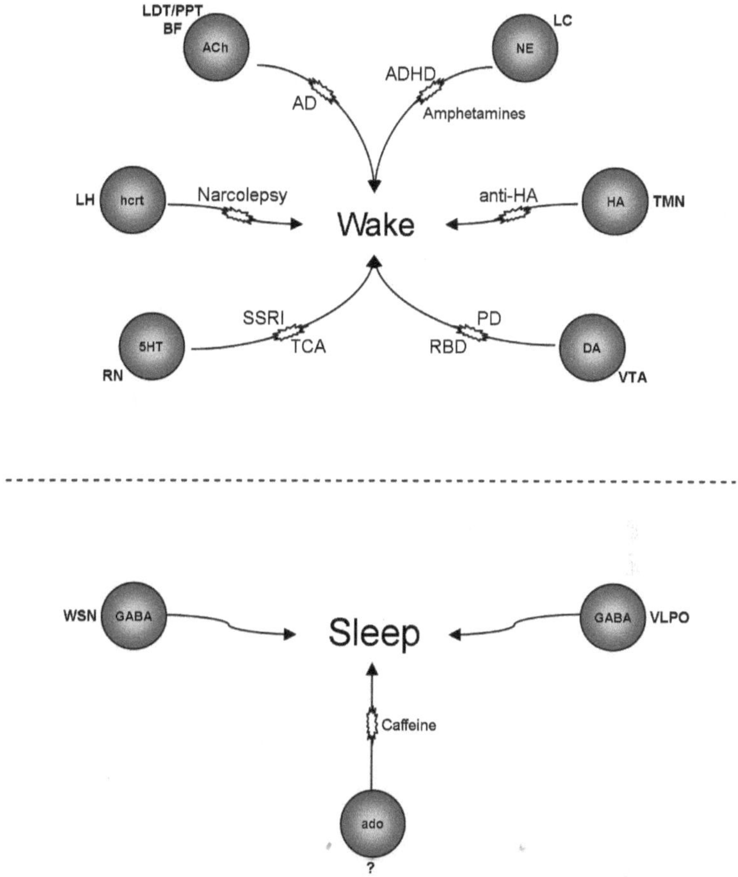

Figure 6.1 Influence of neuromodulatory systems on wake (upper panel) and sleep (lower panel). Each circle represents a different, anatomically distinct neurotransmitter system that promotes sleep or wake, with the neurotransmitter encircled and the name of the neuron cluster on the outside of the circle. Drugs or conditions that commonly disrupt the normal influence of these pathways on wake or sleep are highlighted within the pathways. Abbreviations: 5HT, serotonin; ACh, acetylcholine; AD, Alzheimer disease; ADHD, attention deficit/hyperactivity disorder; ado, adenosine; BF, basal forebrain; DA, dopamine; GABA, gamma-aminobutyric acid; HA, histamine; hcrt, hypocretin; LC, locus coeruleus; LDT, laterodorsal tegmentum; LH, lateral hypothalamus; NE, norepinephrine; PD, Parkinson's disease; PPT, pendunculopontine tegmentum; RBD, rapid eye movement behavior disorder; RN, raphe nuclei; SSRI, selective serotonin reuptake inhibitor; TCA, tricyclic antidepressant; TMN, tuberomamillary nucleus; VLPO, ventrolateral preoptic nucleus; VTA, ventral tegmental area; WSN, warm-sensitive neurons.

membrane potential of thalamocortical neurons to such a degree as to inactivate the T- and H-currents and return these neurons to single-spike firing and the EEG to its desynchronized state.[3] This is the general manner by which the activity of large numbers of cortical neurons may be synchronized during sleep.

The primary neuromodulatory systems in the brain stem that are involved with EEG synchronization include the dorsal raphe (serotonin), locus coeruleus (LC; noradrenaline), ventral periaqueductal gray/ventral tegmental area (dopamine), and pedunculopontine/laterodorsal tegmentum (acetylcholine). Each of these systems exhibits state-based activity. For example, the dorsal raphe in a cat fires progressively less frequently as the cat proceeds from active wake to quiet wake to NREMS to REMS.[4] This pattern is reflected by differential release of serotonin during wake, NREMS, and REMS, in humans as well.[5] The influence of serotonin on sleep can also be observed in both the pathological state of depression and the effects of antidepressants that act to increase serotonergic tone (e.g., selective serotonin reuptake inhibitors, serotonin/norepinephrine reuptake inhibitors, tricyclic antidepressants). Depression itself is associated with sleep disruption,[6] with sleep disruption often preceding onset of an episode of depression in individuals with recurrent depression. Much debate has centered on whether the depression causes the disruption in sleep or if the disruption in sleep leads to the depression. The neurobiology indicates that it may perhaps be that upstream disruption of serotonin networks leads to both sleep disruption and depression, but that sleep disruption is more easily observed at lower levels of serotonin disruption. The change in serotonin signaling that is hypothesized to accompany the use of many antidepressants is often associated with a disruption of sleep, specifically REMS.[7] As serotonin release is progressively reduced with depth of NREMS and nearly absent during REMS, the pharmacologic increase of serotonin tone, such as might follow the use of a serotonin transport inhibitor, would be expected to shift sleep into lighter states, away from REMS and stage 3 of NREMS. This is somewhat borne out by sleep studies conducted in individuals who chronically use serotonin-acting antidepressants.[7] The balance, of course, is that given the highly associated nature, and possibly etiology, of mood and sleep, normalization of mood despite disruption of sleep may be a more clinically favorable outcome.

The LC provides noradrenergic input to both the thalamus and cortex. LC-derived noradrenaline is involved not only in thalamic desynchronization of cortical EEG but also in general mechanisms of attention. Based on recordings done in primates, it has been postulated that the relationship between LC firing and attention is like an inverted-U,[8] low firing (low noradrenergic tone) being associated with inattentiveness, elevated firing (very high noradrenergic tone) with highly labile attention, and firing between the extremes with normal, focused attention. Attention deficit/hyperactivity disorder (ADHD) is often treated with medications that increase noradrenergic tone (e.g., methylphenidate, D-/L-amphetamine), effectively moving people from inattentiveness to normal attention. It has been hypothesized that a significant number of children who are diagnosed with ADHD are misdiagnosed as such and are actually suffering from a disruption to their sleep as caused by obstructive sleep apnea or insufficient sleep opportunity.[9] Reduction in sleep can cause the same type of inattentiveness and hyperactivity (often referred to as paradoxical hyperactivity, as it is thought to be a method by which the brain is attempting to rectify low noradrenergic tone in the presence of or despite excessive sleepiness). Narcolepsy, a disease characterized by a loss of hypocretin neurons (see below), is characterized by cataplexy (loss of skeletal muscle tone in response to positive emotions such as laughter) and a fundamental disruption of both wake and sleep consolidation.[10] This disease is currently treated symptomatically. Individuals with narcolepsy who are untreated have difficulty maintaining episodes of wake for more than a few hours.[11] Current treatment involves the use of amphetamines and their derivatives, which primarily act by stimulating noradrenergic pathways.[12] Thus, when exogenously stimulated, the noradrenaline system has the capacity to provide a substantial wake-promoting signal, though the relative degree of contribution of endogenous noradrenaline to wake promotion is likely much less.[13]

Substantially, less research has been done into the role of dopamine in the regulation of sleep and wake. Dopamine is generally considered a modulator of motor function, but it is also involved in diverse brain functions such as memory acquisition[14] and regulation of wake[15] among others. While an initial report indicated that mesencephalic dopamine neurons do not vary their activity with sleep or wake state,[16] a more recent study indicates that at least some of these neurons, specifically in the ventral periaqueductal gray matter, are specifically wake active.[17] The loss of these dopamine-expressing neurons in Parkinson's disease (PD) may help to explain the commonly observed disruption of daytime wake in individuals with PD.[18] The disruption of wake in PD may precede observable motor disruption,[19] implying either a progressive movement of the disease from wake-promoting dopamine neurons to motor-modulating dopamine neurons or a varied threshold at which wake and motor symptoms become evident. Dopamine is also likely involved in the control of movement during sleep itself. Dysfunction of dopamine is thought to underlie REM behavior disorder (RBD). RBD is characterized by unusual gross motor activity during REMS, imagined as people "acting out their dreams." Normally during sleep, descending spinal motor pathways are inhibited via descending glycinergic axons from the ventromedial medullary reticular formation.[20] The locale of the effect of dopamine on spinal motoneuron excitability is unknown and might occur at the level of the pons, medulla, or the spinal cord itself.[21] Individuals with RBD are more likely to develop PD, with about one-third converting to PD within 5 years of initial presentation to a sleep clinic.[22] This too implies either a progression of the disease or varying thresholds for different pathological phenotypes. It has been suggested that daytime sleepiness or RBD may be useful as prodromal markers in PD, allowing for earlier treatment when, perhaps, slowing the disease might prove more effective than current treatments that begin after the onset of motor symptoms.

There are two separate loci of cholinergic neurons that appear to contribute to the regulation of sleep and wake—basal forebrain and brain stem pedunculopontine and laterodorsal tegmentum (PPT/LDT). Basal forebrain cholinergic neurons mainly innervate the cortex, while those of the PPT/LDT primarily innervate the thalamus.[23,24] These cholinergic neurons provide a major source of excitation to both cortical and thalamic neurons and are mainly responsible for the control of cortical synchronization that varies between wake, NREMS, and REM.[25] As described earlier, it is the return of cholinergic tone during REMS that appears to drive thalamocortical neurons from bursting to single-spike mode, resulting in a desynchronization of cortical neurons. A dramatic loss of cholinergic neurons is the hallmark of Alzheimer disease (AD). Not unexpectedly, there is often severe disruption of sleep in individuals with AD.[26] The common form of sleep disruption found in those with AD is a deterioration of the diurnal organization of sleep and wake, so much so that in many cases, sleep and wake occur randomly across the day and night. This type of sleep disturbance is so disruptive that it often causes institutionalization of individuals with AD long before memory disturbances become a bother to the primary caregiver.[27,28] It has been surmised that this breakdown in sleep/wake organization is secondary to a reduction in the influence of the circadian clock (see below) due to degeneration of this nucleus.[29] There have been no reported studies in humans, however, that have demonstrated a reduced amplitude of the circadian clock in individuals with AD. Likewise, efforts to increase circadian amplitude via administration of light during the daytime have had mixed results.[30] It is possible, however, that the loss of cholinergic tone that accompanies AD is directly responsible for the observed loss of temporal organization of sleep and wake. In healthy older individuals without AD, there is an increased likelihood of napping[31] and an increased rate of nocturnal awakenings.[32] The reduction in cholinergic tone in individuals with AD could further increase the likelihood of daytime sleep, which would reduce the pressure for sleep to occur at night. The increased wake at night would then lead to increased sleep during the day and so on. Given the role of sleep in memory consolidation, this type of sleep disruption has also been hypothesized to compound AD-associated dementia.[33] Amelioration of sleep disruption, therefore, could be a target to improve memory in individuals with AD, especially early in the course of the disease.

2. THE HYPOTHALAMUS

Although the hypothalamus was identified as a key contributor to the modulation of sleep and wake nearly a century ago by pioneering work of Von Economo,[34] hypothalamic mechanisms went largely understudied until the 1980s. Hypothalamic mechanisms can be grossly separated into five groups of neurons—the hypocretin-expressing neurons of the lateral hypothalamus, the suprachiasmatic nucleus (SCN, location of the circadian pacemaker), histamine-expressing neurons, the ventrolateral preoptic nucleus (VLPO), and temperature-sensitive neurons in the anterior hypothalamus.

Hypocretin (also known as orexin) is a neuropeptide expressed in a discreet group of hypothalamic neurons. The loss of these neurons results in the sleep disorder narcolepsy (see earlier).[35] There is strong evidence to suggest that narcolepsy is an autoimmune-mediated destruction of hypocretin-expressing neurons.[36] The onset of this autoimmune attack may be a confluence of genetic background, age, and an appropriately timed infection. Antistreptococcal antibodies have been found in the serum of individuals with a recent onset of narcolepsy, implying that a recent infection by streptococcus bacteria might have the capacity to trigger narcolepsy.[37] More recently, it has been reported that there was a substantial increase in new cases of narcolepsy following H1N1 influenza infection and after receiving the adjuvant-coupled version of the H1N1 vaccine.[38] Whether the H1N1 virus or vaccine directly induces the autoimmune attack on hypocretin-expressing neurons or whether they create a permissive environment (e.g., upregulated immune activity, leakiness of the blood–brain barrier) is not known.

Under normal circumstances, hypocretin release is elevated during wake[39,40] and has been hypothesized to be critical in the consolidation of sleep and wake into single, daily episodes[41] as well as suppression of motor activity during REMS.[42] Hypocretin is likely involved in the consolidation of sleep and wake in humans. This consolidation process has been modeled as the interaction of two distinct, but highly interactive systems—the circadian clock and the sleep/wake homeostat.[43,44] The circadian clock, located in the SCN (see Chapter 1), is a near 24-h oscillator that can be entrained (synchronized) with the geophysical day. The circadian clock is considered a predictive mechanism as it modifies endogenous physiology in advance of changes actually occurring. The human circadian clock has been modeled as providing two distinct controls of sleep and wake consolidation—a wake-promoting signal that increases throughout the normal waking day, peaking just before habitual bedtime, and a sleep-promoting signal that increases throughout the normal time of the sleep, peaking just before habitual waketime.[44] Hypocretin signaling has been proposed to be a neurological correlate of the circadian wake drive, though this remains to be proven.[40] This circadian drive for sleep and wake interacts with the homeostatic drive for sleep and wake. Homeostasis, a term popularized by Walter Cannon in the 1930s, refers to a system that tends toward equilibrium.[45] In essence, the longer you stay awake, the more tired you get and the longer you sleep, the less tired you become. It has been hypothesized that extracellular adenosine may be the neurological correlate of the homeostatic drive for sleep. As wake progresses in duration, the rate of release of adenosine by neurons in the basal forebrain increases, resulting in elevations of extracellular adenosine.[46] This rate of release of adenosine may be secondary to cellular metabolism of adenosine triphosphate (ATP) or may be specifically stimulated by prostaglandin D2 (PGD2) via the DP1 receptor.[47] PGD2 is a naturally occurring somnogenic agent that is produced in the brain by the meninges and choroid plexus.[47] PGD2 production is also increased in response to inflammatory stimuli and may be part of the cellular link between inflammatory processes and the increase in sleepiness that often accompanies such.[48]

Concentrations of adenosine in the basal forebrain rapidly decrease during sleep.[46] This extracellular adenosine is hypothesized to inhibit basal forebrain cholinergic neurons, decreasing cerebral cholinergic tone, and decreasing wakefulness.[49] It is likely that this modulation of adenosine is tissue specific as other areas of cortical and subcortical tissue, both in rodent models and in humans, show little if any increase with increased time awake.[50,51] The role of adenosine in wake regulation is well established by the ubiquitous use of caffeine as a wake-promoting agent—at physiologic concentrations, the main effect of caffeine is to block adenosine receptors.[52] A recent study has proposed that the wake-promoting effects of caffeine are due to specific blockade of adenosine A2A receptors in the shell of the nucleus accumbens.[53] The role of the nucleus accumbens in sleep regulation and how adenosine regulation in this nucleus compares with adenosine regulation in the basal forebrain are yet to be discovered.

The homeostatic and circadian systems interact to create single, daily periods of stable sleep and wake. As the waking day goes on, the homeostatic drive for sleep increases and is offset by an increasing circadian drive for wake, resulting in stable wakefulness for about 16 h. Around habitual bedtime, the homeostatic drive for sleep begins to overwhelm the circadian drive for wake, which has now begun to dissipate, and sleep begins. As sleep continues, the homeostatic drive for sleep is quenched, but the circadian drive for sleep offsets this decline and sleep is maintained for about 8 h. When there is a mismatch between the timing of homeostasis and the circadian system, such as occurs during jet lag or shift work, sleep and wake disruptions become evident (see Chapter 7) with potentially significant effects on physical and mental well-being (see Chapters 7, 9, 10, 11). This type of sleep and wake fragmentation is caused by controllable behavioral influences. In normal aging, there appears to be a decline in the strength of the consolidation of both sleep and wake. Healthy older adults will often nap during the daytime and have fragmented sleep at night. Of course, these two behaviors are somewhat reciprocal in that reduced sleep at night will cause tiredness and napping during the day, which decreases homeostatic pressure for sleep and increases awakening at night and so on. Even when kept awake all day, there is an increased sleep fragmentation in healthy older adults.[54] In pathological states of aging, such as AD, there can be even greater fragmentation of sleep and wake. About half of individuals with AD will have nearly a complete loss of sleep and wake consolidation.[55] In other words, the occurrence of sleep and wake becomes nearly random across the day and night. In many epidemiologic studies, this sleep/wake fragmentation is the top- or second-most cited reasons for institutionalization of older adults with AD (with incontinence being similarly ascribed).[27,56] It is possible that a decline of hypocretin release (i.e., modeled as a decline in the circadian wake drive) within the normal physiologic range may be partially responsible for this decline in wake consolidation and that repletion of this peptide might reverse this effect.[57]

The tuberomamillary nucleus (TMN) in the hypothalamus is the site of neurons that express histamine—the only neural source of histamine in the brain. TMN histamine-expressing neurons project widely, innervating many of the wake-promoting brain stem neuromodulatory systems.[58] These TMN neurons are wake-active[59] and histamine itself is wake-promoting.[60] Over-the-counter antihistamines are very commonly used medications to induce sleep, providing *prime facie* evidence for the important role of histamine in wake regulation. Commonly used antihistamines (e.g., diphenhydramine) not only block the effects of histamine on TMN H1 histamine receptors but also have effects on cholinergic receptors, leading to side effects such as dry mouth and urinary retention, as well as peripheral H1 histamine receptors that are involved in response to histamine released by mast cells and basophils. As such, much research has been done to develop molecules that act as agonists or antagonists at the H3 histamine receptor.[61] The H3 histamine receptor is an autoreceptor present only on TMN histamine-expressing neurons, thereby obviating any drug effect on peripheral histamine transduction.[62] As part of an autoinhibitory regulatory feedback loop, activation of the H3 histamine receptor leads to decreased activity of TMN neurons, while inhibition of the H3 histamine receptor leads to increased activity of TMN neurons and increased release of histamine.[63] As such, molecules that agonize the H3 histamine receptor are predicted to decrease brain histamine release and result in increased sleepiness, while molecules that antagonize this receptor would increase brain histamine release and increase alertness. Early studies have confirmed this conjecture[64] and may lead to a new method of treating insomnia and hypersomnia, respectively.

Unlike most of the systems thus far discussed, the VLPO and the temperature-sensitive anterior hypothalamic area are regions involved in the promotion of sleep, rather than wake. The VLPO is a GABA-expressing group of cells that inhibit the firing of most wake-promoting neurons, including the TMN, LC, raphe, and periaqueductal gray.[65] The VLPO is also preferentially active during sleep.[66] It has been hypothesized that the VLPO is a key player in the determination whether a brain is in the wake or sleep state. This has been modeled as a "flip-flop" switch in which the VLPO drives the presence of sleep and is itself controlled by the reciprocal activity of the TMN, raphe, and LC.[67] It is the reciprocal inhibition between the wake-promoting neurons and those of the VLPO that are hypothesized to create a binary environment (either wake or sleep) in which ambiguous states are neurochemically selected against.

Close to the VLPO are a distributed group of neurons in the hypothalamic preoptic area that increase firing in response to warming (so-called warm-sensitive neurons or WSNs).[68] WSNs are preferentially active during sleep, and they exhibit an increase in firing in anticipation of the transition from wake to sleep.[69] Activation of these neurons leads to inhibition of wake-promoting neurons in the lateral hypothalamus and basal forebrain and a subsequent increase in NREMS.[70] Unlike the VLPO, however, these neurons do not appear to have direct, reciprocal innervation from wake-promoting areas of the brain and are, therefore, likely involved in the modulation but not stability of the sleep state. These WSN are also likely not involved in the circadian- and sleep-modulation of body temperature[71] or changes in sleep in response to fever,[72] but may be involved in modulating sleep in response to changes in environmental temperatures.

Robust changes in temperature often accompany immune responses, which are themselves linked to changes in sleep. Molecules nominally linked to the immune system, such as the cytokines interleukin-1β (IL1β) and tumor necrosis factor-α (TNFα), may also play a role under non-pathophysiologic conditions. When given exogenously at nonpyrogenic concentrations, both IL1β and TNFα can increase NREMs.[73] When animals are sleep deprived, both of these cytokines increase in concentration in the cerebrospinal fluid.[74] These and other data have led to the suggestion that levels of IL1β and TNFα may act as a downstream signal of a homeostatic drive for sleep, possible being a stable, biochemical expression of extracellular ATP concentrations.[75] IL1β and TNFα might mediate their hypnogenic effect through activation of sleep-promoting neurons in the preoptic area and inhibition of wake-promoting neurons in the hypothalamus,[76] though their effects may be more widespread. Under conditions of infection, there is a further, immune-stimulated increase of IL1β and TNFα. This may be responsible for the commonly observed increase in sleep during many active and some chronic infections.

3. THE CORTEX

Most of this chapter has discussed the fairly well-characterized effects of subcortical structures on sleep and wake. Vis-à-vis the cortex, these structures can be simply described as coordinating the movement of cortical neurons into "wake" and "sleep" states. There is growing evidence that cortical neurons may exhibit "local" patterns of sleep. One of the first examples in the human literature of sleep occurring at a local level was the demonstration that a motor task that increased wake EEG activity in a specific area of the cortex had a similar increase in slow wave activity in the same area of cortex during sleep.[77] This implies that, to a certain degree, the amount of activity occurring in a group of cortical neurons during sleep is due to their activity during the daytime. It has been hypothesized that the increased activity at night may be part of a local learning process, with synaptic paring and enhancement occurring simultaneously.[78] Relatively little is understood about local networks of sleep, especially the many different anatomical levels that it might occur (e.g., functional region, cortical column, single neuron). Pathophysiologically, the loss of coordinated sleep may be best exemplified by parasomnias such as hypnagogic hallucinations, RBD, and sleep walking. More work in the coming years may discover more subtle forms of desynchronized local cortical sleep, which may lead to some of the memory deficits that occur in a variety of different dementias (e.g., AD, Pick disease).

4. THE ENDOCRINE SYSTEM

In the recent literature, there are a plethora of studies that have examined the effects of sleep on the output of the endocrine system, especially the relationship between disrupted sleep and changes in hormones that are critical to metabolism.[79] While it is well established that many hormones are directly impacted by sleep, there is less literature examining the potentially reciprocal relationship of the impact of these hormones on sleep. One of the most commonly studied hormones is melatonin, most of which is derived from the pineal gland. Melatonin has a simple, near square-wave pattern with blood concentrations rising a few hours before habitual sleep time and falling within an hour or so after habitual wake time. The production of melatonin is tightly controlled by the SCN and modulated by light input such that light acutely suppresses melatonin production in a ose-dependent manner.[80] In many nonmammalian vertebrates (e.g., zebrafish,[81] chickens[82]), melatonin is an important regulator of sleep. In many seasonally breeding mammals, it encodes day length.[83] Its importance in the regulation of human sleep, however, has been extensively debated. Exogenous administration of melatonin has mild hypnotic properties mainly associated with a decrease in sleep latency.[84] The loss of endogenous melatonin, however, secondary to either surgical removal of the pineal or neurologically complete damage to the cervical spinal cord leads to few if any changes in sleep architecture.[85–87] An acute reduction in melatonin concentrations through ocular light exposure is associated with a decrease in alertness, but no causal mechanisms have been established.[88] Low melatonin production also may naturally occurs in many older adults (more as a consequence of being old as opposed to the aging process itself).[89] Supplementation of this low endogenous melatonin with exogenous melatonin has had mixed results, but may have a small impact on sleep onset mechanisms.[84] If anything, endogenous melatonin is likely to evoke a mild hypnotic effect predominantly at sleep onset. Mechanistically, this might occur through suppression of a SCN-derived wake signal[90] or through modulation of thermoregulatory processes that are important for active heat loss at sleep onset.[91] It must be noted, however, that the importance of melatonin to the production of normal sleep in humans is not well established.

Another hormone that has been studied in the context of sleep regulation is cortisol. Cortisol is produced in a highly pulsatile fashion, with low levels in the evening and early sleep period and a rise late in the sleep period that peaks just after habitual wake time. Cortisol, as with melatonin, is under strong circadian control with a minor influence of sleep.[92] At physiologic levels, there is little evidence for the relevance of cortisol to sleep regulation. Individuals with Addison's disease, characterized by a severely diminished capacity to produce cortisol, have relatively normal sleep.[93,94] At elevated concentrations, cortisol may enhance the occurrence of slow wave sleep and inhibit REMS.[93,95] Individuals with Cushing's disease, characterized by constitutively elevated cortisol, have markedly disturbed sleep, exhibiting increased sleep fragmentation and reduced latency to onset of REMS.[96,97] Thus, short-term elevations in cortisol may result in an increase in slow wave sleep while, a chronic elevation in cortisol may result in a form of down-regulation and a reduction in NREMS.

5. THE REST

There are, of course, many other components of sleep/wake regulation, including other immune molecules (e.g., IL1α, IL4, IL6), hormones (e.g., prolactin, epinephrine, thyrotropin-releasing hormone, vasopressin, growth hormone), peptides (e.g., cholecystokinin, vasoactive intestinal peptide), and neuropeptides (e.g., neuropeptide S, cortistatin, somatostatin, corticotropin-releasing factor). The data for these, however, are mixed and less substantial on the role that they have in the regulation of sleep and wake. Given that hypocretin was discovered about 15 years ago and now considered a major regulator of sleep and wake, the roles of these various modulators and how they interact to produce both normal and pathological forms of sleep must be considered a work in progress.

REFERENCES

1. Bremer F. Cerveau isole et physiologie du sommeil. *C R Soc Biol (Paris)*. 1935;118:1235–1241.
2. Moruzzi G, Magoun HW. Brain stem reticular formation and activation of the EEG. *Electroencephalogr Clin Neurophysiol*. 1949;1:455–473.
3. Steriade M, McCormick DA, Sejnowski TJ. Thalamocortical oscillations in the sleeping and aroused brain. *Science*. 1993;262:679–685.
4. Trulson ME, Jacobs BL. Raphe unit activity in freely moving cats: correlation with level of behavioral arousal. *Brain Res*. 1979;163:135–150.
5. Zeitzer JM, Maidment NT, Behnke EJ, et al. Ultradian sleep-cycle variaiton of serotonin in the human lateral ventricle. *Neurology*. 2002;59:1272–1274.
6. Riemann D. Insomnia and comorbid psychiatric disorders. *Sleep Med*. 2007;8(Suppl. 4): S15–S20.
7. Sharpley AL, Cowen PJ. Effect of pharmacologic treatments on the sleep of depressed patients. *Biol Psychiatry*. 1995;37:85–98.
8. Aston-Jones G, Rajkowski J, Cohen J. Role of locus coeruleus in attention and behavioral flexibility. *Biol Psychiatry*. 1999;46:1309–1320.
9. Tsai MH, Huang YS. Attention-deficit/hyperactivity disorder and sleep disorders in children. *Med Clin North Am*. 2010;94:615–632.
10. Mignot E. A hundred years of narcolepsy research. *Arch Ital Biol*. 2001;139:207–220.
11. Broughton R, Valley V, Aguirre M, Roberts J, Suwalski W, Dunham W. Excessive daytime sleepiness and the pathophysiology of narcolepsy-cataplexy: a laboratory perspective. *Sleep*. 1986;9:205–215.
12. Zeitzer JM, Nishino S, Mignot E. The neurobiology of hypocretins (orexins), narcolepsy and therapeutic interventions. *Trends Pharmacol Sci*. 2006;27:368–374.
13. Blanco-Centurion C, Gerashchenko D, Shiromani PJ. Effects of saporin-induced lesions of three arousal populations on daily levels of sleep and wake. *J Neurosci*. 2007;27:14041–14048.
14. Fried I, Wilson CL, Morrow JW, et al. Increased dopamine release in the human amygdala during performance of cognitive tasks. *Nat Neurosci*. 2001;4:201–206.
15. Rye DB, Jankovic J. Emerging views of dopamine in modulating sleep/wake state from an unlikely source: PD. *Neurology*. 2002;58:341–346.
16. Trulson ME, Preussler DW, Howell GA. Activity of substantia nigra units across the sleep-waking cycle in freely moving cats. *Neurosci Lett*. 1981;26:183–188.
17. Lu J, Jhou TC, Saper CB. Identification of wake-active dopaminergic neurons in the ventral periaqueductal gray matter. *J Neurosci*. 2006;26:193–202.
18. Tandberg E, Larsen JP, Karlsen K. A community-based study of sleep disorders in patients with Parkinson's disease. *Mov Disord*. 1998;13:895–899.
19. Abbott RD, Ross GW, White LR, et al. Excessive daytime sleepiness and subsequent development of Parkinson disease. *Neurology*. 2005;65:1442–1446.
20. Siegel JM. The neurobiology of sleep. *Semin Neurol*. 2009;29:277–296.
21. Taepavarapruk N, Taepavarapruk P, John J, et al. State-dependent changes in glutamate, glycine, GABA, and dopamine levels in cat lumbar spinal cord. *J Neurophysiol*. 2008;100:598–608.
22. Schenck CH, Bundlie SR, Mahowald MW. Delayed emergence of a parkinsonian disorder in 38% of 29 older men initially diagnosed with idiopathic rapid eye movement sleep behaviour disorder. *Neurology*. 1996;46:388–393.

23. Hallanger AE, Levey AI, Lee HJ, Rye DB, Wainer BH. The origins of cholinergic and other subcortical afferents to the thalamus in the rat. *J Comp Neurol.* 1987;262:105–124.
24. Rye DB, Wainer BH, Mesulam MM, Mufson EJ, Saper CB. Cortical projections arising from the basal forebrain: a study of cholinergic and noncholinergic components employing combined retrograde tracing and immunohistochemical localization of choline acetyltransferase. *Neuroscience.* 1984;13:627–643.
25. Jones BE. Modulation of cortical activation and behavioral arousal by cholinergic and orexinergic systems. *Ann N Y Acad Sci.* 2008;1129:26–34.
26. Vitiello MV, Borson S. Sleep disturbances in patients with Alzheimer's disease: epidemiology, pathophysiology and treatment. *CNS Drugs.* 2001;15:777–796.
27. Hope T, Keene J, Gedling K, Fairburn CG, Jacoby R. Predictors of institutionalization for people with dementia living at home with a carer. *Int J Geriatr Psychiatry.* 1998;13:682–690.
28. Yaffe K, Fox P, Newcomer R, et al. Patient and caregiver characteristics and nursing home placement in patients with dementia. *JAMA.* 2002;287:2090–2097.
29. Swaab DF, Fliers E, Partiman TS. The suprachiasmatic nucleus of the human brain in relation to sex, age and senile dementia. *Brain Res.* 1985;342:37–44.
30. Friedman L, Zeitzer JM, Kushida CA, et al. Scheduled bright light for treatment of insomnia in older adults. *J Am Geriatr Soc.* 2009;57:441–452.
31. Buysse DJ, Browman KE, Monk TH, Reynolds III CF, Fasiczka AL, Kupfer DJ. Napping and 24-hour sleep/wake patterns in healthy elderly and young adults. *J Am Geriatr Soc.* 1992;40:779–786.
32. Klerman EB, Dijk DJ. Age-related reduction in the maximal capacity for sleep—implications for insomnia. *Curr Biol.* 2008;18:1118–1123.
33. Westerberg CE, Lundgren EM, Florczak SM, et al. Sleep influences the severity of memory disruption in amnestic mild cognitive impairment: results from sleep self-assessment and continuous activity monitoring. *Alzheimer Dis Assoc Disord.* 2010;24:325–333.
34. Von Economo C. *Encephalitis Lethargica: Its Sequelae and Treatment.* London: Oxford University Press; 1931.
35. Peyron C, Faraco J, Rogers W, et al. A mutation in a case of early onset narcolepsy and a generalized absence of hypocretin peptides in human narcoleptic brains. *Nat Med.* 2000;6:991–997.
36. Kornum BR, Faraco J, Mignot E. Narcolepsy with hypocretin/orexin deficiency, infections and autoimmunity of the brain. *Curr Opin Neurobiol.* 2011;21:897–903.
37. Aran A, Lin L, Nevsimalova S, et al. Elevated anti-streptococcal antibodies in patients with recent narcolepsy onset. *Sleep.* 2009;32:979–983.
38. Dauvilliers Y, Montplaisir J, Cochen V, et al. Post-H1N1 narcolepsy-cataplexy. *Sleep.* 2010;33:1428–1430.
39. Salomon RM, Ripley B, Kennedy JS, et al. Diurnal variation of CSF hypocretin-1 (orexin-A) levels in control and depressed subjects. *Biol Psychiatry.* 2003;54:96–104.
40. Zeitzer JM, Buckmaster CL, Parker KJ, Hauck CM, Lyons DM, Mignot E. Circadian and homeostatic regulation of hypocretin in a primate model: implications for the consolidation of wakefulness. *J Neurosci.* 2003;23:3555–3560.
41. Zeitzer JM, Mignot E. Role of hypocretin/orexin in the neurobiology of sleep and alertness. In: Bassetti C, Billiard M, Mignot E, eds. *Narcolepsy and Hypersomnia.* New York: Informa Healthcare; 2006:359–374.
42. Yamuy J, Fung SJ, Xi M, Chase MH. Hypocretinergic control of spinal cord motoneurons. *J Neurosci.* 2004;24:5336–5345.
43. Borbély AA. A two process model of sleep regulation. *Hum Neurobiol.* 1982;1:195–204.
44. Dijk D-J, Czeisler CA. Paradoxical timing of the circadian rhythm of sleep propensity serves to consolidate sleep and wakefulness in humans. *Neurosci Lett.* 1994;166:63–68.
45. Cooper SJ. From Claude Bernard to Walter Cannon. Emergence of the concept of homeostasis. *Appetite.* 2008;51:419–427.

46. Porkka-Heiskanen T, Strecker RE, Thakkar M, Bjorkum AA, Greene RW, McCarley RW. Adenosine: a mediator of the sleep-inducing effects of prolonged wakefulness. *Science*. 1997;276:1265–1268.
47. Urade Y, Hayaishi O. Prostaglandin D2 and sleep/wake regulation. *Sleep Med Rev*. 2011;15:411–418.
48. Saper CB, Romanovsky AA, Scammell TE. Neural circuitry engaged by prostaglandins during the sickness syndrome. *Nat Neurosci*. 2012;15:1088–1095.
49. McCarley RW. Neurobiology of REM and NREM sleep. *Sleep Med*. 2007;8:302–330.
50. Porkka-Heiskanen T, Strecker RE, McCarley RW. Brain site-specificity of extracellular adenosine concentration changes during sleep deprivation and spontaneous sleep: an *in vivo* microdialysis study. *Neuroscience*. 2000;99:507–517.
51. Zeitzer JM, Moreales-Villagran A, Maidment NT, et al. Extracellular adenosine in the human brain during sleep and sleep deprivation: an *in vivo* microdialysis study. *Sleep*. 2006;29:455–461.
52. Fredholm BB, Bättig K, Holmén J, Nehlig A, Zvartau EE. Actions of caffeine in the brain with special reference to factors that contribute to its widespread use. *Pharmcol Rev*. 1999;51:83–133.
53. Lazarus M, Shen HY, Cherasse Y, et al. Arousal effect of caffeine depends on adenosine A2A receptors in the shell of the nucleus accumbens. *J Neurosci*. 2011;31:10067–10075.
54. Klerman EB, Davis JB, Duffy JF, Dijk D-J, Kronauer RE. Older people awaken more frequently but fall back asleep at the same rate as younger people. *Sleep*. 2004;27:793–798.
55. Prinz PN, Peskind ER, Vitaliano PP, et al. Changes in the sleep and waking EEGs of nondemented and demented elderly subjects. *J Am Geriatr Soc*. 1982;30:86–93.
56. Bianchetti A, Scuratti A, Zanetti O, et al. Predictors of mortality and institutionalization in Alzheimer disease patients 1 year after discharge from an Alzheimer dementia unit. *Dementia*. 1995;6:108–112.
57. Friedman L, Zeitzer JM, Lin L, et al. Hypocretin-1 is inversely associated with wake fragmentation in Alzheimer's disease. *Neurology*. 2007;68:793–794.
58. Panula P, Pirvola U, Auvinen S, Airaksinen MS. Histamine-immunoreactive nerve fibers in the rat brain. *Neuroscience*. 1989;28:585–610.
59. Steininger TL, Alam MN, Gong H, Szymusiak R, McGinty D. Sleep-waking discharge of neurons in the posterior lateral hypothalamus of the albino rat. *Brain Res*. 1999;840:138–147.
60. Lin JS. Brain structures and mechanisms involved in the control of cortical activation and wakefulness, with emphasis on the posterior hypothalamus and histaminergic neurons. *Sleep Med Rev*. 2000;4:471–503.
61. Passani MB, Lin J-S, Hancock A, Crochet S, Blandina P. The histamine H3 receptor as a novel therapeutic target for cognitive and sleep disorders. *Trends Pharmacol Sci*. 2004;25:618–625.
62. Arrang JM, Garbarg M, Schwartz JC. Auto-inhibition of brain histamine release mediated by a novel class (H3) of histamine receptor. *Nature*. 1983;302:832–837.
63. Itoh Y, Oishi R, Nishibori M, Saeki K. Characterization of histamine release from the rat hypothalamus as measured by in vivo microdialysis. *J Neurochem*. 1991;56:769–774.
64. Lin J-S, Dauvilliers Y, Arnulf I, et al. An inverse agonist of the histamine H3 receptor improves wakefulness in narcolepsy: studies in orexin-/- mice and patients. *Neurobiol Dis*. 2008;30:74–83.
65. Sherin JE, Elmquist JK, Torrealba F, Saper CB. Innervation of histaminergic tuberomammillary neurons by GABAergic and galaninergic neurons in the ventrolateral preoptic nucleus of the rat. *J Neurosci*. 1998;18:4705–4721.
66. Sherin JE, Shiromani PJ, McCarley RW, Saper CB. Activation of ventrolateral preoptic neurons during sleep. *Science*. 1996;271:216–219.

67. Saper CB, Cano G, Scammell TE. Homeostatic, circadian, and emotional regulation of sleep. *J Comp Neurol.* 2005;493:92–98.
68. Boulant JA. Hypothalamic mechanisms in thermoregulation. *Fed Proc.* 1981;40: 2843–2850.
69. Alam MN, McGinty D, Szymusiak R. Neuronal discharge of preoptic/anterior hypothalamic thermosensitive neurons: relation to NREM sleep. *Am J Physiol.* 1995;269: R1240–R1249.
70. McGinty D, Szymusiak R, Thomson D. Preoptic/anterior hypothalamic warming increases EEG delta frequency activity within non-rapid eye movement sleep. *Brain Res.* 1994;667:273–277.
71. Lu J, Greco MA, Shiromani P, Saper CB. Effect of lesions of the ventrolateral preoptic nucleus on NREM and REM sleep. *J Neurosci.* 2000;20:3830–3842.
72. Krueger JM, Takahashi S. Thermoregulation and sleep. Closely linked but separable. *Ann N Y Acad Sci.* 1997;813:281–286.
73. Shoham S, Davenne D, Cady AB, Dinarello CA, Krueger JM. Recombinant tumor necrosis factor and interleukin 1 enhance slow-wave sleep. *Am J Physiol.* 1987;253: R142–R149.
74. Lue FA, Bail M, Jephthah-Ochola J, Carayanniotis K, Gorczynski R, Moldofsky H. Sleep and cerebrospinal fluid interleukin-1-like activity in the cat. *Int J Neurosci.* 1988;42:179–183.
75. Krueger JM, Clinton JM, Winters BD, et al. Involvement of cytokines in slow wave sleep. *Prog Brain Res.* 2011;193:39–47.
76. Alam MN, McGinty D, Bashir T, et al. Interleukin-1beta modulates state-dependent discharge activity of preoptic area and basal forebrain neurons: role in sleep regulation. *Eur J Neurosci.* 2004;20:207–216.
77. Huber R, Ghilardi MF, Massimini M, Tononi G. Local sleep and learning. *Nature.* 2004;430:78–81.
78. Tononi G, Cirelli C. Sleep function and synaptic homeostasis. *Sleep Med Rev.* 2006;10:49–62.
79. Spiegel K, Tasali E, Leproult R, van Cauter E. Effects of poor and short sleep on glucose metabolism and obesity risk. *Nat Rev Endocrinol.* 2009;5:253–261.
80. Zeitzer JM, Dijk D-J, Kronauer RE, Brown EN, Czeisler CA. Sensitivity of the human circadian pacemaker to nocturnal light: melatonin phase resetting and suppression. *J Physiol.* 2000;526:695–702.
81. Zhdanova IV, Wang SY, Leclair OU, Danilova NP. Melatonin promotes sleep-like state in zebrafish. *Brain Res.* 2001;903:263–268.
82. Hishikawa Y, Cramer H, Kuhlo W. Natural and melatonin-induced sleep in young chickens—a behavioral and electrographic study. *Exp Brain Res.* 1969;7:84–94.
83. Goldman BD. Mammalian photoperiodic system: formal properties and neuroendocrine mechanisms of photoperiodic time measurement. *J Biol Rhythms.* 2001;16: 283–301.
84. Buscemi N, Vandermeer B, Hooton N, et al. Efficacy and safety of exogenous melatonin for secondary sleep disorders and sleep disorders accompanying sleep restriction: meta-analysis. *Br Med J.* 2006;332:385–393.
85. Kocher L, Brun J, Borson-Chazot F, Gonnaud P-M, Claustrat B. Increased REM sleep associated with melatonin deficiency after pinealectomy: a case study. *Chronobiol Intl.* 2006;23:889–901.
86. Scheer FAJL, Zeitzer JM, Ayas NT, Brown R, Czeisler CA, Shea SA. Reduced sleep efficiency in cervical spinal cord injury; association with abolished night time melatonin secretion. *Spinal Cord.* 2006;44:78–81.
87. Krieg SM, Slawik H, Meyer B, Wiegand M, Stoffel M. Sleep disturbance after pinealectomy in patients with pineocytoma WHO°I. *Acta Neurochir (Wien).* 2012;154:1399–1405.

88. Cajochen C, Zeitzer JM, Czeisler CA, Dijk D-J. Dose-response relationship for light intensity and ocular and electroencephalographic correlates of alertness in humans. *Behav Brain Res*. 2000;115:75–83.
89. Zeitzer JM, Daniels JE, Duffy JF, et al. Plasma melatonin concentration: does it decline with age? *Am J Med*. 1999;107:432–436.
90. von Gall C, Stehle JH, Weaver DR. Mammalian melatonin receptors: molecular biology and signal transduction. *Cell Tissue Res*. 2002;309:151–162.
91. Viswanathan M, Laitinen JT, Saavedra JM. Expression of melatonin receptors in arteries involved in thermoregulation. *Proc Natl Acad Sci USA*. 1990;87:6200–6203.
92. Weitzman ED, Zimmerman JC, Czeisler CA, Ronda J. Cortisol secretion is inhibited during sleep in normal man. *J Clin Endocrinol Metab*. 1983;56:352–358.
93. Gillin JC, Jacobs LS, Snyder F, Henkin RI. Effects of decreased adrenal corticosteroids: changes in sleep in normal subjects and patients with adrenal cortical insufficiency. *Electroencephalogr Clin Neurophysiol*. 1974;36:283–289.
94. Lovas K, Husebye ES, Holsten F, Bjorvatn B. Sleep disturbances in patients with Addison's disease. *Eur J Endocrinol*. 2003;148:449–456.
95. Born J, DeKloet ER, Wenz H, Kern W, Fehm HL. Gluco- and antimineralocorticoid effects on human sleep: a role of central corticosteroid receptors. *Am J Physiol*. 1991;260: E183–E188.
96. Shipley JE, Schteingart DE, Tandon R, Starkman MN. Sleep architecture and sleep apnea in patients with Cushing's disease. *Sleep*. 1992;15:514–518.
97. Krieger DT, Glick SM. Sleep EEG stages and plasma growth hormone concentration in states of endogenous and exogenous hypercortisolemia or ACTH elevation. *J Clin Endocrinol Metab*. 1974;39:986–1000.

CHAPTER SEVEN

Circadian Rhythms, Sleep Deprivation, and Human Performance

Namni Goel[*], Mathias Basner[*], Hengyi Rao[*,†], David F. Dinges[*]
[*]Division of Sleep and Chronobiology, Department of Psychiatry, Perelman School of Medicine, University of Pennsylvania, Philadelphia, Pennsylvania, USA
[†]Department of Neurology, Center for Functional Neuroimaging, Perelman School of Medicine, University of Pennsylvania, Philadelphia, Pennsylvania, USA

Contents

Abstract

Much of the current science on, and mathematical modeling of, dynamic changes in human performance within and between days is dominated by the two-process model of sleep–wake regulation, which posits a neurobiological drive for sleep that varies homeostatically (increasing as a saturating exponential during wakefulness and decreasing in a like manner during sleep), and a circadian process that neurobiologically modulates both the homeostatic drive for sleep and waking alertness and performance.

Progress in Molecular Biology and Translational Science, Volume 119
ISSN 1877-1173
http://dx.doi.org/10.1016/B978-0-12-396971-2.00007-5

Endogenous circadian rhythms in neurobehavioral functions, including physiological alertness and cognitive performance, have been demonstrated using special laboratory protocols that reveal the interaction of the biological clock with the sleep homeostatic drive. Individual differences in circadian rhythms and genetic and other components underlying such differences also influence waking neurobehavioral functions. Both acute total sleep deprivation and chronic sleep restriction increase homeostatic sleep drive and degrade waking neurobehavioral functions as reflected in sleepiness, attention, cognitive speed, and memory. Recent evidence indicating a high degree of stability in neurobehavioral responses to sleep loss suggests that these trait-like individual differences are phenotypic and likely involve genetic components, including circadian genes. Recent experiments have revealed both sleep homeostatic and circadian effects on brain metabolism and neural activation. Investigation of the neural and genetic mechanisms underlying the dynamically complex interaction between sleep homeostasis and circadian systems is beginning. A key goal of this work is to identify biomarkers that accurately predict human performance in situations in which the circadian and sleep homeostatic systems are perturbed.

1. INTRODUCTION

Sleep is a ubiquitous biological imperative that appears to be evolutionarily conserved across species.[1] Sleep of sufficient duration, continuity, and intensity (depth) without circadian disruption is necessary to promote high levels of attention and cognitive performance during the wake period, and to prevent physiological changes that may predispose individuals to adverse health outcomes.[2] The evidence linking habitually short sleep or circadian desynchrony to conditions such as weight gain,[3,4] obesity,[5] diabetes,[6] and hypertension,[7] as well as to increased mortality,[8] has accumulated over the past decade. These negative cognitive and health consequences of sleep restriction are provocative, given that current representative surveys indicate 35–40% of the adult US population report sleeping less than 7 h on weekday nights,[9] which has been experimentally demonstrated to result in cumulative deficits in behavioral alertness and vigilant attention.[10]

A lifestyle of chronic partial sleep loss that is often paired with chronic stimulant use (e.g., caffeine)[11] may at least in part be explained by the fact that humans frequently alter the timing and duration of sleep in exchange for other activities. This altered behavior appears to be prevalent in current industrialized societies, where the biological imperative to sleep adequately often opposes the cultural imperative to spend more time awake.[12] Sleep may be perceived as a flexible commodity that is traded for other activities considered more pressing or of greater value.[13] Analyses of the American Time Use Survey (ATUS) revealed that paid work time and commuting to and from work were the two waking activities most often exchanged for sleep time.[14] Sleep time was lowest in the 45- to 54-year-old respondents, shorter in men than in women, and shorter on weekdays compared to weekends. An ATUS analysis on waking activities in the 2-h period before retiring in the evening and after waking up in the morning showed that watching TV was the dominant (>50%) activity in the 2 h before retiring.[15] Long work hours were associated with progressively earlier wake-up times in the morning, while long-hour workers, short-hour workers, and those who did not work did not differ in the times when they retired at night.[15] We speculate that some of this sleep-restriction behavior may be explained by respondents with a late evening circadian phase preference, who awaken early by alarm clock to commute for paid work. These individuals cannot easily advance their sleep onset, but they can use an alarm clock to advance their sleep offset (for commuting and paid work), resulting in a restricted sleep period. This misalignment of biological and social time has been termed "social jet lag" by Roenneberg and colleagues.[16] Individuals with a late circadian preference thus often engage in chronic sleep restriction during the work week, and try to pay off their sleep debt on the weekend. Furthermore, shift work affects sleep and alertness of approximately one out of five working Americans, with 15% of full-time salaried workers usually working shifts that include nights.[17] Shift work includes working evenings, nights, or rotating shifts and is often associated with shorter-than-normal and disrupted sleep periods at an adverse circadian phase.[18] The International Agency for Research on Cancer concluded in 2007 that shift work involving circadian disruption is probably carcinogenic to humans.[17,19]

2. SLEEP–WAKE AND CIRCADIAN REGULATION: TWO-PROCESS MODEL

The two-process model of sleep–wake regulation has been applied to the temporal profiles of sleep[20,21] and daytime vigilance.[22] The model consists of a homeostatic process (S) and a circadian process (C), which combine to determine the timing of sleep onset and offset. The homeostatic process represents the drive for sleep that increases as a saturating exponential during wakefulness (as can be observed when wakefulness is maintained beyond habitual bedtime into the night and subsequent day) and decreases as a saturating exponential during sleep (which represents recuperation obtained from sleep). When the homeostat increases above a certain threshold, sleep is triggered; when it decreases below a different threshold, wakefulness occurs. The circadian process represents daily oscillatory modulation of these threshold levels. It has been suggested that the circadian system actively promotes wakefulness more than sleep.[23] The circadian drive for wakefulness may be manifested as spontaneously enhanced alertness and better cognitive performance in the early evening after one night or multiple nights without sleep[24,25] (Figs. 7.1 and 7.2).

The endogenous circadian regulating system (i.e., biological clock) that modulates the timing of both sleep and wakefulness is located in the suprachiasmatic nuclei (SCN) of the anterior hypothalamus. Beyond gating the timing of sleep onset and offset, the SCN modulates waking behavior in a circadian manner, as reflected in subjective and physiological sleepiness, behavioral alertness, and a number of fundamental cognitive functions, including vigilant attention, psychomotor and perceptual cognitive speed, and working memory.

Alertness and performance, sleep and sleeplessness are neurobehavioral outputs that involve dynamic circadian variation. Recent forced desynchrony protocols, which serve to experimentally reveal the variance in neurobehavioral functions attributable primarily to endogenous circadian control and the variance attributable primarily to the sleep homeostatic drive, have revealed that circadian dynamics can expose large neurobehavioral vulnerability during chronic sleep restriction.[26,27] These studies demonstrated that sleep restriction induced decreased vigilant attention, as measured by the Psychomotor Vigilance Test (PVT),[25] most prominently during circadian night, even with short prior wake duration. Another study found that time of day modulated the effects of chronic sleep restriction, whereby the build-up rate of cumulative neurobehavioral deficits across days was largest at 0800 h and became progressively smaller across the hours of the day, especially between 1600 and 2000 h, indicating a late afternoon/early evening period of relatively protected alertness.[28]

Thus, while the two-process model has been very successful in explaining changes in neurobehavioral performance in acute total sleep deprivation paradigms, it fails to adequately predict the escalating declines in vigilant attention observed under chronic sleep-restriction conditions.

The two-process model has proved to be most useful for generating mathematical predictions of the dynamics of human alertness and performance under varying conditions of sleep loss and circadian misalignment.[29]

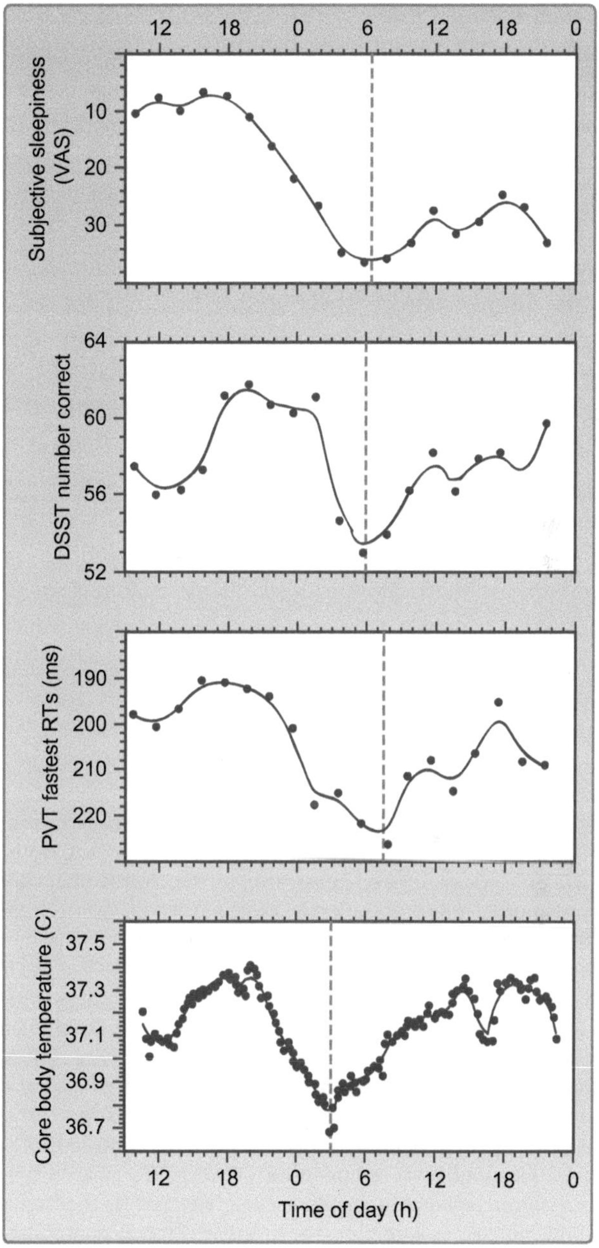

Figure 7.1 See legend on next page

When these mathematical models are compared to emerging experimental data on performance relative to sleep–wake dynamics, they often reveal new deficiencies in the two-process model.[30] An excellent example of this need to continually improve the predictions of the two-process model can be found in a mathematical modeling paper by McCauley and colleagues,[31] who showed that the two-process model belongs to a broader class of models formulated in terms of coupled nonhomogeneous first-order ordinary differential equations. Their new model includes an additional component modulating the homeostatic process across days and weeks, and better reflects the neurobehavioral changes observed under both acute total and chronic partial sleep loss than the original two-process model. The authors speculate that adenosine receptor upregulation (wakefulness) and downregulation (sleep) constitute the underlying neurobiological mechanism. Importantly, the model predicts a critical amount of daily wake duration of 20.2 h. If daily wake duration is above ca. 15.8 h[32] but below 20.2 h (corresponding to a total sleep time of 3.8–8.2 h), the model, over a period of weeks, converges to an asymptotically stable equilibrium (i.e., performance deficits will stabilize at a certain level). If daily wake duration is above 20.2 h, the model diverges and, similar to acute total sleep deprivation, performance impairments escalate.[31] The model of McCauley *et al.*[31] also

Figure 7.1 Circadian variation across a 40-h period of wakefulness in measures of subjective sleepiness as assessed by visual analogue scale (VAS, note reversed scale direction); in cognitive performance speed as assessed by the digit symbol substitution task (DSST); in psychomotor speed as reflected in the 10% fastest reaction times (RT) assessed by the Psychomotor Vigilance Test (PVT); and in core body temperature (CBT) as assessed by a rectal thermistor. Data shown are the mean values from five subjects who remained awake in dim light, in bed, in a constant routine protocol, for 36 h consecutively (a distance-weighted least-squares function was fitted to each variable). The circadian trough is evident in each variable (marked by vertical broken lines). A phase difference is also apparent such that all three neurobehavioral variables had their average minimum between 3.0 and 4.5 h after the time of the body temperature minimum. This phase delay in neurobehavioral functions relative to CBT has been consistently observed. Although body temperature reflects predominantly the endogenous circadian clock, neurobehavioral functions are also affected by the homeostatic pressure for sleep, which escalates with time awake and which may contribute to the phase delay through interaction with the circadian clock. Neurobehavioral functions usually show a circadian decline at night as is observed in CBT, but they continue their decline after CBT begins to rise, making the subsequent 2–6 h period (clock time approximately 0600–1000 h) a zone of maximum vulnerability to loss of alertness and to performance failure. *Reprinted with permission from Ref. 256.* (See Color Insert.)

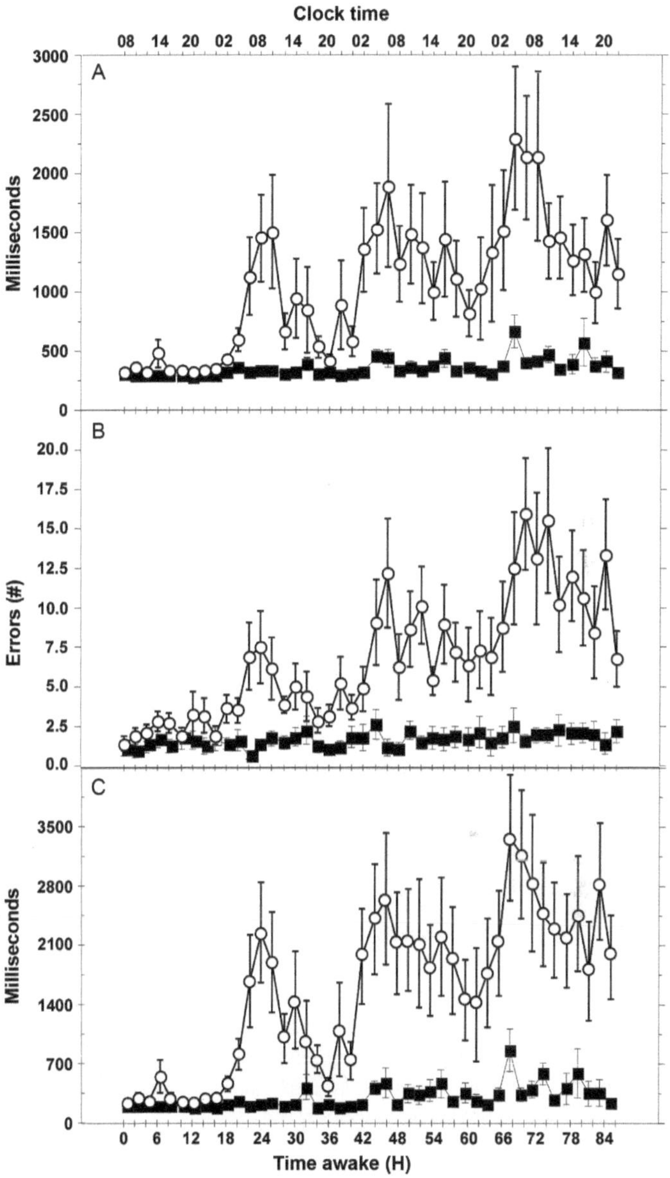

Figure 7.2 See legend on next page

predicts that a single night of recovery sleep is inadequate to recover from a prolonged period of sleep restriction, a finding we recently experimentally confirmed.[33] It is recognized that further model development is needed to integrate more comprehensive mathematical models of the circadian component and to account for sleep inertia and trait-like individual differences in vulnerability to sleep loss.[31]

3. CIRCADIAN RHYTHMS OF PERFORMANCE

3.1. Subjective measures of sleepiness and alertness

A variety of subjective measures of sleepiness and alertness reflect circadian variation, as long as the scale requests ratings about the near immediate state of the subject. These include visual analogue scales,[34] Likert-type rating scales such as the Stanford Sleepiness Scale[35] and the Karolinska Sleepiness Scale,[36] and certain fatigue-related subscales of standard adjective checklists such as the Activation–Deactivation Adjective Check List[37] and Profile of Mood States.[38] Despite structural differences among these scales, all self-reports of sleepiness are highly intercorrelated and because they are relative psychometrics, they are subject to a number of sources of variance, including different uses of the scale response range by different subjects. The effects of cognitive performance testing on subsequent posttest subjective alertness ratings are evident only when sleep loss has commenced and this effect is modulated by circadian variation.[28]

3.2. Objective measures of cognitive performance

Many studies rely on objective performance measures to track the temporal dynamics of endogenous circadian rhythmicity. Circadian variation in

Figure 7.2 Psychomotor Vigilance Test (PVT) performance parameters of healthy adults during an 88 h period of limited to no sleep in the laboratory. The open circles represent 13 subjects undergoing 88 h of total sleep deprivation, and the filled squares represent 15 control subjects given a 2-h time in bed nap opportunity once every 12 h (0245–0445 h and 1445–1645 h) throughout the 88 h period (nap times are not shown in the figure). Graph A: mean (SEM) PVT reaction times (RT), which as RTs > 500 ms are frank errors of omission and referred to as lapses of attention (i.e., responding too slowly). Graph B: mean (SEM) PVT errors of commission, which result from premature responses and reflect impulsiveness (i.e., responding too fast). Graph C: mean (SEM) of PVT RT standard deviations for each test bout, reflecting the magnitude of interindividual differences in performance. The subjects who underwent 88 h without sleep showed clear circadian variation in both lapses of attention (A) and premature responses (B), as well as interindividual differences in these effects (C). *Figure adapted and modified with permission from Ref. 24.*

performance is most evident when sleep loss is present, and sleep loss has its largest effects on attention, working memory, and cognitive throughput.[39] Examples of such cognitive performance measures that have historically been reported to display circadian variation include the following: search-and-detection tasks[40] and simple and choice reaction time tasks,[41] sorting,[42] logical reasoning,[43] memory access,[44] and real-world tasks such as meter reading accuracy[45] and school performance.[46] Typically, response speed and accuracy to a series of repetitive stimuli are analyzed, although the sensitivity of the performance metric used to track circadian variation depends on whether the task is work-paced versus subject-paced, on speed versus accuracy trade-offs in performance metrics,[47] on the rate and number of responses acquired during the task, on whether the task metrics reflect performance variability or mean performance, and on the overall technical precision of the measurement. Even short-duration, work-paced tasks that precisely measure variability in performance can be used to demonstrate circadian variation.[48] It is likely that the modulatory effects the circadian system has on speed and accuracy make many tasks sensitive to process C, more so than any unique aspect of task demand.

Earlier studies concluded that different tasks[49,50] and different task outcomes[51,52] may yield distinct peak phases of circadian rhythmicity, suggesting that many distinct circadian rhythms utilizing different clock mechanisms exist.[53,54] Under strictly controlled laboratory conditions, most intertask differences in circadian variation disappear.[55,56] As illustrated in Fig. 7.1, under controlled sleep deprivation conditions, the circadian rhythms of neurobehavioral performance variables covary with each other and with subjective sleepiness. Importantly, these rhythms mimic the circadian profile of core body temperature, one conventional marker of the biological clock.[57,58] Under entrained conditions, higher and lower core body temperature values typically correspond to good and poor performance, respectively.[56,59,60]

3.3. Masking factors

Subjective measures of sleepiness and alertness are vulnerable to numerous confounding influences that can "mask" their circadian rhythmicity. Masking refers to the evoked effects of noncircadian factors on measurements of circadian rhythmicity. The context in which such measurements are taken (i.e., the environmental and experimental conditions) is a major source of masking effects. Masking can alter or obscure a circadian rhythm, or create the appearance of a circadian rhythm. Masking factors specifically affecting sleepiness and alertness include the following demand characteristics of the experiment,[61] distractions by environmental stimuli and noise,[62] boredom and motivational factors,[63–65] stress,[66] food intake,[67,68] posture and activity,[69,70] ambient temperature,[64] lighting conditions,[71,72] and stimulant drug intake (e.g., caffeine, modafinil, amphetamine).[73–75]

Physical, mental, and social activities can represent masking factors that interact with endogenous circadian rhythms in neurobehavioral functions. The effects of performing cognitive tests on subjective estimates of alertness are apparent at certain circadian phases during sleep deprivation. Subjects report feeling less alert after they are challenged to perform. Thus, prior activity can influence subjective estimates, and can interact with circadian effects if not properly controlled when measuring the rhythmicity of subjective states.

Sleep and sleep loss can also be considered masking factors when measuring circadian rhythmicity in certain neurobehavioral variables. Therefore, neurobehavioral measures reflect to varying degrees a combination of endogenous circadian rhythmicity, sleep homeostatic drive, and masking effects interacting to produce behavioral outcomes.

4. PROTOCOLS TO ASSESS CIRCADIAN VARIATION IN NEUROBEHAVIORAL FUNCTIONS

Considerable research has been devoted to unmasking circadian rhythms, that is, eliminating sources of extraneous variance to expose the endogenous circadian rhythms of variables of interest, including alertness and cognitive performance. Two such experimental approaches are the use of a constant routine protocol and the use of a forced desynchrony protocol.

4.1. Constant routine

The constant routine procedure[76] is generally regarded as the gold standard for measuring circadian rhythmicity. By keeping subjects awake with a fixed posture in a constant laboratory environment for at least 24 h, circadian rhythms in a variety of physiologic and neurobehavioral variables can be recorded without biases (Fig. 7.1). Indeed, the circadian rhythm of body temperature is believed to be largely free of masking effects when derived under a constant routine.

However, when neurobehavioral variables are considered, sleep deprivation and the stimuli used to sustain wakefulness can constitute masking factors. In constant routine experiments, these masking effects are evident in subjective measures of sleepiness and alertness.[57,77] Figure 7.1 shows the somewhat reduced values for subjective alertness as well as cognitive and psychomotor performance after 30 h awake in a constant routine, compared with the values of these variables 24 h earlier (i.e., at the same circadian phase but without sleep deprivation).

Recently, the constant routine protocol has been used to examine metabolites in saliva and plasma at different times of day to identify those that are under circadian control and are independent of sleep.[78,79] Remarkably, one study found that metabolites from blood taken every 2 h, which were used to form a circadian timetable, could subsequently be used to predict internal time within a 3-h interval using only two blood samples.[80] More recently, a constant routine was used to examine the effects of chronic sleep restriction on circadian rhythmicity and amplitude of genes that were upregulated or downregulated using a transcriptome analysis, highlighting the critical interaction between sleep homeostasis and circadian rhythms at the mRNA level.[81]

A progressive change associated with the time spent awake is typically superimposed on the circadian rhythm of neurobehavioral variables.[82,83] When total sleep deprivation is continued for several days (whether in a constant routine procedure or an experimental design involving ambulation), the detrimental effects on alertness and performance increase, and although the circadian process can be exposed,[84] it is overlaid on a continuing (nearly linear) change reflecting increasing homeostatic pressure for sleep.[85] This is illustrated in Fig. 7.2 for PVT performance lapses—perhaps the most sensitive waking measure of homeostatic sleep drive and circadian phase, and the least masked by aptitude and learning.[25,86] It is noteworthy that decreased alertness during the circadian trough is associated with increased intra-individual variability in performance. This is evidenced by intermittent lapsing (reaction times >500 ms)[87] which reflects wake state instability.[24,25] The wake state instability hypothesis posits that sleep-initiating mechanisms may interfere with wakefulness, making sustained performance unstable and dependent on compensatory mechanisms.[25]

4.2. Forced desynchrony

The forced desynchrony protocol[88,89] conducted in temporally and environmentally isolated conditions, is an experimental procedure particularly suitable for studying the interaction of the circadian and homeostatic processes.[55,90,91] In this protocol, a subject's imposed timing and duration of wake and sleep (typically maintained in a 2:1 ratio) deviate from the normal 24-h day (e.g., 20- or 28-h days), such that the subject's biological clock is unable to entrain to this schedule. The subject experiences two distinct influences simultaneously—the schedule of predetermined sleep and waking times representing the homeostatic system and the rhythm of the subject's unsynchronized (i.e., free-running) circadian system. Neurobehavioral functions are assayed during the waking periods. By folding the data over either the free-running circadian rhythm or the imposed sleep–wake cycle, the other component can be balanced out. Thus, the separate effects of circadian rhythms and wake duration (i.e., homeostatic drive for sleep) on neurobehavioral variables can be assessed.

Forced desynchrony studies have found that both the circadian and homeostatic processes influence sleepiness and performance.[26,27] The interaction of the two systems is oppositional during diurnal wake periods (from approximately 0700 h until 2300 h), such that a relatively stable level of alertness and performance can be maintained throughout the day.[89,90] This explains why in many studies of alertness and performance, very little temporal variation is observed during the waking portion of a normal day, especially when there is no sleep deprivation[24] (Fig. 7.2).

The interaction of the homeostatic and circadian processes is believed to be nonlinear (i.e., nonadditive).[90,92] Therefore, the separation of circadian and homeostatic influences on neurobehavioral variables presents a conceptual and mathematical challenge, and it is difficult, if not impossible, to quantify the relative importance of the two influences on neurobehavioral functions. Moreover, their relative contributions may vary across different experimental conditions[55,90] and among subjects.[93]

5. INTERINDIVIDUAL VARIABILITY IN CIRCADIAN RHYTHMS

Healthy adults show interindividual differences in the free-running circadian period (tau),[94–98] which shows robust stability within individuals.[97] Subjects also demonstrate interindividual differences in circadian amplitude[58,99] and circadian phase[57,58,95,99] which are in part due to genetic influences.[99] There are several standardized methods for assessing interindividual differences in circadian rhythms. One newer method, using molecular techniques, can determine individual differences in tau, amplitude, and phase-resetting, which relate to diurnal phase preference, using cultured human fibroblasts from skin biopsies or blood samples.[100–102] While these *in vitro* skin fibroblasts can determine circadian rhythms and period, they do not necessarily correlate with *in vivo* physiological rhythms,[97] limiting the validity and utility of this technique. Standard physiological estimates of circadian phase include the dim light melatonin onset[103] and core body temperature minimum.[57,58] These methods are important for characterizing interindividual variation in circadian rhythmicity.

5.1. Chronotype (morningness–eveningness)

Chronotype or morningness–eveningness (i.e., the tendency to be an early "lark" or a late "owl") is perhaps the most frequently measured interindividual variation in circadian rhythmicity. Morning- and evening-type individuals differ endogenously in the circadian phase of their biological clocks.[57,58] Self-report questionnaires, such as the Horne–Östberg morningness–eveningness questionnaire[104] and its variants,[105] and the Munich ChronoType Questionnaire,[106,107] differentiate timing of activities on workdays versus free days. They are the most commonly utilized measures of circadian phase preference, because of their convenience and cost effectiveness.

Age affects morningness–eveningness as shown in laboratory studies[108] and more naturalistic population-based settings.[107,109] In addition, gender influences morningness–eveningness with women showing a greater skew toward morningness than men.[107,110–112] Women also have been reported to have a shorter average intrinsic circadian period than men,[113] though not consistently,[114] and blacks have been reported to have a shorter average intrinsic circadian period than whites.[114] These differences in circadian phase preference (and possibly in circadian period) appear to be enduring traits, with a significant genetic basis across various diverse populations.[115–120] As such, chronotype is a phenotypic aspect of circadian rhythmicity in humans.[121]

In line with the two-process model, the relationship of chronotype to the regulation of sleep and neurobehavioral responses to sleep deprivation has been investigated in laboratory studies. Chronotypes showed differences in homeostatic sleep regulation[122–124] and in homeostatic response to sleep fragmentation.[125] Moreover, chronotypes showed differences in neurobehavioral responses to sleep fragmentation[126] and to total sleep deprivation,[127] and to risk-taking propensity at baseline and following total sleep deprivation.[128]

5.2. Genetics of individual differences in chronotype and circadian rhythms

Morningness–eveningness is estimated to be about 50% heritable.[129] The genetic basis of morningness–eveningness in the general population has been investigated by targeting several core circadian genes, producing inconsistent results.[130] For example, the 3111C allele of the *CLOCK* gene 5′-UTR region has been associated with eveningness and delayed sleep timing in some studies[131–133] but not others.[98,134–138] Similarly, the variable number tandem repeat (VNTR) polymorphism in *PERIOD3* (*PER3*), another core clock gene, has been linked to diurnal preference, but not consistently,[135,139–148] thereby underscoring the need for further investigation on this topic. Both the 111G polymorphism in the 5′-untranslated region of *PERIOD2* (*PER2*) and the T2434C polymorphism of *PERIOD1* have been associated with morning preference[149,150] though not consistently.[134] Since morningness–eveningness represents a continuum, it is likely this trait is polygenic, influenced by several genes, each contributing to the determination of circadian phase preference. Thus, further studies investigating other clock genes, as well as replication of the *PER* and *CLOCK* findings, are needed to establish precisely the molecular components of behavioral circadian phase preference.

Interindividual differences in morningness–eveningness are believed to manifest into extreme cases classified as primary circadian rhythm sleep disorders (CRSDs), with altered phase relationships of the biological clock to the light–dark cycle, including alterations in sleep timing.[151,152] Thus, extreme eveningness is thought to result in CRSD, delayed sleep phase type (usually referred to as a disorder and abbreviated as DSPD[152]), while extreme morningness can manifest as CRSD, advanced sleep phase type (usually referred to as a disorder and abbreviated as ASPD).[151,152] The extent to which these phase-displacement disorders reflect differences in endogenous circadian period, amplitude, coupling, entrainment, or other aspects of clock neurobiology has been the focus of recent research.

The genetic basis of DSPD and ASPD related to phenotypic chronotype has been investigated in recent years, both demonstrating links to core clock genes.[130,153,154] DSPD, the most common CRSD in the general population, is characterized by an inability to fall asleep at the desired and "normal" time of day; the average onset of sleep in DSPD occurs in the early morning (0300–0600 h), and the average wake-up time occurs in the late morning to early afternoon (1100–1400 h).[152] DSPD also may be characterized by a longer than normal tau (e.g., $\geq$25 h).[155] The VNTR polymorphism in *PER3* is associated with DSPD in large sample studies,[139,140,142] and the

3111C allele of the *CLOCK* gene 5′-UTR region also has been related to DSPD.[131] In addition, a specific haplotype of *PER3*, which includes the polymorphism G647, is associated positively with DSPD,[142] while the N408 allele of casein kinase I epsilon protects against DSPD in a Japanese population[156] but not in a Brazilian population.[157]

ASPD is a rarer disorder than DSPD and is characterized by 3- to 4-h advanced sleep onsets and wake times relative to desired, normal times.[152,158] It may be associated with a shorter-than-normal tau (e.g., <24 h).[159] In one study, ASPD was shown to be associated with a mutation in *PER2*,[160] although a later study failed to replicate this finding.[161] Another report implicated mutations in casein kinase I delta in ASPD.[162] Future studies on additional core clock genes are needed to determine other mutations that may underlie this disorder.

Morningness–eveningness and differences in circadian phase preference are reflected in the diurnal time course of neurobehavioral variables[163]—some people perform consistently better in the morning, whereas others are more alert and perform better in the evening.

How genetic variants underlying morningness–eveningness and chronotype disorders affect performance and alertness under normal and sleep-deprived conditions remains an emerging and important field of investigation. Two studies have shown that the longer, 5-repeat allele of the VNTR polymorphism in *PER3*, a clock gene linked to diurnal preference and DSPD, may be associated with higher sleep propensity both at baseline and after total sleep deprivation, and worse cognitive performance following sleep loss.[143,144] However, a study from our laboratory found that this polymorphism related to individual differences in sleep homeostatic responses, but not performance responses to chronic sleep restriction.[148] The role of other core clock gene polymorphisms in response to total sleep deprivation or chronic sleep restriction remains unknown.

A number of core clock genes have been associated with interindividual differences in diurnal preference or its extreme variants. This area of research has promising implications for objectively detecting differential vulnerability to circadian disorders and lifestyles that adversely affect alertness, performance and sleep duration, and homeostasis.

6. SLEEP DEPRIVATION AND PERFORMANCE

Sleep deprivation induces a variety of physiological and neurobehavioral changes.[164] Both objective and subjective measures of sleep propensity increase with sleep deprivation. Sleep deprivation affects a wide range of cognitive domains (including attention, working memory, abstraction, and decision making) and results in decreases in both the encoding of new information and memory consolidation.[165] Vigilant attention performance and psychomotor speed, as assessed with the PVT, are affected early and progressively more severely by sleep deprivation.[86,166] Although sustained attention seems a prerequisite for high levels of performance on more complex cognitive tasks, several studies have shown that the latter are less affected by sleep loss than attention, probably because they are more challenging and engaging than sustained attention tasks that unmask fatigue by their limited evocation of additional neural processing areas.[39,167] In addition, some of the differences among tasks in sensitivity to sleep deprivation may be explained by practice effects confounding the effects of sleep deprivation on more complex tasks. At the same time, the ability of stimulants to counteract the effects of sleep deprivation seems to depend on the cognitive domain studied.[168]

The neurobehavioral effects of chronic sleep restriction are less severe than those observed after acute total sleep deprivation, but the former can reach levels of deficit equivalent to total sleep loss when the sleep restriction is severe enough (i.e., the consecutive days of restricted sleep continue long enough).[10,32] Chronic sleep-restriction experiments suggest that the neurobiology underlying the neurobehavioral deficits can continue to undergo long-term changes. This is supported by a study investigating recovery after a period of chronic sleep restriction that suggests a single recovery night of up to 10 h time in bed is insufficient for some behavioral functions to return to prerestriction levels.[33] Evidence of longer time constants in homeostatic sleep pressure manifesting in waking neurobehavioral functions was reported by Rupp *et al.*[169] who varied the amount of baseline nightly sleep prior to chronic sleep restriction and found that it affected both the rate at which alertness was degraded and the rate at which deficits were reversed by repeated nights of recovery sleep.

6.1. Phenotypic and genotypic differences in response to sleep deprivation

We have repeatedly demonstrated that there are large and highly replicable, trait-like individual differences in the magnitude of fatigue, sleepiness, sleep homeostatic, and cognitive performance vulnerability to acute total sleep deprivation[170,171] and to chronic sleep restriction.[32,148,172,173] Some individuals are highly vulnerable to neurobehavioral performance deficits when sleep deprived, others demonstrate remarkable levels of neurobehavioral resistance to sleep loss, and others show intermediate responses.[171,174] Thus far, studies from our laboratory and others indicate these phenotypic responses occur as a normal distribution,[170,175] which suggests the phenotype, like chronotype, may be polygenetic.

It remains unclear, however, whether the same individuals vulnerable to the adverse neurobehavioral effects of chronic sleep restriction are also vulnerable to acute total sleep deprivation. Some studies have reported differences in behavioral, sleep homeostatic and/or physiological responses to both types of deprivation.[32,176,177] Moreover, only a few experiments have systematically examined the same subjects in both types of deprivation.[167,175,178–180] These studies reported inconsistent results, likely due to small sample sizes, different populations, varying doses of sleep restriction, and different outcome measures.

The reasons for differential neurobehavioral vulnerabilities to sleep loss are unknown, and thus far have not been accounted for by demographic factors, IQ, or sleep need. Moreover, psychometric scales have not reliably identified cognitively vulnerable individuals.[181] The stable, trait-like interindividual differences observed in response to acute total sleep deprivation—with intraclass correlation coefficients accounting for 50–90% of the variance in neurobehavioral measures[170,171]—point to underlying genetic components. In support of this statement, a recent study by Kuna *et al.*[182] conducted in monozygotic and dizygotic twin pairs, found substantial differences in individual neurobehavioral responses to total sleep deprivation—56.2% of the total variance in the monozygotic twins was due to variance between pairs whereas only 14.5% of the total variance in dizygotic twins was due to variance between pairs (Fig. 7.3), indicating that the response to acute total sleep deprivation is a highly stable, genetically determined trait. Indeed, data from unrelated individuals further indicate that common genetic polymorphisms involved in sleep–wake, circadian, and cognitive regulation may underlie these large interindividual differences in neurobehavioral vulnerability to sleep deprivation in healthy adults.[164,181,183]

Because of reported differences in behavioral, sleep homeostatic, and physiological responses to chronic sleep restriction and acute total sleep deprivation, specific candidate genes may play different roles in the degree of vulnerability and/or resilience to the neurobehavioral and homeostatic effects of acute total sleep deprivation and chronic sleep restriction. Two examples—one from a genetic variation involved in circadian regulation and one from a genetic variation involved in a cognitive regulation—

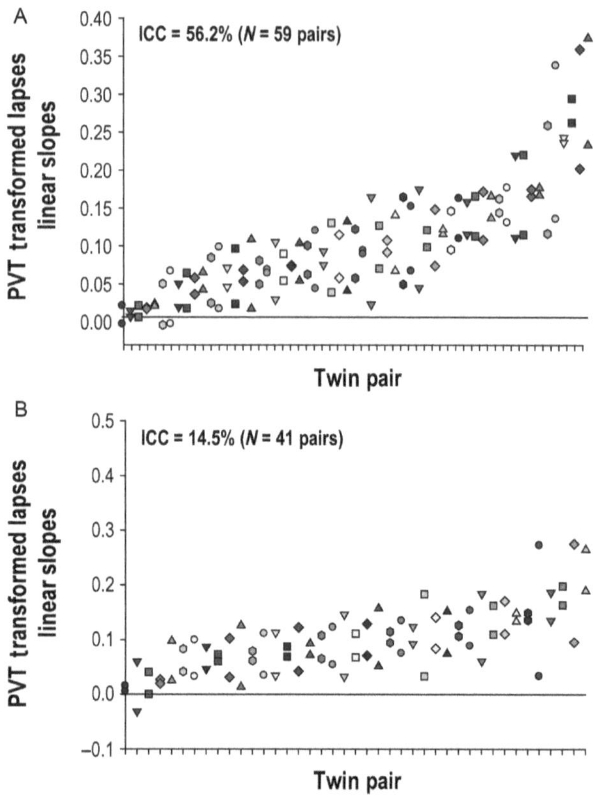

Figure 7.3 The individual linear slopes of the change in Psychomotor Vigilance Task (PVT) transformed lapses during 38 h of total sleep deprivation in monozygotic (MZ; A) and dizygotic (DZ; B) twin pairs. Data for each MZ and DZ twin pair are plotted together on the abscissa. In each panel, the pairs are ordered by the magnitude of their impairment (averaged over each pair), with the most resistant twin pair on the left and the most vulnerable twin pair on the right. The panels reveal substantial differences in individual responses to sleep deprivation. The intraclass correlation (ICC) revealed greater similarity within MZ twin pairs than within DZ twin pairs. There was 56.2% of the total variance in the MZ twins due to variance between pairs whereas only 14.5% of the total variance in DZ twins was due to variance between pairs. *Reprinted with permission from Ref. 182.*

illustrate this point. As mentioned previously, the *PER3* VNTR polymorphism has been associated with individual differences in sleep homeostatic and executive performance responses to acute total sleep deprivation.[143,144] We showed that this polymorphism related to individual differences in sleep homeostatic responses, but not cognitive performance responses to chronic sleep restriction.[148] By contrast, two recent studies,[167,184] which used different sleep-restriction paradigms than that of Goel *et al.*,[148] claimed that the *PER3* VNTR polymorphism was related to individual differences in neurobehavioral responses to chronic sleep restriction. Notably, one of these[184] failed to include subjects from the critical $PER3^{5/5}$ putatively vulnerable genotype, and thus its findings must be interpreted cautiously and replicated in the appropriately inclusive genotypes. As another example, we found that the *catechol-O-methyltransferase* Val158Met polymorphism predicted individual differences in sleep homeostatic responses to chronic sleep restriction,[173] but such prediction has not been shown to acute total sleep deprivation.[185] Clearly, more studies are warranted to investigate potential genotypic markers of phenotypic vulnerability to sleep loss and the differential role they might play in response to different types of sleep loss.

6.2. Neuroimaging of sleep deprivation and circadian variations in brain metabolism and neural activity

With few exceptions, the influences of sleep deprivation and circadian variations on brain metabolism and neural activity have been studied separately in the past two decades using various neuroimaging methods, particularly positron emission tomography (PET) and functional magnetic resonance imaging (fMRI).

PET studies of sleep deprivation have consistently reported significant reductions in metabolic rates in the thalamic, parietal, and prefrontal regions after sleep loss, which correlated with declines of cognitive performance and alertness.[186–189] An early PET study examined the effects of time of day (a surrogate for circadian phase) on the cerebral metabolic rate of glucose and observed a trend toward increased whole brain glucose metabolism from the morning to the afternoon scans.[190] A more recent PET study found increased relative glucose metabolism in the brainstem and hypothalamic arousal systems and decreased relative metabolism in the posterior cortical regions during evening wakefulness compared with morning wakefulness.[191] Moreover, variations in regional brain glucose metabolism have been reported to differ across morning and evening scans in depressed and healthy adult subjects.[192] New PET studies on neurotransmitter receptors have shown downregulation of striatal dopamine receptors[193] but increases in cerebral serotonin receptor binding with sleep loss,[194] which may reflect a complex adaptive brain response to sleep deprivation. However, due to its invasiveness and the rapid decay of radioactive tracers, further utility of PET in imaging human brain metabolism variations associated with sleep deprivation and time-of-day effects is limited.

The vast majority of sleep deprivation and time-of-day or circadian neuroimaging studies are based on the blood oxygenation level-dependent (BOLD) fMRI. Compared with PET, BOLD fMRI is noninvasive, more cost effective, and easier to apply, thus making it the most widely used imaging method for localizing regional brain function. BOLD studies typically compare fMRI signals during a specific cognitive task with those during a control or baseline condition to obtain task-related brain activation. A large number of BOLD studies have investigated the effects of acute total or partial sleep deprivation on brain activation during performance on a broad range of neurocognitive tasks, including arithmetic calculation,[195] attention,[196–208] decision making,[209–211] emotional processing,[212] episodic memory,[213–215] inhibition control,[216] logical reasoning,[217] spatial navigation,[218] verbal learning,[219,220] visuomotor adaptation memory,[221] and working memory tasks.[222–230] Many BOLD fMRI studies have found changes in neural activity after sleep deprivation. For example, a reproducibility study showed that brain activation patterns were highly correlated across test–retest sessions and the magnitude of decreased activation in parietal regions was preserved and reproducibly correlated with behavioral decline after acute total sleep deprivation.[228] Reduced frontoparietal activation was found during lapses on a visual, selective attention task in addition to decreased overall activation after total sleep deprivation.[199] However, robust interindividual differences in brain responses to sleep loss have also been reported. Individuals cognitively vulnerable to sleep deprivation showed reduced frontoparietal activation, while resilient individuals showed increased parietal activation associated with lapses of attention during total sleep deprivation[201] suggesting a potential neurobiological compensatory mechanism in some individuals.

Far fewer neuroimaging studies have been conducted to examine either time-of-day or circadian phase effects on brain activation. One study used functional near-infrared spectroscopy to examine circadian variability of the hemodynamic response in visual cortex throughout the day from 0800–1800 h, reporting no significant time-of-day influences on visual activation.[231] However, BOLD fMRI studies have shown significant time-of--day effects on brain activation when subjects performed various neurocognitive tasks. For example, Gorfine and Zisapel[232] found that left hippocampal activation was reduced during an autobiographic memory task at 2200 h compared with 1600 h, indicative of diurnal variation. Vimal and colleagues[233] showed significantly increased BOLD activation in response to light stimuli in the suprachiasmatic nucleus at night compared with midday, while Peres and colleagues[234] demonstrated systematic BOLD signal differences across the day in the motor areas during a self-paced finger-tapping task. Significant time-of-day effects were also observed in the brain orienting attentional system including the inferior parietal and frontal eye field regions during a Stroop-like task, suggesting that bottom-up attention orientation may be vulnerable to circadian factors.[235]

Importantly, a few recent BOLD fMRI studies have demonstrated significant interindividual differences in circadian variation of brain activation and the complex interactions between sleep homeostasis, circadian phase, and genotype. For example, using an auditory 3-back working memory task, Vandewalle and colleagues[236] showed no changes in brain responses during the normal sleep–wake cycle for the putatively less-vulnerable $PER3^{4/4}$ genotype, while reduced activation in the posterior prefrontal area was found in the putatively vulnerable $PER3^{5/5}$ genotype when comparing evening and morning activation during a normal sleep–wake cycle. These authors also reported that blue light increased brain responses in the frontoparietal regions only in $PER3^{4/4}$ individuals in the morning after one night of normal sleep, while blue light increased brain responses in the thalamic and frontoparietal regions only in $PER3^{5/5}$ individuals in the morning after one night of total sleep deprivation.[237] In addition, Schmidt and colleagues[238] showed that morning and evening chronotypes differed in brain activation in the suprachiasmatic area at night during PVT performance. They further found that brain activation associated with conflict processing and inhibition function were maintained or increased in evening chronotypes from the subjective morning to the subjective evening but decreased in morning chronotypes under the same conditions.[239]

While the above findings are informative, one major limitation of task-related BOLD fMRI is that it can only measure *relative* signal changes and lacks absolute quantification of brain activity. Therefore, it is difficult to determine whether the observed BOLD activation changes are due to changes at baseline or changes during performance of specific tasks, or both.

It is also difficult for task-related BOLD fMRI to dissociate the effects of sleep loss, time-of-day, or circadian phase on brain function *per se* and on behavioral performance that subsequently confounds brain activation. In contrast to BOLD, arterial spin-labeled (ASL) perfusion fMRI—a relatively new imaging technique—can noninvasively measure absolute cerebral blood flow (CBF) that is tightly coupled to regional brain function,[240,241] providing a method for imaging variations of brain function at different time of day or circadian phases or after sleep loss. ASL has been increasingly used to assess waking brain function at task-free resting states as well as during different cognitive tasks.[242,243]

We successfully used ASL to quantify CBF changes after prolonged cognitive workload without sleep loss.[244] Currently, only one published study has used ASL and measured resting CBF changes after one night of sleep restriction.[245] This study reported significantly reduced frontoparietal CBF following sleep loss, but only in participants with significant signs of drowsiness, while nondrowsy participants maintained CBF in the frontoparietal regions and increased CBF in basal forebrain and cingulate regions. These findings also suggest a potential neurobiological mechanism to compensate for drowsiness after sleep loss. Ongoing studies in our group as well as others are using ASL to quantify regional neural activity changes associated with time-of-day variation and sleep deprivation.[246] Our preliminary data from scans during PVT performance in the morning and afternoon in two independent groups also showed significant time-of-day effects. Both morning and afternoon scans showed similar sensorimotor, cingulate, and frontoparietal activation while subjects performed the PVT. However, thalamic activation was observed only in the morning PVT scan, while increased activation in the right frontal eye field was observed in the afternoon PVT scan (Fig. 7.4).

Another emerging imaging method for studying sleep deprivation and time-of-day or circadian phase effects on brain activity is resting-state functional connectivity fMRI (FC-fMRI), which usually uses low frequency fluctuations of resting-state BOLD signal to examine intrinsic and spontaneous neural activity in the absence of external stimuli or tasks.[247,248] Converging evidence from resting-state fMRI studies has indicated an organized mode of resting brain function and identified a number of brain networks associated with different domains of neurocognitive functioning.[249–252] Two recent studies have used FC-fMRI to investigate the effect of one night of either total or partial sleep deprivation on functional connectivity.[253,254] Both studies found that sleep deprivation reduced resting functional connectivity within the default mode network (DMN) and between DMN and its anticorrelated network, suggesting that reduced brain functional connectivity may be a precursor to behavioral impairments from sleep loss. In addition, one recent study used FC-fMRI to

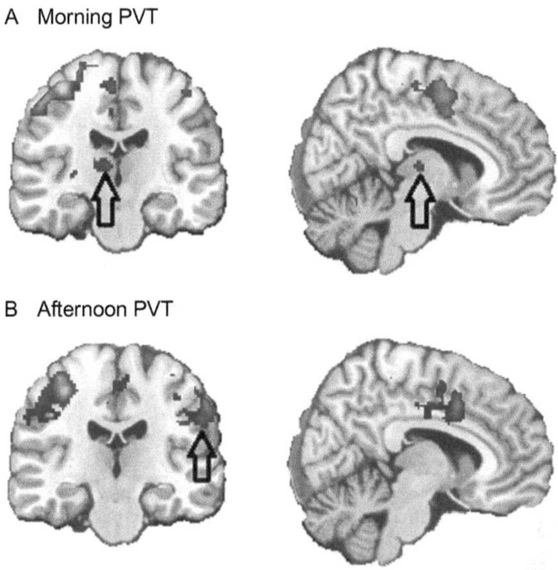

Figure 7.4 Time-of-day effects on absolute cerebral blood flow (CBF) activation during the Psychomotor Vigilance Test (PVT). Twenty healthy adults performed the PVT in the morning (between 0700– and 0900 h) and a separate group of 15 healthy adults performed the PVT in the afternoon (between 1400– and 1700 h)—both groups did so during ASL perfusion fMRI scanning. Brain scans at both times of day showed significant activation in the sensorimotor, cingulate, and frontoparietal regions. However, thalamic activation (indicated by the arrows in A) was only observed in the morning scan while increased activation in the right frontal eye field (indicated by the arrow in B) was observed in the afternoon scan (Hengyi Rao, unpublished data). (See Color Insert.)

examine daily variations in resting brain functional connectivity and found that the DMN and sensorimotor network showed highly rhythmic connectivity patterns while the executive control network was most stable across the day.[255]

Almost all published neuroimaging studies to date have focused on acute total sleep deprivation or time-of-day effects—very few studies have examined the dynamic effects of chronic partial sleep loss and recovery on brain function and their interactions with circadian timing. Findings from the few available ASL and resting-state FC-fMRI studies already provide some important new insights. However, application of these new methods to sleep deprivation and circadian research is still in the early stages and studies are needed to further elucidate the dynamic effects of both acute and chronic sleep loss as well as circadian timing on neural activity.

7. CONCLUSIONS

The circadian drive for wakefulness, the homeostatic drive for sleep, and masking factors simultaneously interact to affect neurobehavioral functioning. Moreover, interindividual differences in circadian parameters, especially phase, and differential vulnerability to sleep loss also markedly affect neurobehavioral responses, suggesting genetic underpinnings. The sleep homeostat and neurobehavioral performance are affected by acute total sleep deprivation and chronic sleep restriction, although the two forms of sleep loss likely differentially affect neural and behavioral responses. Identification of biomarkers that accurately predict alertness and performance via the complex interactions of the sleep homeostatic and circadian systems is of high priority and will aid in predicting performance deficits and implementing countermeasures in a variety of situations in which these two processes are dynamically covarying, such as shift work, jet lag, and imposed acute, chronic, or intermittent sleep loss.

ACKNOWLEDGMENTS

Preparation of this chapter was supported by ONR N00014-11-1-0361 (NG), National Space Biomedical Research Institute through NASA NCC 9-58 (MB, DFD), NIH HL102119 (HR), and NIH NR004281 (DFD).

REFERENCES

1. Siegel JM. Sleep viewed as a state of adaptive inactivity. *Nat Rev Neurosci.* 2009;10:747–753.
2. Buxton OM, Cain SW, O'Connor SP, et al. Adverse metabolic consequences in humans of prolonged sleep restriction combined with circadian disruption. *Sci Transl Med.* 2012;4:129ra143.
3. Arble DM, Bass J, Laposky AD, Vitaterna MH, Turek FW. Circadian timing of food intake contributes to weight gain. *Obesity.* 2009;17:2100–2102.
4. Goel N, Stunkard AJ, Rogers NL, et al. Circadian rhythm profiles in women with night eating syndrome. *J Biol Rhythms.* 2009;24:85–94.
5. Kobayashi D, Takahashi O, Deshpande GA, Shimbo T, Fukui T. Association between weight gain, obesity, and sleep duration: a large-scale 3-year cohort study. *Sleep Breath.* 2012;16:829–833.
6. Chao CY, Wu JS, Yang YC, et al. Sleep duration is a potential risk factor for newly diagnosed type 2 diabetes mellitus. *Metabolism.* 2011;60:799–804.
7. Wang Q, Xi B, Liu M, Zhang Y, Fu M. Short sleep duration is associated with hypertension risk among adults: a systematic review and meta-analysis. *Hypertens Res.* 2012;35:1012–1018.
8. Cappuccio FP, D'Elia L, Strazzullo P, Miller MA. Sleep duration and all-cause mortality: a systematic review and meta-analysis of prospective studies. *Sleep.* 2010;33:585–592.

9. Centers for Disease Control, Prevention. Effect of short sleep duration on daily activities–United States, 2005–2008. *MMWR Morb Mortal Wkly Rep*. 2011;60: 239–242.
10. Belenky G, Wesensten NJ, Thorne DR, et al. Patterns of performance degradation and restoration during sleep restriction and subsequent recovery: a sleep dose–response study. *J Sleep Res*. 2003;12:1–12.
11. Basner M. Sleep duration and chronic sleep debt: are 6 hours enough? *Biol Psychol*. 2011;87:15–16.
12. Foster RG, Wulff K. The rhythm of rest and excess. *Nat Rev Neurosci*. 2005;6:407–414.
13. Biddle JE, Hamermesh DS. Sleep and the allocation of time. *J Polit Econ*. 1990;98:922–943.
14. Basner M, Fomberstein KM, Razavi FM, et al. American time use survey: sleep time and its relationship to waking activities. *Sleep*. 2007;30:1085–1095.
15. Basner M, Dinges DF. Dubious bargain: trading sleep for Leno and Letterman. *Sleep*. 2009;32:747–752.
16. Wittmann M, Dinich J, Merrow M, Roenneberg T. Social jetlag: misalignment of biological and social time. *Chronobiol Int*. 2006;23:497–509.
17. Costa G, Haus E, Stevens R. Shift work and cancer—considerations on rationale, mechanisms, and epidemiology. *Scand J Work Environ Health*. 2010;36:163–179.
18. Driscoll TR, Grunstein RR, Rogers NL. A systematic review of the neurobehavioural and physiological effects of shiftwork systems. *Sleep Med Rev*. 2007;11:179–194.
19. Straif K, Baan R, Grosse Y, et al. Carcinogenicity of shift-work, painting, and firefighting. *Lancet Oncol*. 2007;8:1065–1066.
20. Borbély AA. A two-process model of sleep regulation. *Hum Neurobiol*. 1982;1:195–204.
21. Daan S, Beersma DGM, Borbély AA. Timing of human sleep: recovery process gated by a circadian pacemaker. *Am J Physiol*. 1984;246:R161–R178.
22. Achermann P, Borbély AA. Simulation of daytime vigilance by the additive interaction of a homeostatic and a circadian process. *Biol Cybern*. 1994;71:115–121.
23. Edgar DM, Dement WC, Fuller CA. Effect of SCN lesions on sleep in squirrel monkeys: evidence for opponent processes in sleep-wake regulation. *J Neurosci*. 1993;13:1065–1079.
24. Doran SM, Van Dongen HP, Dinges DF. Sustained attention performance during sleep deprivation: evidence of state instability. *Arch Ital Biol*. 2001;139:253–267.
25. Lim J, Dinges DF. Sleep deprivation and vigilant attention. *Ann N Y Acad Sci*. 2008;1129:305–322.
26. Cohen DA, Wang W, Wyatt JK, et al. Uncovering residual effects of chronic sleep loss on human performance. *Sci Transl Med*. 2010;2:14ra13.
27. Zhou X, Ferguson SA, Matthews RW, et al. Sleep, wake and phase dependent changes in neurobehavioral function under forced desynchrony. *Sleep*. 2011;34:931–941.
28. Mollicone DJ, Van Dongen HPA, Rogers NL, Dinges DF. Response surface mapping of neurobehavioral performance: testing the feasibility of split sleep schedules for space operations. *Acta Astronaut*. 2008;63:833–840.
29. Mallis MM, Mejdal S, Nguyen TT, Dinges DF. Summary of the key features of seven biomathematical models of human fatigue and performance. *Aviat Space Environ Med*. 2004;75:A4–A14.
30. Van Dongen HPA. Comparison of mathematical model predictions to experimental data of fatigue and performance. *Aviat Space Environ Med*. 2004;75:A15–A36.
31. McCauley P, Kalachev LV, Smith AD, Belenky G, Dinges DF, Van Dongen HPA. A new mathematical model for the homeostatic effects of sleep loss on neurobehavioral performance. *J Theor Biol*. 2009;256:227–239.

32. Van Dongen HP, Maislin G, Mullington JM, Dinges DF. The cumulative cost of additional wakefulness: dose–response effects on neurobehavioral functions and sleep physiology from chronic sleep restriction and total sleep deprivation. *Sleep.* 2003;26:117–126.
33. Banks S, Van Dongen HP, Maislin G, Dinges DF. Neurobehavioral dynamics following chronic sleep restriction: dose–response effects of one night of recovery. *Sleep.* 2010;33:1013–1026.
34. Monk TH. A visual analogue scale technique to measure global vigor and affect. *Psychiatry Res.* 1989;27:89–99.
35. Hoddes E, Zarcone V, Smythe H, Phillips R, Dement WC. Quantification of sleepiness: a new approach. *Psychophysiology.* 1973;10:431–436.
36. Åkerstedt T, Gillberg M. Subjective and objective sleepiness in the active individual. *Int J Neurosci.* 1990;52:29–37.
37. Thayer RE. Factor analytic and reliability studies on the Activation-Deactivation Adjective Check List. *Psychol Rep.* 1978;42:747–756.
38. McNair DM, Lorr M, Druppleman LF. *EITS Manual for the Profile of Mood States.* San Diego, CA: Educational and Industrial Test Services; 1971.
39. Lim J, Dinges DF. A meta-analysis of the impact of short-term sleep deprivation on cognitive variables. *Psychol Bull.* 2010;136:375–389.
40. Colquhoun WP. Circadian variations in mental efficiency. In: Colquhoun WP, ed. *Biological Rhythms in Human Performance.* London: Academic Press; 1971:39–107.
41. Kleitman N. *Sleep and Wakefulness.* Chicago, IL: University of Chicago Press; 1963.
42. Kleitman N. Diurnal variation in performance. *Am J Physiol.* 1933;104:449–456.
43. Folkard S. Diurnal variation in logical reasoning. *Br J Psychol.* 1975;66:1–8.
44. Folkard S, Monk TH. Circadian rhythms in human memory. *Br J Psychol.* 1980;71:295–307.
45. Bjerner B, Holm A, Swensson A. Diurnal variation in mental performance: a study of three-shift workers. *Br J Ind Med.* 1955;12:103–110.
46. Laird DA. Relative performance of college students as conditioned by time of day and day of week. *J Exp Psychol.* 1925;8:50–63.
47. Osman A, Lou L, Muller-Gethmann H, Rinkenauer G, Mattes S, Ulrich R. Mechanisms of speed-accuracy tradeoff: evidence from covert motor processes. *Biol Psychol.* 2000;51:173–199.
48. Dorrian J, Rogers NL, Dinges DF. Psychomotor vigilance performance: a neurocognitive assay sensitive to sleep loss. In: Kushida C, ed. *Sleep Deprivation: Clinical Issues, Pharmacology and Sleep Loss Effects.* New York, NY: Marcel Dekker Inc.; 2005:39–70.
49. Hockey GRJ, Colquhoun WP. Diurnal variation in human performance: a review. In: Colquhoun WP, ed. *Aspects of Human Efficiency: Diurnal Rhythm and Loss of Sleep.* London: English Universities Press; 1972:1–23.
50. Gillooly PB, Smolensky MH, Albright DL, Hsi B, Thorne DR. Circadian variation in human performance evaluated by the Walter Reed performance assessment battery. *Chronobiol Int.* 1990;7:143–153.
51. Monk TH, Leng VC. Time of day effects in simple repetitive tasks: some possible mechanisms. *Acta Psychol.* 1982;51:207–221.
52. Kirkcaldy BD. Performance and circadian rhythms. *Eur J Appl Physiol Occup Physiol.* 1984;52:375–379.
53. Folkard S, Wever RA, Wildgruber CM. Multi-oscillatory control of circadian rhythms in human performance. *Nature.* 1983;305:223–226.
54. Monk TH, Weitzman ED, Fookson JE, Moline ML, Kronauer RE, Gander PH. Task variables determine which biological clock controls circadian rhythms in human performance. *Nature.* 1983;304:543–545.

55. Johnson MP, Duffy JF, Dijk DJ, Ronda JM, Dyal CM, Czeisler CA. Short-term memory, alertness and performance: a reappraisal of their relationship to body temperature. *J Sleep Res.* 1992;1:24–29.
56. Monk TH, Buysse DJ, Reynolds III CF, et al. Circadian rhythms in human performance and mood under constant conditions. *J Sleep Res.* 1997;6:9–18.
57. Kerkhof GA, Van Dongen HPA. Morning-type and evening-type individuals differ in the phase position of their endogenous circadian oscillator. *Neurosci Lett.* 1996;218:153–156.
58. Baehr EK, Revelle W, Eastman CI. Individual differences in the phase and amplitude of the human circadian temperature rhythm: with an emphasis on morningness-eveningness. *J Sleep Res.* 2000;9:117–127.
59. Kleitman N, Jackson DP. Body temperature and performance under different routines. *J Appl Physiol.* 1950;51:309–328.
60. Blake MJF. Time of day effects on performance in a range of tasks. *Psychon Sci.* 1967;9:349–350.
61. Orne MT. On the social psychology of the psychological experiment: with particular reference to demand characteristics and their implications. *Am Psychol.* 1962;17:776–783.
62. Landström U, Englund K, Nordström B, Åström A. Laboratory studies of a sound system that maintains wakefulness. *Percept Mot Skills.* 1998;86:147–161.
63. Minors DS, Waterhouse JM. Circadian rhythm amplitude: is it related to rhythm adjustment and/or worker motivation? *Ergonomics.* 1983;26:229–241.
64. Mavjee V, Horne JA. Boredom effects on sleepiness/alertness in the early afternoon vs. early evening and interactions with warm ambient temperature. *Br J Psychol.* 1994;85:317–333.
65. Hull JT, Wright Jr KP, Czeisler CA. The influence of subjective alertness and motivation on human performance independent of circadian and homeostatic regulation. *J Biol Rhythms.* 2003;18:329–338.
66. Orr WC, Hoffman HJ, Hegge FW. The assessment of time-dependent changes in human performance. *Chronobiologia.* 1976;3:293–305.
67. Paz A, Berry EM. Effect of meal composition on alertness and performance of hospital night-shift workers. *Ann Nutr Metab.* 1997;41:291–298.
68. Wells AS, Read NW, Idzikowski C, Jones J. Effects of meals on objective and subjective measures of daytime sleepiness. *J Appl Physiol.* 1998;84:507–515.
69. Kräuchi K, Cajochen C, Wirz-Justice A. A relationship between heat loss and sleepiness: effects of postural change and melatonin administration. *J Appl Physiol.* 1997;83:134–139.
70. Johns MW. Sleep propensity varies with behaviour and the situation in which it is measured: the concept of somnificity. *J Sleep Res.* 2002;11:61–67.
71. Cajochen C, Zeitzer JM, Czeisler CA, Dijk DJ. Dose–response relationship for light intensity and ocular and electroencephalographic correlates of human alertness. *Behav Brain Res.* 2000;115:75–83.
72. Phipps-Nelson J, Redman JR, Dijk DJ, Rajaratnam SMW. Daytime exposure to bright light, as compared to dim light, decreases sleepiness and improves psychomotor vigilance performance. *Sleep.* 2003;26:695–700.
73. Åkerstedt T, Ficca G. Alertness-enhancing drugs as a countermeasure to fatigue in irregular work hours. *Chronobiol Int.* 1997;14:145–158.
74. Makris AP, Rush CR, Frederich RC, Taylor AC, Kelly TH. Behavioral and subjective effects of d-amphetamine and modafinil in healthy adults. *Exp Clin Psychopharmacol.* 2007;15:123–133.
75. Dagan Y, Doljansky JT. Cognitive performance during sustained wakefulness: a low dose of caffeine is equally effective as modafinil in alleviating the nocturnal decline. *Chronobiol Int.* 2006;23:973–983.

76. Mills JN, Minors DS, Waterhouse JM. Adaptation to abrupt time shifts of the oscillator(s) controlling human circadian rhythms. *J Physiol.* 1978;285:455–470.
77. Monk TH, Carrier J. Speed of mental processing in the middle of the night. *Sleep.* 1997;20:399–401.
78. Ang JE, Revell V, Mann A, et al. Identification of human plasma metabolites exhibiting time-of-day variation using an untargeted liquid chromatography-mass spectrometry metabolomic approach. *Chronobiol Int.* 2012;29:868–881.
79. Dallmann R, Viola AU, Tarokh L, Cajochen C, Brown SA. The human circadian metabolome. *Proc Natl Acad Sci USA.* 2012;109:2625–2629.
80. Kasukawa T, Sugimoto M, Hida A, et al. Human blood metabolite timetable indicates internal body time. *Proc Natl Acad Sci USA.* 2012;109:15036–15041.
81. Möller-Levet CS, Archer SN, Bucca G, et al. Effects of insufficient sleep on circadian rhythmicity and expression amplitude of the human blood transcriptome. *Proc Natl Acad Sci USA.* 2013;110:E1132–E1141.
82. Åkerstedt T, Gillberg M, Wetterberg L. The circadian covariation of fatigue and urinary melatonin. *Biol Psychiatry.* 1982;17:547–554.
83. Van Dongen HPA, Dinges DF. Sleep, circadian rhythms, and psychomotor vigilance performance. *Clin Sports Med.* 2005;24:237–249.
84. Babkoff H, Mikulincer M, Caspy T, Carasso RL, Sing HC. The implications of sleep loss for circadian performance accuracy. *Work Stress.* 1989;3:3–14.
85. Bohlin G, Kjellberg A. Self-reported arousal during sleep deprivation and its relation to performance and physiological variables. *Scand J Psychol.* 1973;14:78–86.
86. Basner M, Dinges DF. Maximizing sensitivity of the Psychomotor Vigilance Test (PVT) to sleep loss. *Sleep.* 2011;34:581–591.
87. Williams HL, Lubin A, Goodnow JJ. Impaired performance with acute sleep loss. *Psychol Monogr Gen Appl.* 1959;73:1–26.
88. Kleitman N, Kleitman E. Effect of non-twenty-four-hour routines of living on oral temperature and heart rate. *J Appl Physiol.* 1953;6:283–291.
89. Dijk DJ, Czeisler CA. Paradoxical timing of the circadian rhythm of sleep propensity serves to consolidate sleep and wakefulness in humans. *Neurosci Lett.* 1994;166:63–68.
90. Dijk DJ, Duffy JF, Czeisler CA. Circadian and sleep/wake dependent aspects of subjective alertness and cognitive performance. *J Sleep Res.* 1992;1:112–117.
91. Monk TH, Moline ML, Fookson JE, Peetz SM. Circadian determinants of subjective alertness. *J Biol Rhythms.* 1989;4:393–404.
92. Van Dongen HPA, Dinges DF. Investigating the interaction between the homeostatic and circadian processes of sleep-wake regulation for the prediction of waking neurobehavioural performance. *J Sleep Res.* 2003;12:181–187.
93. Lenné MG, Triggs TJ, Redman JR. Interactive effects of sleep deprivation, time of day, and driving experience on a driving task. *Sleep.* 1998;21:38–44.
94. Czeisler CA, Duffy JF, Shanahan TL, et al. Stability, precision, and near-24-hour period of the human circadian pacemaker. *Science.* 1999;284:2177–2181.
95. Smith MR, Burgess HJ, Fogg LF, Eastman CI. Racial differences in the human endogenous circadian period. *PLoS One.* 2009;4:e6014.
96. Lázár AS, Santhi N, Hasan S, et al. Circadian period and the timing of melatonin onset in men and women: predictors of sleep during the weekend and in the laboratory. *J Sleep Res.* 2013;22:155–159.
97. Hasan S, Santhi N, Lázár AS, et al. Assessment of circadian rhythms in humans: comparison of real-time fibroblast reporter imaging with plasma melatonin. *FASEB J.* 2012;26:2414–2423.
98. Chang AM, Buch AM, Bradstreet DS, Klements DJ, Duffy JF. Human diurnal preference and circadian rhythmicity are not associated with the CLOCK 3111C/T gene polymorphism. *J Biol Rhythms.* 2011;26:276–279.

99. Burgess HJ, Fogg LF. Individual differences in the amount and timing of salivary melatonin secretion. *PLoS One*. 2008;3:e3055.
100. Brown SA, Fleury-Olela F, Nagoshi E, et al. The period length of fibroblast circadian gene expression varies widely among human individuals. *PLoS Biol*. 2005;3:e338.
101. Brown SA, Kunz D, Dumas A, et al. Molecular insights into human daily behavior. *Proc Natl Acad Sci USA*. 2008;105:1602–1607.
102. Yang S, Van Dongen HPA, Wang K, Berrettini W, Bućan M. Assessment of circadian function in fibroblasts of patients with bipolar disorder. *Mol Psychiatry*. 2009;14:143–155.
103. Lewy AJ, Sack RL. The dim light melatonin onset as a marker for circadian phase position. *Chronobiol Int*. 1989;6:93–102.
104. Horne JA, Östberg O. A self-assessment questionnaire to determine morningness–eveningness in human circadian rhythms. *Int J Chronobiol*. 1976;4:97–110.
105. Smith CS, Reilly D, Midkiff K. Evaluation of three circadian rhythm questionnaires with suggestions for an improved measure of morningness. *J Appl Psychol*. 1989;74:728–738.
106. Roenneberg T, Wirz-Justice A, Merrow M. Life between clocks: daily temporal patterns of human chronotypes. *J Biol Rhythms*. 2003;18:80–90.
107. Roenneberg T, Kuehnle T, Juda M, et al. Epidemiology of the human circadian clock. *Sleep Med Rev*. 2007;11:429–438.
108. Duffy JF, Dijk DJ, Klerman EB, Czeisler CA. Later endogenous circadian temperature nadir relative to an earlier wake time in older people. *Am J Physiol*. 1998;275: R1478–R1487.
109. Foster RG, Roenneberg T. Human responses to the geophysical daily, annual and lunar cycles. *Curr Biol*. 2008;18:R784–R794.
110. Adan A, Natale V. Gender differences in morningness–eveningness preference. *Chronobiol Int*. 2002;19:709–720.
111. Randler C. Gender differences in morningness–eveningness assessed by self-report questionnaires: a meta-analysis. *Pers Individ Dif*. 2007;43:1667–1675.
112. Mongrain V, Paquet J, Dumont M. Contribution of the photoperiod at birth to the association between season of birth and diurnal preference. *Neurosci Lett*. 2006;406:113–116.
113. Duffy JF, Cain SW, Chang AM, et al. Sex difference in the near-24-hour intrinsic period of the human circadian timing system. *Proc Natl Acad Sci USA*. 2001;108:15602–15608.
114. Eastman CI, Molina TA, Dziepak ME, Smith MR. Blacks (African Americans) have shorter free-running circadian periods than whites (Caucasian Americans). *Chronobiol Int*. 2012;29:1072–1077.
115. Hur YM. Stability of genetic influence on morningness-eveningness: a cross-sectional examination of South Korean twins from preadolescence to young adulthood. *J Sleep Res*. 2007;16:17–23.
116. Hur YM, Bouchard Jr TJ, Lykken DT. Genetic and environmental influence on morningness–eveningness. *Pers Individ Dif*. 1998;25:917–925.
117. Vink JM, Groot AS, Kerkhof GA, Boomsma DI. Genetic analysis of morningness and eveningness. *Chronobiol Int*. 2001;18:809–822.
118. Barclay NL, Eley TC, Parsons MJ, Willis TA, Gregory AM. Monozygotic twin differences in non-shared environmental factors associated with chronotype. *J Biol Rhythms*. 2013;28:51–61.
119. Klei L, Reitz P, Miller M, et al. Heritability of morningness-eveningness and self-report sleep measures in a family-based sample of 521 hutterites. *Chronobiol Int*. 2005;22:1041–1054.
120. Watson NF, Buchwald D, Noonan C, Vitiello MV, Pack AI, Goldberg J. Is circadian type associated with sleep duration in twins? *Sleep Biol Rhythms*. 2012;10:61–68.

121. Van Dongen HPA, Kerkhof GA, Dinges DF. Human circadian rhythms. In: Sehgal A, ed. *Molecular Biology of Circadian Rhythms*. New York, NY: John Wiley & Sons; 2004:255–269.
122. Mongrain V, Carrier J, Dumont M. Chronotype and sex effects on sleep architecture and quantitative sleep EEG in healthy young adults. *Sleep*. 2005;28:819–827.
123. Mongrain V, Carrier J, Dumont M. Difference in sleep regulation between morning and evening circadian types as indexed by antero–posterior analyses of the sleep EEG. *Eur J Neurosci*. 2006;23:497–504.
124. Mongrain V, Carrier J, Paquet J, Bélanger-Nelson E, Dumont M. Morning and evening-type differences in slow waves during NREM sleep reveal both trait and state-dependent phenotypes. *PLoS One*. 2011;6:e22679.
125. Mongrain V, Dumont M. Increased homeostatic response to behavioral sleep fragmentation in morning types compared to evening types. *Sleep*. 2007;30:773–780.
126. Mongrain V, Noujaim J, Blais H, Dumont M. Daytime vigilance in chronotypes: diurnal variations and effects of behavioral sleep fragmentation. *Behav Brain Res*. 2008;190:105–111.
127. Taillard J, Philip P, Claustrat B, et al. Time course of neurobehavioral alertness during extended wakefulness in morning- and evening-type healthy sleepers. *Chronobiol Int*. 2011;28:520–527.
128. Killgore WD. Effects of sleep deprivation and morningness-eveningness traits on risk-taking. *Psychol Rep*. 2007;100:613–626.
129. Barclay NL, Gregory AM. Quantitative genetic research on sleep: a review of normal sleep, sleep disturbances and associated emotional, behavioural, and health-related difficulties. *Sleep Med Rev*. 2013;17:29–40.
130. Takahashi JS, Hong HK, Ko CH, McDearmon EL. The genetics of mammalian circadian order and disorder: implications for physiology and disease. *Nat Rev Genet*. 2008;9:764–775.
131. Katzenberg D, Young T, Finn L, Lin L, King DP, Takahashi JS. A CLOCK polymorphism associated with human diurnal preference. *Sleep*. 1998;21:569–576.
132. Mishima K, Tozawa T, Satoh K, Saitoh H, Mishima Y. The 3111T/C polymorphism of hClock is associated with evening preference and delayed sleep timing in a Japanese population sample. *Am J Med Genet B Neuropsychiatr Genet*. 2005;133:101–104.
133. Garaulet M, Sánchez-Moreno C, Smith CE, Lee YC, Nicolás F, Ordovás JM. Ghrelin, sleep reduction and evening preference: relationships to CLOCK 3111T/C SNP and weight loss. *PLoS One*. 2011;6:e17435.
134. Choub A, Mancuso M, Coppedè F, et al. Clock T3111C and Per2 C111G SNPs do not influence circadian rhythmicity in healthy Italian population. *Neurol Sci*. 2011;32:89–93.
135. Barclay NL, Eley TC, Mill J, et al. Sleep quality and diurnal preference in a sample of young adults: associations with 5HTTLPR, PER3, and CLOCK 3111. *Am J Med Genet B Neuropsychiatr Genet*. 2011;156B:681–690.
136. Robilliard DL, Archer SN, Arendt J, et al. The 3111 Clock gene polymorphism is not associated with sleep and circadian rhythmicity in phenotypically characterized human subjects. *J Sleep Res*. 2002;11:305–312.
137. Iwase T, Kajimura N, Uchiyama M, et al. Mutation screening of the human Clock gene in circadian rhythm sleep disorders. *Psychiatry Res*. 2002;109:121–128.
138. Pedrazzoli M, Louzada FM, Pereira DS, et al. Clock polymorphisms and circadian rhythms phenotypes in a sample of the Brazilian population. *Chronobiol Int*. 2007;24:1–8.
139. Archer SN, Robilliard DL, Skene DJ, et al. A length polymorphism in the circadian clock gene PER3 is linked to delayed sleep phase syndrome and extreme diurnal preference. *Sleep*. 2003;26:413–415.

140. Pereira DS, Tufik S, Louzada FM, et al. Association of the length polymorphism in the human PER3 gene with the delayed sleep-phase syndrome: does latitude have an influence upon it? *Sleep*. 2005;28:29–32.
141. Jones KH, Ellis J, von Schantz M, Skene DJ, Dijk DJ, Archer SN. Age-related change in the association between a polymorphism in the PER3 gene and preferred timing of sleep and waking activities. *J Sleep Res*. 2007;16:12–16.
142. Ebisawa T, Uchiyama M, Kajimura N, et al. Association of structural polymorphisms in the human period3 gene with delayed sleep phase syndrome. *EMBO Rep*. 2001;2:342–346.
143. Viola AU, Archer SN, James LM, et al. PER3 polymorphism predicts sleep structure and waking performance. *Curr Biol*. 2007;17:613–618.
144. Groeger JA, Viola AU, Lo JC, von Schantz M, Archer SN, Dijk DJ. Early morning executive functioning during sleep deprivation is compromised by a PERIOD3 polymorphism. *Sleep*. 2008;31:1159–1167.
145. Lázár AS, Slak A, Lo JC, et al. Sleep, diurnal preference, health, and psychological well-being: a prospective single-allelic-variation study. *Chronobiol Int*. 2012;29:131–146.
146. Osland TM, Bjorvatn BR, Steen VM, Pallesen S. Association study of a variable-number tandem repeat polymorphism in the clock gene PERIOD3 and chronotype in Norwegian university students. *Chronobiol Int*. 2011;28:764–770.
147. Voinescu BI, Coogan AN. A variable-number tandem repeat polymorphism in PER3 is not associated with chronotype in a population with self-reported sleep problems. *Sleep Biol Rhythms*. 2012;10:23–26.
148. Goel N, Banks S, Mignot E, Dinges DF. *PER3* polymorphism predicts cumulative sleep homeostatic but not neurobehavioral changes to chronic partial sleep deprivation. *PLoS One*. 2009;4:e5874.
149. Carpen JD, Archer SN, Skene DJ, Smits M, von Schantz M. A single-nucleotide polymorphism in the 5′-untranslated region of the hPER2 gene is associated with diurnal preference. *J Sleep Res*. 2005;14:293–297.
150. Carpen JD, von Schantz M, Smits M, Skene DJ, Archer SN. A silent polymorphism in the PER1 gene associates with extreme diurnal preference in humans. *J Hum Genet*. 2006;51:1122–1125.
151. The International Classification of Sleep Disorders. *Diagnostic and Coding Manual*. 2nd ed. Westchester, IL: American Academy of Sleep Medicine; 2005.
152. Sack RL, Auckley D, Auger RR, et al. Circadian rhythm sleep disorders: part II, advanced sleep phase disorder, delayed sleep phase disorder, free-running disorder, and irregular sleep-wake rhythm. An American Academy of Sleep Medicine review. *Sleep*. 2007;30:1484–1501.
153. Albrecht U. Circadian clocks in mood-related behaviors. *Ann Med*. 2010;42:241–251.
154. Lamont EW, James FO, Boivin DB, Cermakian N. From circadian clock gene expression to pathologies. *Sleep Med*. 2007;8:547–556.
155. Campbell SS, Murphy PJ. Delayed sleep phase disorder in temporal isolation. *Sleep*. 2007;30:1225–1228.
156. Takano A, Uchiyama M, Kajimura N, et al. A missense variation in human casein kinase I epsilon gene that induces functional alteration and shows an inverse association with circadian rhythm sleep disorders. *Neuropsychopharmacology*. 2004;29:1901–1909.
157. Castro RM, Barbosa AA, Pedrazzoli M, Tufik S. Casein kinase I epsilon (CKIvarepsilon) N408 allele is very rare in the Brazilian population and is not involved in susceptibility to circadian rhythm sleep disorders. *Behav Brain Res*. 2008;193:156–157.
158. Reid KJ, Chang AM, Dubocovich ML, Turek FW, Takahashi JS, Zee PC. Familial advanced sleep phase syndrome. *Arch Neurol*. 2001;58:1089–1094.
159. Jones CR, Campbell SS, Zone SE, et al. Familial advanced sleep-phase syndrome: a short-period circadian rhythm variant in humans. *Nat Med*. 1999;5:1062–1065.

160. Toh KL, Jones CR, He Y, et al. An hPer2 phosphorylation site mutation in familial advanced sleep phase syndrome. *Science*. 2001;291:1040–1043.
161. Satoh K, Mishima K, Inoue Y, Ebisawa T, Shimizu T. Two pedigrees of familial advanced sleep phase syndrome in Japan. *Sleep*. 2003;26:416–417.
162. Xu Y, Padiath QS, Shapiro RE, et al. Functional consequences of a CKIdelta mutation causing familial advanced sleep phase syndrome. *Nature*. 2005;434:640–644.
163. Kerkhof GA. Inter-individual differences in the human circadian system: a review. *Biol Psychol*. 1985;20:83–112.
164. Goel N, Rao H, Durmer JS, Dinges DF. Neurocognitive consequences of sleep deprivation. *Semin Neurol*. 2009;29:320–339.
165. Diekelmann S, Born J. The memory function of sleep. *Nat Rev Neurosci*. 2010;11:114–126.
166. Basner M, Mollicone DJ, Dinges DF. Validity and sensitivity of a brief Psychomotor Vigilance Test (PVT-B) to total and partial sleep deprivation. *Acta Astronaut*. 2011;69:949–959.
167. Lo JC, Groeger JA, Santhi N, et al. Effects of partial and acute total sleep deprivation on performance across cognitive domains, individuals and circadian phase. *PLoS One*. 2012;7:e45987.
168. Killgore WDS, Grugle NL, Balkin TJ. Gambling when sleep deprived: don't bet on stimulants. *Chronobiol Int*. 2012;29:43–54.
169. Rupp TL, Wesensten NJ, Bliese PD, Balkin TJ. Banking sleep: realization of benefits during subsequent sleep restriction and recovery. *Sleep*. 2009;32:311–321.
170. Van Dongen HPA, Baynard MD, Maislin G, Dinges DF. Systematic interindividual differences in neurobehavioral impairment from sleep loss: evidence of trait-like differential vulnerability. *Sleep*. 2004;27:423–433.
171. Van Dongen HP, Maislin G, Dinges DF. Dealing with interindividual differences in the temporal dynamics of fatigue and performance: importance and techniques. *Aviat Space Environ Med*. 2004;75:A147–A154.
172. Goel N, Banks S, Mignot E, Dinges DF. *DQB1**0602* predicts interindividual differences in physiologic sleep, sleepiness and fatigue. *Neurology*. 2010;75: 1509–1519.
173. Goel N, Banks S, Lin L, Mignot E, Dinges DF. Catechol-O-methyltransferase Val158Met polymorphism associates with individual differences in sleep physiologic responses to chronic sleep loss. *PLoS One*. 2011;6:e29283.
174. Goel N, Dinges DF. Behavioral and genetic markers of sleepiness. *J Clin Sleep Med*. 2011;7:S19–S21.
175. Rupp TL, Wesensten NJ, Balkin TJ. Trait-like vulnerability to total and partial sleep loss. *Sleep*. 2012;35:1163–1172.
176. Rowland LM, Thomas ML, Thorne DR, et al. Oculomotor responses during partial and total sleep deprivation. *Aviat Space Environ Med*. 2005;76:C104–C113.
177. Drummond SP, Anderson DE, Straus LD, Vogel EK, Perez VB. The effects of two types of sleep deprivation on visual working memory capacity and filtering efficiency. *PLoS One*. 2012;7:e35653.
178. Tassi P, Schimchowitsch S, Rohmer O, et al. Effects of acute and chronic sleep deprivation on daytime alertness and cognitive performance of healthy snorers and non-snorers. *Sleep Med*. 2012;13:29–35.
179. Philip P, Sagaspe P, Prague M, et al. Acute versus chronic partial sleep deprivation in middle-aged people: differential effect on performance and sleepiness. *Sleep*. 2012;35:997–1002.
180. Drake CL, Roehrs TA, Burduvali E, Bonahoom A, Rosekind M, Roth T. Effects of rapid versus slow accumulation of eight hours of sleep loss. *Psychophysiology*. 2001;38:979–987.

181. Goel N. Genetics of sleep timing, duration and homeostasis in humans. *Sleep Med Clin.* 2011;6:171–182.
182. Kuna ST, Maislin G, Pack FM, et al. Heritability of performance deficit accumulation during acute sleep deprivation in twins. *Sleep.* 2012;35:1223–1233.
183. Goel N, Dinges DF. Predicting risk in space: genetic markers for differential vulnerability to sleep restriction. *Acta Astronaut.* 2012;77:207–213.
184. Rupp TL, Wesensten NJ, Newman R, Balkin TJ. PER3 and ADORA2A polymorphisms impact neurobehavioral performance during sleep restriction. *J Sleep Res.* 2013;22:160–165.
185. Bodenmann S, Xu S, Luhmann U, et al. Pharmacogenetics of modafinil after sleep loss: catechol-*O*-methyltransferase genotype modulates waking functions but not recovery sleep. *Clin Pharmacol Ther.* 2009;85:296–304.
186. Wu JC, Gillin JC, Buchsbaum MS, et al. The effect of sleep deprivation on cerebral glucose metabolic rate in normal humans assessed with positron emission tomography. *Sleep.* 1991;14:155–162.
187. Wu JC, Gillin JC, Buchsbaum MS, et al. Frontal lobe metabolic decreases with sleep deprivation not totally reversed by recovery sleep. *Neuropsychopharmacology.* 2006;31:2783–2792.
188. Thomas M, Sing H, Belenky G, et al. Neural basis of alertness and cognitive performance impairments during sleepiness. I. Effects of 24 h of sleep deprivation on waking human regional brain activity. *J Sleep Res.* 2000;9:335–352.
189. Thomas M, Sing H, Belenky G, et al. Neural basis of alertness and cognitive performance impairments during sleepiness II. Effects of 48 and 72 h of sleep deprivation on waking human regional brain activity. *Thalamus Relat Syst.* 2003;2:199–229.
190. Bartlett EJ, Brodie JD, Wolf AP, Christman DR, Laska E, Meissner M. Reproducibility of cerebral glucose metabolic measurements in resting human subjects. *J Cereb Blood Flow Metab.* 1988;8:502–512.
191. Buysse DJ, Nofzinger EA, Germain A, et al. Regional brain glucose metabolism during morning and evening wakefulness in humans: preliminary findings. *Sleep.* 2004;27:1245–1254.
192. Germain A, Nofzinger EA, Meltzer CC, et al. Diurnal variation in regional brain glucose metabolism in depression. *Biol Psychiatry.* 2007;62:438–445.
193. Volkow ND, Tomasi D, Wang GJ, et al. Evidence that sleep deprivation downregulates dopamine D2R in ventral striatum in the human brain. *J Neurosci.* 2002;32:6711–6717.
194. Elmenhorst D, Kroll T, Matusch A, Bauer A. Sleep deprivation increases cerebral serotonin 2A receptor binding in humans. *Sleep.* 2012;35:1615–1623.
195. Drummond SPA, Brown GG, Stricker JL, Buxton RB, Wong EC, Gillin JC. Sleep deprivation-induced reduction in cortical functional response to serial subtraction. *Neuroreport.* 1999;10:3745–3748.
196. Drummond SPA, Gillin JC, Brown GG. Increased cerebral response during a divided attention task following sleep deprivation. *J Sleep Res.* 2001;10:85–92.
197. Mander BA, Reid KJ, Davuluri VK, et al. Sleep deprivation alters functioning within the neural network underlying the covert orienting of attention. *Brain Res.* 2008;1217:148–156.
198. Tomasi D, Wang RL, Telang F, et al. Impairment of attentional networks after 1 night of sleep deprivation. *Cereb Cortex.* 2009;19:233–240.
199. Chee MW, Tan JC, Zheng H, et al. Lapsing during sleep deprivation is associated with distributed changes in brain activation. *J Neurosci.* 2008;28:5519–5528.
200. Chee MW, Tan JC, Parimal S, Zagorodnov V. Sleep deprivation and its effects on object-selective attention. *Neuroimage.* 2010;49:1903–1910.
201. Chee MWL, Tan JC. Lapsing when sleep deprived: neural activation characteristics of resistant and vulnerable individuals. *Neuroimage.* 2010;51:835–843.

202. Chee MW, Goh CS, Namburi P, Parimal S, Seidl KN, Kastner S. Effects of sleep deprivation on cortical activation during directed attention in the absence and presence of visual stimuli. *Neuroimage*. 2011;58:595–604.
203. Lim J, Tan JC, Parimal S, Dinges DF, Chee MW. Sleep deprivation impairs object-selective attention: a view from the ventral visual cortex. *PLoS One*. 2010;5:e9087.
204. Jackson ML, Hughes ME, Croft RJ, et al. The effect of sleep deprivation on BOLD activity elicited by a divided attention task. *Brain Imaging Behav*. 2011;5:97–108.
205. Kong D, Soon CS, Chee MW. Reduced visual processing capacity in sleep deprived persons. *Neuroimage*. 2011;55:629–634.
206. Kong D, Soon CS, Chee MW. Functional imaging correlates of impaired distractor suppression following sleep deprivation. *Neuroimage*. 2012;61:50–55.
207. Czisch M, Wehrle R, Harsay HA, et al. On the need of objective vigilance monitoring: effects of sleep loss on target detection and task-negative activity using combined EEG/fMRI. *Front Neurol*. 2012;3:67.
208. Muto V, Shaffii-le Bourdiec A, Matarazzo L, et al. Influence of acute sleep loss on the neural correlates of alerting, orientating and executive attention components. *J Sleep Res*. 2012;21:648–658.
209. Venkatraman V, Chuah YM, Huettel SA, Chee MW. Sleep deprivation elevates expectation of gains and attenuates response to losses following risky decisions. *Sleep*. 2007;30:603–609.
210. Venkatraman V, Huettel SA, Chuah LY, Payne JW, Chee MW. Sleep deprivation biases the neural mechanisms underlying economic preferences. *J Neurosci*. 2011;31:3712–3718.
211. Libedinsky C, Smith DV, Teng CS, et al. Sleep deprivation alters valuation signals in the ventromedial prefrontal cortex. *Front Behav Neurosci*. 2011;5:70.
212. Yoo SS, Gujar N, Hu P, Jolesz FA, Walker MP. The human emotional brain without sleep—a prefrontal amygdale disconnect. *Curr Biol*. 2007;17:R877–R878.
213. Yoo SS, Hu PT, Gujar N, Jolesz FA, Walker MP. A deficit in the ability to form new human memories without sleep. *Nat Neurosci*. 2007;10:385–392.
214. Chuah LY, Chong DL, Chen AK, et al. Donepezil improves episodic memory in young individuals vulnerable to the effects of sleep deprivation. *Sleep*. 2009;32:999–1010.
215. Darsaud A, Dehon H, Lahl O, et al. Does sleep promote false memories? *J Cogn Neurosci*. 2011;23:26–40.
216. Chuah YM, Venkatraman V, Dinges DF, Chee MW. The neural basis of inter-individual variability in inhibitory efficiency after sleep deprivation. *J Neurosci*. 2006;26:7156–7162.
217. Drummond SP, Brown GG, Salamat JS, Gillin JC. Increasing task difficulty facilitates the cerebral compensatory response to total sleep deprivation. *Sleep*. 2004;27:445–451.
218. Strangman G, Thompson JH, Strauss MM, Marshburn TH, Sutton JP. Functional brain imaging of a complex navigation task following one night of total sleep deprivation: a preliminary study. *J Sleep Res*. 2005;14:369–375.
219. Drummond SPA, Brown GG, Gillin JC, Stricker JL, Wong EC, Buxton RB. Altered brain response to verbal learning following sleep deprivation. *Nature*. 2000;403:655–657.
220. Drummond SP, Meloy MJ, Yanagi MA, Orff HJ, Brown GG. Compensatory recruitment after sleep deprivation and the relationship with performance. *Psychiatry Res*. 2005;140:211–223.
221. Albouy G, Vandewalle G, Sterpenich V, et al. Sleep stabilizes visuomotor adaptation memory: a functional magnetic resonance imaging study. *J Sleep Res*. 2013;22:144–154.
222. Chee MW, Choo WC. Functional imaging of working memory after 24 hr of total sleep deprivation. *J Neurosci*. 2004;24:4560–4567.

223. Chee MW, Chuah LY, Venkatraman V, Chan WY, Philip P, Dinges DF. Functional imaging of working memory following normal sleep and after 24 and 35 h of sleep deprivation: correlations of fronto-parietal activation with performance. *Neuroimage*. 2006;31:419–428.
224. Chee MW, Chuah YM. Functional neuroimaging and behavioral correlates of capacity decline in visual short-term memory after sleep deprivation. *Proc Natl Acad Sci USA*. 2007;104:9487–9492.
225. Choo WC, Lee WW, Venkatraman V, Sheu FS, Chee MW. Dissociation of cortical regions modulated by both working memory load and sleep deprivation and by sleep deprivation alone. *Neuroimage*. 2005;25:579–587.
226. Mu Q, Nahas Z, Johnson KA, et al. Decreased cortical response to verbal working memory following sleep deprivation. *Sleep*. 2005;28:55–67.
227. Mu Q, Mishory A, Johnson KA, et al. Decreased brain activation during a working memory task at rested baseline is associated with vulnerability to sleep deprivation. *Sleep*. 2005;28:433–446.
228. Lim J, Choo WC, Chee MW. Reproducibility of changes in behaviour and fMRI activation associated with sleep deprivation in a working memory task. *Sleep*. 2007;30:61–70.
229. Chuah LY, Dolcos F, Chen AK, Zheng H, Parimal S, Chee MW. Sleep deprivation and interference by emotional distracters. *Sleep*. 2010;33:1305–1313.
230. Lythe KE, Williams SC, Anderson C, Libri V, Mehta MA. Frontal and parietal activity after sleep deprivation is dependent on task difficulty and can be predicted by the fMRI response after normal sleep. *Behav Brain Res*. 2012;233:62–70.
231. Schroeter ML, Bücheler MM, Scheid R. Circadian variability is negligible in primary visual cortices as measured by fNIRS. *Int J Psychophysiol*. 2006;62:9–13.
232. Gorfine T, Zisapel N. Melatonin and the human hippocampus, a time dependent interplay. *J Pineal Res*. 2007;43:80–86.
233. Vimal RL, Pandey-Vimal MU, Vimal LS, et al. Activation of suprachiasmatic nuclei and primary visual cortex depends upon time of day. *Eur J Neurosci*. 2009;29:399–410.
234. Peres I, Vetter C, Blautzik J, et al. Chronotype predicts activity patterns in the neural underpinnings of the motor system during the day. *Chronobiol Int*. 2011;28:883–889.
235. Marek T, Fafrowicz M, Golonka K, et al. Diurnal patterns of activity of the orienting and executive attention neuronal networks in subjects performing a Stroop-like task: a functional magnetic resonance imaging study. *Chronobiol Int*. 2010;27:945–958.
236. Vandewalle G, Archer SN, Wuillaume C, et al. Functional magnetic resonance imaging-assessed brain responses during an executive task depend on interaction of sleep homeostasis, circadian phase, and PER3 genotype. *J Neurosci*. 2009;29:7948–7956.
237. Vandewalle G, Archer SN, Wuillaume C, et al. Effects of light on cognitive brain responses depend on circadian phase and sleep homeostasis. *J Biol Rhythms*. 2011;26:249–259.
238. Schmidt C, Collette F, Leclercq Y, et al. Homeostatic sleep pressure and responses to sustained attention in the suprachiasmatic area. *Science*. 2009;324:516–519.
239. Schmidt C, Peigneux P, Leclercq Y, et al. Circadian preference modulates the neural substrate of conflict processing across the day. *PLoS One*. 2012;7:e29658.
240. Detre JA, Leigh JS, Williams DS, Koretsky AP. Perfusion imaging. *Magn Reson Med*. 1992;23:37–45.
241. Detre JA, Alsop DC. Perfusion magnetic resonance imaging with continuous arterial spin labeling: methods and clinical applications in the central nervous system. *Eur J Radiol*. 1999;30:115–124.
242. Detre JA, Wang J, Wang Z, Rao H. ASL perfusion MRI in basic and clinical neuroscience. *Curr Opin Neurol*. 2009;22:348–355.

243. Detre JA, Rao H, Wang DJ, Chen YF, Wang Z. Applications of arterial spin labeled MRI in the brain. *J Magn Reson Imaging*. 2012;35:1026–1037.
244. Lim J, Wu WC, Wang J, Detre JA, Dinges DF, Rao H. Imaging brain fatigue from sustained mental workload: an ASL perfusion study of the time-on-task effect. *Neuroimage*. 2010;49:3426–3435.
245. Poudel GR, Innes CRH, Jones RD. Cerebral perfusion differences between drowsy and non-drowsy individuals following acute sleep restriction. *Sleep*. 2012;35:1085–1096.
246. Rao H, Lim J, Zhu S, Hu S, Detre JA, Dinges DF. Differential effects of total sleep deprivation on regional cortical activity at rest and during Psychomotor Vigilance Test performance. *Sleep*. 2011;34:A97.
247. Biswal B, Yetkin FZ, Haughton VM, Hyde JS. Functional connectivity in the motor cortex of resting human brain using echo-planar MRI. *Magn Reson Med*. 1995;34:537–541.
248. Fox MD, Raichle ME. Spontaneous fluctuations in brain activity observed with functional magnetic resonance imaging. *Nat Rev Neurosci*. 2007;8:700–711.
249. Fox MD, Snyder AZ, Vincent JL, Corbetta M, Van Essen DC, Raichle ME. The human brain is intrinsically organized into dynamic, anticorrelated functional networks. *Proc Natl Acad Sci USA*. 2005;102:9673–9678.
250. Damoiseaux JS, Rombouts SA, Barkhof F, et al. Consistent resting-state networks across healthy subjects. *Proc Natl Acad Sci USA*. 2006;103:13848–13853.
251. Laird AR, Fox PM, Eickhoff SB, et al. Behavioral interpretations of intrinsic connectivity networks. *J Cogn Neurosci*. 2011;23:4022–4037.
252. Raichle ME. The restless brain. *Brain Connect*. 2011;1:3–12.
253. Sämann PG, Tully C, Spoormaker VI, et al. Increased sleep pressure reduces resting state functional connectivity. *MAGMA*. 2010;23:375–389.
254. De Havas JA, Parimal S, Soon CS, Chee MW. Sleep deprivation reduces default mode network connectivity and anti-correlation during rest and task performance. *Neuroimage*. 2012;59:1745–1751.
255. Blautzik J, Vetter C, Peres I, et al. Classifying fMRI-derived resting-state connectivity patterns according to their daily rhythmicity. *Neuroimage*. 2012;71C:298–306.
256. Goel N, Van Dongen HPA, Dinges DF. Circadian rhythms in sleepiness, alertness, and performance. In: Kryger MH, Roth T, Dement WC, eds. *Principles and Practice of Sleep Medicine*. 5th ed. Philadelphia, PA: Elsevier; 2011:445–455.

CHAPTER EIGHT

Timing, Sleep, and Respiration in Health and Disease

Gordon F. Buchanan[*,†]

[*]Department of Neurology, Yale University School of Medicine, New Haven, Connecticut, USA
[†]Veteran's Affairs Medical Center, West Haven, Connecticut, USA

Contents

Abstract

Breathing is perhaps the physiological function that is most vital to human survival. Without breathing and adequate oxygenation of tissues, life ceases. As would be expected for such a vital function, breathing occurs automatically, without the requirement of conscious input. Breathing is subject to regulation by a variety of factors including circadian rhythms and vigilance state. Given the need for breathing to occur continuously with little tolerance for interruption, it is not surprising that breathing is subject to both circadian phase-dependent and vigilance-state-dependent regulation. Similarly, the information regarding respiratory state, including blood–gas concentrations, can affect circadian timing and sleep–wake state. The exact nature of the interactions between breathing, circadian phase, and vigilance state can vary depending

Progress in Molecular Biology and Translational Science, Volume 119
ISSN 1877-1173
http://dx.doi.org/10.1016/B978-0-12-396971-2.00008-7

upon the species studied and the methodologies employed. These interactions between breathing, circadian phase, and vigilance state may have important implications for a variety of human diseases, including sleep apnea, asthma, sudden unexpected death in epilepsy, and sudden infant death syndrome.

1. INTRODUCTION

The physiological function most vital to mammalian existence is breathing. Breathing provides the vital oxygen (O_2) that most mammalian tissues require to function properly and expels the dangerous carbon dioxide (CO_2), thus regulating serum pH within a tight range. This regulation maintains respiratory drive to ensure continued O_2/CO_2 exchange and pH regulation. Breathing is something we, as mammals, do automatically. We can consciously alter our breathing rate and depth, but still breathe even when higher cortical function is isolated from the brainstem respiratory centers. Breathing can be altered by a wide variety of factors including feedback from airway muscles; feedback from pulmonary stretch receptors in the lungs, body temperature, limbic factors (e.g., stress, excitement), physical exertion, cardiac factors, and conscious modulation; and chemosensory feedback regarding oxygen and carbon dioxide concentrations in the blood. In addition to these factors, breathing is also subject to circadian control and varies as a function of vigilance state. Conversely, as the regulator of O_2 and CO_2 concentrations in the blood, breathing can also modify the phase and amplitude of circadian rhythms and modulate vigilance state. The focus of this review will be on the influences of circadian factors and vigilance state in governing respiratory control, mechanisms by which respiratory control can in turn modulate circadian rhythms and sleep–wake regulation; these factors together can influence human disease.

2. NORMAL BREATHING

Breathing occurs automatically without the need for conscious input. A major purpose of breathing is to allow exchange of oxygen (O_2) inspired from the environment for carbon dioxide (CO_2) produced by the body. This occurs at the level of the alveolar capillaries and serves to both oxygenate blood for delivery to tissues and maintain arterial partial pressure of CO_2 (P_{CO_2}) which is the primary regulator of respiratory drive. The mammalian respiratory tract is comprised of the nose, mouth, pharynx, larynx, trachea, bronchi, bronchioles, and alveoli. The latter three components are housed within the lungs and along with the trachea constitute the lower airway. The remaining components comprise the upper airway. The mechanical act of breathing is accomplished via rhythmic contraction of primary and accessory respiratory muscles including the diaphragm and intercostal muscles. Contraction of the diaphragm creates negative pressure with the chest cavity allowing O_2-rich air to pass into the respiratory tract. Following gas exchange within the alveoli, diaphragmatic relaxation increases pressure within the chest cavity and allows CO_2-laden air to be expelled from the respiratory tract through the mouth and nose. Rhythm-dictating input to these muscles originates in the respiratory pattern generator (RPG) within the caudal medulla.[1] The RPG will be described in more detail shortly. Output from the RPG is transmitted to anterior horn motor neurons of the cervical and thoracic spinal cord, which in turn contract the diaphragm and intercostal muscles, respectively.[2] While breathing occurs automatically—from birth to death for most animals including humans—several factors can influence respiratory rate (RR) and rhythm. The most notable among these are the respiratory chemical stimuli, hypoxia, and hypercapnia.[3,4] A schematic overview of the respiratory system is given in Fig. 8.1. Respiratory pattern generation, airway control, and mechanisms of respiratory rhythm modulation will now be discussed in greater detail.

2.1. Respiratory pattern generation

Unlike the heart that has intrinsic pacemakers,[5] the lungs do not have self-contained pacemaking capabilities. For the respiratory system, generation of the respiratory rhythm is relegated to the medullary brainstem.[1,6] Within the medulla are two bilateral neuronal populations, the dorsal and ventral respiratory groups (DRG and VRG, respectively), which are involved in respiration. The DRG is primarily comprised of neurons that are most active during inspiration and that are situated within the ventrolateral subgroup of the nucleus tractus solitarius (NTS).[7] The VRG is subdivided into a rostral part comprised largely of inspiratory premotor neurons[8] and a caudal part containing primarily expiratory premotor neurons.[9] Rostral to the VRG are the Bötzinger complex (BötC), the pre-Bötzinger complex (pre-BötC), and the parafacial nucleus. The BötC contains inhibitory expiratory neurons and cranial motor neurons.[10] Generation of the basic respiratory rhythm is thought to be the responsibility of the pre-BötC[11] and parafacial respiratory group,[12,13] which includes the retrotrapezoid nucleus (RTN).[6,14] The

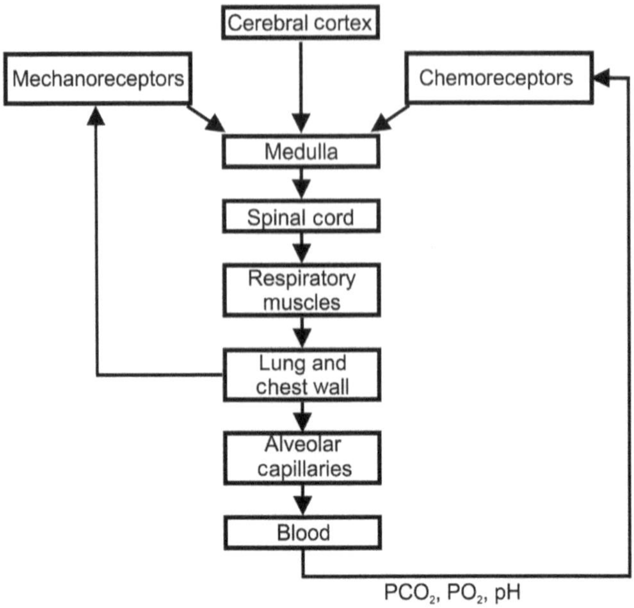

Figure 8.1 Central control of respiration. The respiratory pattern is generated in the medulla. Efferent projections from the respiratory pattern generator synapse on respiratory muscle motor neurons in the spinal cord to control respiratory muscle, diaphragm, and intercostal muscle contraction. This drives expansion and contraction of the lung and chest wall to move air through the respiratory tract and allows O_2 to be exchanged for CO_2 at the level of the alveolar capillaries. The resultant changes in P_{O_2}, (P_{CO_2}), and pH are monitored by chemoreceptors, which then feedback onto the RPG to modulate breathing. Lung, chest wall, and diaphragmatic activity are monitored by mechanoreceptors, which also feed back onto the medullary pattern generator. Connections to the pattern generator from the cortex allow for conscious modulation of breathing.

DRG and VRG both contain neurons that project directly to respiratory motor neurons within the spinal cord, where they can directly drive the RR and rhythm, and to nonmotor medullary respiratory neurons, where they can indirectly influence RR and rhythm.[1]

2.2. Airway control

Airway resistance is modulated in an inverse relationship with airway tone, whereby increasing airway tone decreases airway resistance and vice versa. Adequate tone of the upper airway muscles is required for airway patency. A major purpose of the upper airway is to condition air as it passes deeper into the respiratory tract; however, reduction of tone in laryngeal and pharyngeal muscles, for instance, contributes to increased resistance and airway obstruction. Control of upper airway tone comes from the trigeminal (CN V), facial (CN VII), and hypoglossal (CN XII) motor neurons. Excitatory drive to these pathways comes from cortical and aminergic systems (i.e., serotonergic, histaminergic, adrenergic, cholinergic systems), is most robust during wakefulness, and progressively decreases through the different stages of sleep. Appropriate activity within respiratory muscles (i.e., diaphragm and intercostals) is required to pump air through the respiratory tract. The state dependence of airway control, respiratory muscle activity, and ventilation will be discussed in more detail below.

2.3. Modulation of breathing

2.3.1 Chemical control of breathing

Arguably, chemical stimuli are the most important regulators of breathing. The partial pressures of oxygen (P_{O_2}) and carbon dioxide (P_{CO_2}) are monitored by chemoreceptors, which function to maintain blood concentrations of O_2, CO_2, and serum pH within a narrow range to ensure normal body tissue function. For the most part, peripheral O_2 chemoreceptors sense changes in P_{O_2},[3] while central CO_2 chemoreceptors sense changes in P_{CO_2}.[15] Chemical aberrations, most typically hypoxia or hypercapnia, lead to activation of the respective chemoreceptor subtype and subsequent modulation of ventilation to correct the aberration.[15]

2.3.1.1 Oxygen chemoreception

The majority of O_2 chemoreception occurs peripherally in the carotid body by the type I glomus cells.[3] Afferent information regarding P_{O_2} is sent via the carotid sinus nerve branch of the glossopharyngeal nerve to the NTS.[16] Under hyperoxic conditions, carotid sinus nerve fibers fire tonically and increase their discharge rate as P_{O_2} falls below the normal level of 100 mmHg.[17] Though most O_2 chemoreception occurs peripherally, animal models indicate that there is a central component to O_2 chemoreception. Implicated brain loci include the pre-BötC,[18,19] the ventrolateral medulla,[20] and hypothalamus.[21] Notably, adult human patients with carotid body resection do not have a hypoxic ventilatory response,[22] suggesting that in humans, little O_2 chemoreception occurs centrally.

2.3.1.2 Carbon dioxide chemoreception

In contrast to O_2 chemoreception, the majority of CO_2 chemoreception occurs within the central nervous system. Neurons in a number of brainstem nuclei have properties consistent with being central CO_2 chemoreceptors, though the relative importance of the specific sites and neuronal cell types responsible for central respiratory chemoreception have been the subject of debate.[15,23–26] Candidate chemoreceptive nuclei include the medullary raphé,[27,28] NTS,[29] locus coeruleus,[30–32] RTN,[33–35] hypothalamus,[21,36–39] and cerebellar fastigial nucleus.[40] The sensitivity of central chemoreceptors to CO_2 is increased by hypoxia.[41] Just as there is evidence for a central component of O_2 chemoreception, there is evidence for a peripheral component of CO_2 chemoreception.[42,43]

Serotonergic medullary raphé neurons are intrinsically chemosensitive *in vitro*.[27,44–46] While the respiratory stimulus that leads to activation of chemoreceptors is hypercapnia, it is the change in brain tissue pH that results from the (P_{CO_2}) change that stimulates central chemoreceptors and not CO_2 itself. Indeed, medullary 5-HT neurons respond to changes in intracellular pH, but do not respond to stimulation with CO_2.[47,48] *In vivo*, hypercapnia increases the firing rate of 5-HT neurons in the awake, behaving cat,[49] leads to increased c-fos expression in medullary 5-HT neurons[50–53] and causes 5-HT release within the mouse hypoglossal nucleus.[54] Direct application of acetazolamide[55] or CO_2-rich artificial cerebrospinal fluid to the medullary raphé induces a focal acidification of the region and increases breathing.[56,57] Acute 5-HT neuron lesion or inactivation with 5,7-dihydroxytryptamine, lidocaine, ibotenic acid, muscimol, 8-OH-DPAT, or saporin conjugated to a 5-HT transporter antibody reduces the hypercapnic ventilatory response (HCVR).[57–62] Mice with a subtotal (Pet1-KO mice) or nearly complete (*Lmx1b$^{f/f/p}$* mice) genetic deletion of 5-HT neurons within the central nervous system[63,64] have an impaired HCVR compared to WT mice.[65,66]

Hypercapnia stimulates LC neurons *in vivo* and *in vitro*.[30,31,67–69] Likewise, neurons within the NTS are also stimulated by acidosis,[29,70] as are neurons in the RTN.[33,35] Neurons in each of these regions, except the RTN,[71] have been shown to retain their chemosensitivity after chemical synaptic blockade or physical isolation. RTN neurons are strongly stimulated by 5-HT, SP, and TRH,[72] and as recently discussed, it is possible that some of their pH sensitivity is due to synaptic input from 5-HT or other neurons.[73,74]

Neurons in the caudal hypothalamus of rabbits,[75] cats,[76] and rats[21] are stimulated by hypercapnia. There are also chemosensitive neurons in the lateral hypothalamus that contain orexin.[36] Stimulation of orexin neurons with acidosis increases breathing *in vivo*.[37] Mice with genetic deletion of hypothalamic orexin neurons have an attenuated HCVR, and this can be partially restored by exogenous orexin.[77] Treatment with an orexin receptor antagonist attenuates the HCVR in WT mice.[78]

It has been proposed that central chemoreception is a widely distributed function of neurons in many brainstem nuclei.[25] This possibility is supported by studies showing that focal acidosis in many of the nuclei discussed above causes an increase in ventilation *in vivo*.[79] It has been further suggested that these sites may not be equally important under normal physiological conditions, but rather play an important role under specific conditions, such as during development, under anesthesia, during sleep, or in various pathological states.[80]

2.3.2 Influence of other factors on breathing

In addition to being subject to chemical regulation, breathing can also be modulated by a number of other factors including conscious control (e.g., voluntary modulation), limbic factors (e.g., stress), cardiac factors (e.g., stress, exertion), body temperature (e.g., fever), and physical exertion (e.g., increased oxygen demand).

The respiratory control system receives descending input from cortical structures, allowing for voluntary modulation of RR, depth of ventilation, and respiratory pattern. For instance, an individual can voluntarily hold his/her breath when submerging the head in water or encountering a noxious olfactory stimulus. An individual can also voluntarily hyperventilate. There are, of course, built-in failsafe mechanisms to prevent an individual from accidentally injuring themselves from these maneuvers. For instance, if breath is held too long, the individual will pass out, thus, releasing the cortical activation and the RPG will resume its preprogrammed rate and rhythm to maintain breathing and normal blood gases. Similarly, if the individual hyperventilates too long, they will experience dyspnea, and this uncomfortable sensation will be relayed to the cortex and cause it to reduce its excitatory input to the respiratory control system.

Afferent signals from Golgi tendon organs contained within the diaphragm feedback to respiratory centers and inhibit respiratory activity. Afferents from muscle spindle fibers embedded within chest wall intercostal muscles relay information regarding chest wall expansion and contraction to higher centers and are involved with reflex control of respiratory activity. These muscle mechanoreceptors allow sampling of volume (length) and pressure (tension) status to detect changes. The balance between volume and pressure is sent to higher centers as "length–tension appropriateness."[81] Any aberration in this system leads to adjustments in medullary respiratory motor activity.[82] In addition to mechanoreceptors, there are irritant receptors within the respiratory tract that are activated by various exogenous stimuli (e.g., dust, etc.) and send afferent to higher centers to typically induce a cough.

3. CIRCADIAN RHYTHMS AND BREATHING

3.1. Circadian regulation of breathing

All organisms display circadian rhythms.[83] In mammals, these rhythms are governed by the master oscillator in the paired suprachiasmatic nuclei (SCN) of the anterolateral hypothalamus.[84,85] The master SCN oscillator drives rhythms in other regions, both central and peripheral.[86] Consequently, the SCN regulates a variety of behavioral and physiological processes in a near 24-h pattern. Among these is breathing.[87]

Respiratory function in humans is typically assessed with spirometry in which a subject inspires maximally and then expires as quickly and completely as possible. Measurements obtained in this manner include total lung capacity, forced vital capacity (FVC), tidal volume (V_T), residual volume, forced expiratory volume in one second (FEV_1), forced expiratory flow, peak expiratory flow (PEF), and FEV_1/FVC (a more patient-specific indicator of airway tone). In addition, RR can be counted manually or measured along with V_T via plethysmography. These measures are used together to determine minute ventilation (V_E), the functional volume of air moved through the lungs in one minute. (P_{CO_2}) and P_{O_2} can be measured via optical or electrochemical sensors or directly in the serum along with pH in an arterial blood gas assay. Finally, the respiratory responses to manipulations including chemical stimulation with hypoxia and hypercapnia (i.e., the hypoxic and hypercapnic ventilator responses, respectively) can be measured.

Through a variety of studies in a wide range of species including humans,[88] non-human primates,[89] dogs,[90] rats,[91] mice,[92]ducks,[93] snakes,[94] cows,[95] and even worms[96] and bees,[97] it has been shown that many respiratory measures can be subject to circadian regulation, though this varies greatly, depending on the species and experimental conditions.[98]

In one study of 30 male subjects, robust circadian variation was seen in V_T, V_E, and V_T/T_I. These participants were subjected to a constant routine paradigm that included sleep deprivation during the study period.[99] In another study of 10 healthy individuals, there was a trend toward circadian variation with V_E being increased during the daytime, but this failed to reach statistical significance. While the sleep–wake cycle is subject to circadian regulation[100,101] and sleep state can certainly affect many of these measures as will be discussed in detail below, that these studies were performed in a constant routine paradigm[102] suggests that circadian regulation of breathing occurs independently of sleep state.[88,103,104]

Animal studies also provide evidence for circadian regulation of breathing. Using whole-body plethysmography in 21 cynomolgus monkeys housed in a standard 12:12 light–dark cycle, Iizuka and colleagues demonstrated circadian variation in RR, V_T, and V_E minute ventilation with small reductions in RR and V_T during the dark phase and larger reductions in V_E during the same time.[89] Unlike the human studies, the investigators did not control for sleep state in this study.[89] Sprague–Dawley rats housed in a 12:12 light–dark cycle display baseline circadian variation in V_{CO_2} and V_T/T_I, but not V_T or RR. There was a trend toward increased V_T in the dark, active period, but this was not statistically significant. They did however observe a heightened HCVR in the dark phase with CO_2 inducing larger increases in V_T and RR compared to during the light phase. In this study, only one time point was sampled during each lighting condition.[91] Since breathing can vary widely throughout the 24-h day and can be subject to ultradian regulation,[105] sampling so infrequently may not be representative of the overall pattern. Using a whole-body plethysmography method similar to that originally described by Drorbaugh and Fenn,[106] which allows for continuous sampling of RR, V_T, and V_E from undisturbed behaving animals in different environmental gas concentrations, rats were found to display a circadian variation of RR, V_T, and V_E.[98] Being nocturnal rodents, V_E is higher during the dark period, in contrast to the diurnal human in which V_E is higher during the light period (Fig. 8.2A).[98] Interestingly, while there are circadian variations in sleep state, body temperature, activity, and metabolism and all of these certainly affect V_E, observed circadian variations in V_E seem to occur independently of these factors.[87,107]

In addition to circadian variation in RR and V_E, there is circadian variation in airway tone and sensitivity to respiratory stimuli. There are diurnal variations in FEV_1,[108] PEF,[109,110] and specific airway conductance.[111] All of these changes are a direct function of airway tone. In the same study of 10 healthy male participants mentioned above, Spengler and Shea observed circadian variation in FEV_1 and FEV_1/FVC, but not in FVC or PEF. In this

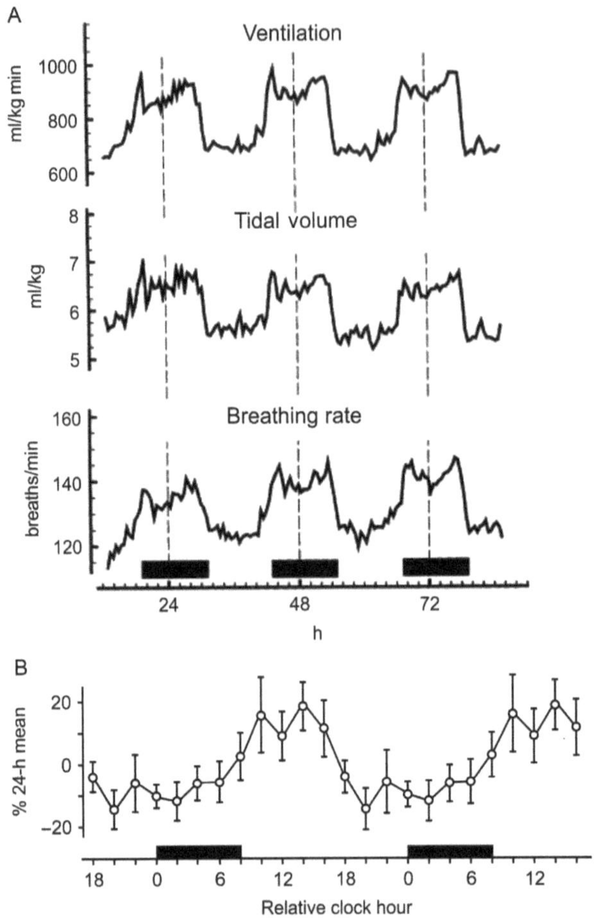

Figure 8.2 Circadian regulation of breathing. (A) 72-h traces depicting average ventilation (top), tidal volume (middle), and breathing rate (bottom) from rats housed in a 12:12 light:dark cycle in room air (21% O_2, balance N_2). Horizontal dark bars indicate time of lights off. (B) 48-h trace depicting the circadian variation in the hypercapnic ventilatory response (HCVR) in humans. *(A) Redrawn with permission from Ref. 98 and (B) from Ref. 103.*

study, the peaks occurred between 10 AM and 2 PM and were phase shifted approximately 4 h from the subjects' circadian rhythm of core body temperature.[88] There are also diurnal variations in the responsiveness of the bronchial musculature to stimuli such as house dust,[112] histamine,[113] or other allergens.[114] For the most part, the airway is most sensitive to these stimuli during the nighttime in humans. As will be seen below, this variation in airway resistance and irritant sensitivity contributes to the nocturnal worsening of symptoms in asthma patients.

Complementary to the master circadian oscillator in the SCN, secondary oscillators exist in a variety of sites and are under the control of the SCN.[86,115] Among these are peripheral oscillators in the respiratory system. Circadian oscillations of mRNA for the clock genes Per1, Per2, Bmal1, and Clock have been detected in the larynx, trachea, bronchus, and lung.[116] Lesion of the bilateral SCN abolishes the circadian respiratory rhythms suggesting that peripheral oscillations are not sufficient to maintain rhythmicity without input from the SCN. Rhythmicity is also lost in arrhythmic $Cry1^{-/-}Cry2^{-/-}$ knockout mice,[116] suggesting that rhythmicity in this peripheral oscillator cannot occur without a normally functioning master oscillator. Vagotomy, but not sympathetectomy, also abolishes the circadian respiratory rhythms, suggesting a vagally mediated role of the parasympathetic nervous system, which is supported by the fact that the rhythms can be attenuated by antimuscarinic agents.[116]

As discussed above, hypercapnia is an important and potent modulator of breathing. The sensitivity of the respiratory system to CO_2 varies in both a circadian and state-dependent manner (Fig. 8.2B). Several studies in humans have demonstrated that the respiratory system is least sensitive to stimulation by hypercapnia in the early morning and most sensitive later in the day. One study demonstrated a similar circadian pattern of sensitivity to hypoxia.[117,118] In another study, the peak sensitivity to CO_2 occurred in the mid-afternoon.[88] Circadian variation in sensitivity to stimulation by hypoxia and hypercapnia is also seen in awake rats[91,119,120] and in ducks.[93]

3.2. Influence of breathing on circadian rhythms

A wide variety of factors have been shown to modulate circadian timing (i.e., reset the circadian clock). The most notable of these is light. This is easy to appreciate, considering that the 24-h day most organisms are subject to is dictated by the rotation of the earth relative to the sun and thus dictated by a light–dark cycle. In addition to light, a number of other phase-shifting stimuli have been identified. Among these are the respiratory stimuli hypercapnia and hypoxia. In human studies, acute hypoxia transiently phase-shifted PEF, temperature, and grip strength,[121] while acute mild hypercapnia caused small and transient phase shift of core body temperature rhythms.[122] Both of these studies were confounded by environmental variables. For golden hamsters, the time of maximal phase shifting of wheel-running activity for both hypoxia and hypercapnia is during the daytime (Fig. 8.3), a time when the nocturnal rodent is less active and in a lower-energy homeostatic state, if not sleeping[123]; however, other time points have not been evaluated. In rats kept in a light–dark cycle, no phase-shifting effect of acute hypoxia is detected.[124] Encountering such powerful homeostatic stimuli during the inactive phase should serve as a profound error signal and would alter the timing of the clock into the active phase in which the animal is more alert and more homeostatically and metabolically able to deal with the stimulus. Most phase-shifting stimuli serve to move the organism from one state of alertness to the other. For instance, when the nocturnal rodent is in its active/dark phase and encounters light, the clock is phase shifted to the less active state usually seen during the light phase (i.e., phase delayed when light is encountered early in the dark period and phase advanced when encountered late in the dark period). Interestingly, both of these stimuli induce arousal from sleep.[125–133]

Just as the reason that these stimuli are able to shift the clock is not clear, the mechanisms are also not clear. The neurotransmitters serotonin (5-HT), neuropeptide Y,[134] pituitary adenylate cyclase activating polypeptide,[135,136] and substance P have all been shown to affect the timing of the circadian clock when presented during the middle of the subjective daytime. Serotonin can advance the timing of the peak in single-unit neuronal firing in rat hypothalamic slices *in vitro*[137] and in the timing of the onset of hamster wheel-running activity when applied into the cerebral ventricles or directly to the SCN *in vivo*.[138] Serotonin neurons are robustly activated by acidosis *in vitro*[23,27,28,44–47,139–142] and by acidosis and elevations in inspired CO_2

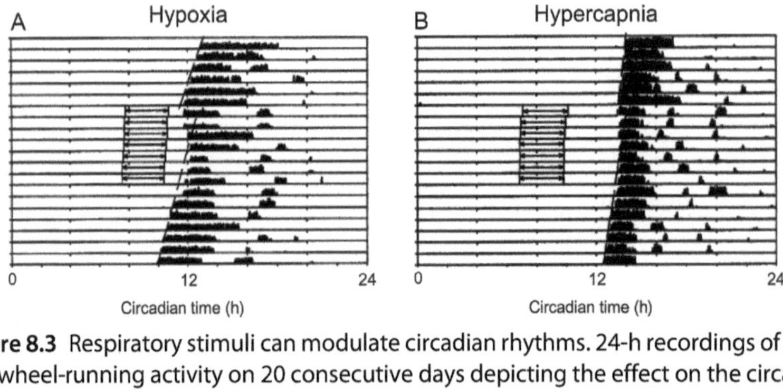

Figure 8.3 Respiratory stimuli can modulate circadian rhythms. 24-h recordings of hamster wheel-running activity on 20 consecutive days depicting the effect on the circadian rhythms following exposure to hypoxia (A) or hypercapnia (B) as indicated. Lines are drawn through the onset of activity for several days before stimulation and after stimulation to aid in visualization of the phase shifts. *Redrawn with permission from Ref. 123.*

concentrations *in vivo*.[49,55,143–145] The timing of sensitivity as well as the direction of the response (i.e., phase advance) for both 5-HT and hypercapnia is similar. Thus it stands to reason that hypercapnia could activate 5-HT neurons leading to 5-HT release onto neurons within the suprachiasmatic nucleus and subsequently the phase shift.

Chronic exposure to hypoxia and hypercapnia can also affect circadian rhythms. Hypoxia has been shown to abolish circadian rhythms of not just respiratory measures, but also of core body temperature, activity, and cortisol secretion.[124,146] As discussed above, hypoxia is sensed primarily via peripheral chemoreceptors in the carotid body. However, sino-aortic node denervation in rats, and thus elimination of output from peripheral chemoreceptors, did not prevent blunting of circadian rhythms of body temperature and locomotor activity. These authors concluded that this suggests direct effect of hypoxia on thermoregulatory centers in the hypothalamus.[147] Certainly, as mentioned above, hypothalamic sites have been shown to be hypoxia chemosensors.[21,148,149] Chronic low-level hypercapnia blunts circadian activity and temperature rhythms in humans[122] but has no effect on rat temperature, activity, or respiratory rhythms.[119] It does, however, increase the baseline and phase amplitude of the circadian rhythms of RR and tidal volume.[119]

4. SLEEP AND BREATHING

4.1. State-dependent control of breathing

Breathing regulation is modulated in a state-dependent manner.[150] As one transitions to light NREM from wakefulness, breathing can be unstable.[151,152] This is termed periodic breathing or Cheyne–Stokes respiration.[153] Breathing instability at the extreme manifests as obstructive, central, or mixed apneas.[154] In deeper NREM, breathing becomes more regular with larger V_T and slower RR. In NREM, these changes are accompanied by a relative increase in arterial (P_{CO_2}) and a concomitant decrease in arterial P_{O_2}. There is an increase in total airway resistance in NREM due largely to the increase in upper airway resistance that results from decreased tonic activity to upper airway muscles.[155–157] Intercostal muscle activity is increased during NREM as evidenced by EMG studies with little or no increase in diaphragmatic muscle activity.[158,159] In contrast to deeper NREM, in REM, respiration is profoundly irregular[151,160] and can be associated with sudden changes in RR and V_T. There can also be apneic intrusions lasting 10–30 s or more. These breathing irregularities are linked to the occurrence of bursts of rapid eye movements for which REM sleep is named.[161–163]

During sleep, there is a reduction in excitatory cortical input to the upper airway muscles resulting in decreased tone and reduced patency.[164–167] This is especially problematic during early NREM (stage N1) and REM, sleep stages in which breathing is particularly unstable.[168] Arousal leads to an increase in excitatory input to the upper airway muscles and improvement in airway patency.[169,170]

Interestingly, there has been some controversy over the effects of sleep deprivation on the HCVR with some studies suggesting that sleep deprivation reduces the HCVR and others showing that it does not. With the aid of the constant routine protocol, Spengler and Shea most convincingly demonstrated that sleep deprivation has no effect on the HCVR.[88]

As discussed, a major purpose of breathing is to maintain CO_2/pH homeostasis in order to preserve respiratory drive and ensure adequate oxygen availability to tissues. This is certainly true during sleep. During sleep, sensitivity of the CO_2 regulating mechanism is reduced. As such, much higher concentrations of CO_2 are required to induce a small increase in ventilation.[171] During sleep, the decreased V_E allows the CO_2 level to rise to the sleep set point which is higher than it is during wakefulness (Fig. 8.4). During brief arousals from sleep, there is an increase in ventilation that serves to reduce the CO_2 level to the wakefulness set point.[172] In a similar manner to hypercapnia, there is also a state-dependent variation in the ventilatory

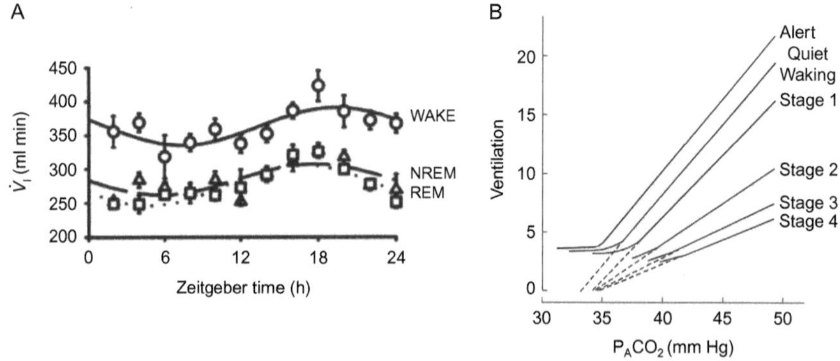

Figure 8.4 Vigilance-state dependence of breathing. (A) Average ventilation over 24 h in Wake, NREM, and REM as indicated in rats housed in a 12:12 light:dark cycle. (B) State-dependent variation in the sensitivity of the hypercapnic ventilatory response. *(A) Redrawn with permission from Ref. 150 and (B) from Ref. 171.*

response to hypoxia, with the responses being less robust in NREM and even lower in REM.[173–176]

Direct stimulation of respiratory control regions including RTN, NTS, lateral hypothalamic orexin neurons, and caudal 5-HT neuronal populations (RPa, ROb, RMag) with acidosis modulates breathing.[34,37,56,143,177,178] These and other sites within the caudal brainstem have been identified as putative respiratory chemoreceptors. The reason for the distributed nature of central respiratory chemoreception has been postulated to revolve around the state dependence of respiratory chemoreception.[25,80] Indeed, there are state-dependent differences in the ability of direct stimulation of the different sites with acidosis to affect breathing.[179] For example, direct stimulation of the RTN with acidosis is now thought to increase tidal volume during wakefulness, whereas stimulation of the rostal medullary raphé (raphé magnus) is most effective during sleep. Stimulation of the caudal medullary raphé (raphé obscuris) is relatively ineffective in sleep and wakefulness, but enhances the chemosensitivity of the RTN during wakefulness.[80]

4.2. Influence of breathing on sleep

Just as sleep state can influence respiratory control, changes in breathing can cause alterations in vigilance state. As discussed above, changes in breathing occur normally as a result of vigilance-state transitions and during dreaming. Breathing changes can also be induced due to untoward circumstances, such as cessation of airflow due to central or obstructive sleep apneas (OSAs) and as a result of noxious stimuli encountered during sleep. The most important chemical modulators of breathing, hypoxia and hypercapnia, are potent arousal inducers (Fig. 8.5).[125–127,129,130,132,133,180]

As discussed earlier, the concentration of CO_2 in the blood is regulated by breathing and is responsible for preserving respiratory drive. While the CO_2 set point is higher during sleep, additional elevations above this threshold might be encountered during sleep due to sleep apnea and poor air exchange or excessive rebreathing of expired CO_2-rich air. The former, of course, is a significant mechanism in OSA, while the latter is postulated to be a significant mechanism in the sudden infant death syndrome (SIDS).[130,181]

Mice with a genetic absence of serotonin neurons in the central nervous system (*Lmx1b*$^{f/f/p}$) have impaired arousal to CO_2 (Fig. 8.5), but display normal arousal to other stimuli, including hypoxia, sounds, and mechanical stimulation, indicating a specific involvement of 5-HT neurons in CO_2-induced

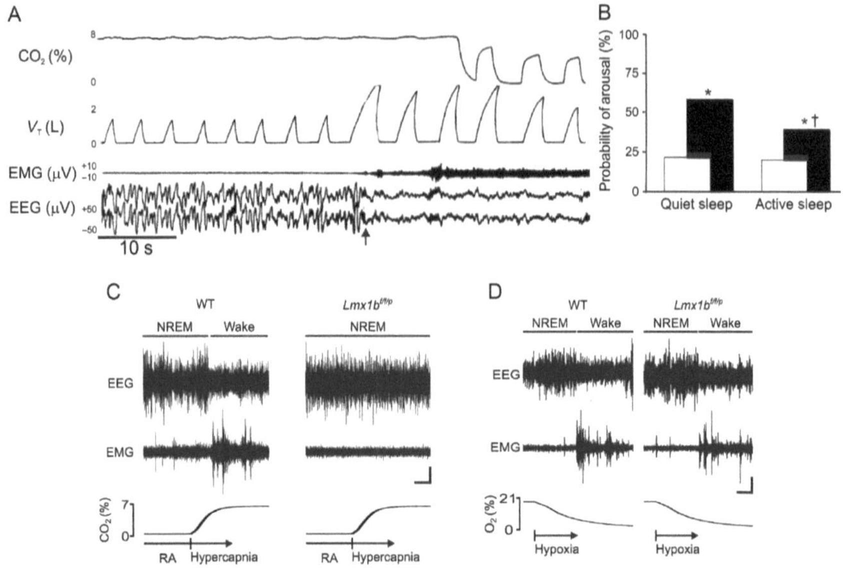

Figure 8.5 Respiratory stimuli can induce arousal from sleep. (A) Tracing depicting human arousal from slow-wave sleep in response to an increase in alveolar CO_2 to ~7%. After a delay of 60 s after elevation of CO_2, there is an increase in tidal volume (V_T) and arousal. Arrow denotes point of arousal. EMG, electromyogram; EEG, electroencephalogram. (B) Bar graph depicting increased waking probability in response to hypercapnic challenge in lambs. Asterisks denote significant difference compared to control ($p<0.05$). Cross denotes significant difference between vigilance states ($p<0.001$). (C) Four-minute EEG (top), EMG (middle), and (P_{CO_2}) (bottom) traces from WT and *Lmx1b*$^{f/f/p}$ mice showing response to 7% CO_2. Arousal in the WT mouse is indicated by a decrease in EEG amplitude (and corresponding increase in EEG frequency) with concomitant increase in EMG amplitude. O_2 level 21% (balance N_2) throughout trace. Scale bars—30 s and 5 μV. (D) Ninety-second EEG (top), EMG (middle), and P_{O_2} (bottom) traces from WT and *Lmx1b*$^{f/f/p}$ mice indicating arousal response to hypoxia (~8% O_2). Arousal indicated as in C. Scale bars—10 s and 5 μV. *(A) Redrawn with permission from Ref. 125, (B) from Ref. 180, and (C and D) from Ref. 130.*

arousal from sleep.[130] 5-HT neurons, including those in the midbrain which are involved in sleep–wake regulation,[182–184] are chemosensitive, increasing their firing rate in response to acidosis *in vitro*[140] and to increased concentration of inspired CO_2 *in vivo*.[145] This likely occurs through a 5-HT_2 receptor-mediated mechanism.[185] It is thought that CO_2-induced arousal is mediated by chemoreceptive neurons in the midbrain since these neurons directly project to thalamus and cortex and are thought to be involved in sleep–wake regulation.[131] A preliminary report suggests that this might be the case.[186] Notably, stimulation of medullary 5-HT neuronal populations can also lead to subcortical arousal as evidenced by brief EEG changes and cardiorespiratory changes in piglets and rat pups.[187–192] This has recently been the subject of extensive review.[132]

5. IMPLICATIONS FOR HUMAN DISEASE

There are a number of diseases in which the interplay between circadian regulation, sleep–wake state, and control of respiration may contribute profoundly to the pathophysiology and associated morbidity of the disease. These include, but are not limited to, sleep apnea, asthma, epilepsy, and sudden unexpected death in epilepsy (SUDEP), and SIDS. In each case, continuing to understand this interplay will undoubtedly lead to improved preventive and therapeutic measures to ultimately reduce associated morbidity and mortality from these diseases.

5.1. Sleep apnea

Perhaps the most obvious interaction between vigilance state and breathing in human disease is seen in sleep apnea. As the name indicates, this is cessation of airflow, or apnea, during sleep. These apneas do not occur during wakefulness. Sleep apnea can be obstructive, central, or mixed. In OSA, there is cessation of airflow due to increased resistance through the airway. Most commonly this is due to the normal decrease in excitatory input to the airway musculature with resultant reduction in airway tone coupled with a small airway which is easily susceptible to obstruction. Central apneas result from loss of central input to the respiratory muscles. In some cases, there can be both an obstructive component and a central component, termed mixed apneas.

Sleep apneas lead to arousals, either through the cessation of respiration or through the elevation in CO_2, or reduction in O_2. The arousals serve to increase excitatory input to the airway muscles thereby improving airway tone and patency and allow resumption of normal breathing. The return of breathing to normal corrects the blood gas aberration. If these are not corrected, the CO_2 will become too high and respiratory drive will be lost. While the arousals ultimately serve to maintain respiratory drive, recurrent sleep disruption leads to excessive daytime sleepiness and contributes profoundly to the morbidity associated with sleep apnea. Many patients with sleep apnea can be treated with a conceptually simple maneuver in which continuous positive pressure is delivered to their airway to maintain its patency and greatly reduce the number of nighttime apneas. While there is clearly an association between apnea and sleep, there may also be a circadian component.[193]

5.2. Asthma

Several factors discussed above are relevant to asthma. First of all, the normal variation in airway tone and airway caliber that occurs in most people is particularly troublesome, even life threatening, in asthmatics since they have a narrowed airway to begin with. Second, the circadian variation in sensitivity to respiratory stimuli can be bothersome. In one study of six asthmatics with known nocturnal worsening and four healthy controls, there was a nighttime increase in lower airway resistance as evidenced by a reduction in FEV_1, in asthmatics compared to controls. This change occurs irrespective of sleep state, but is enhanced during nighttime sleep. Despite the changes in airway resistance, subjects maintained minute ventilation largely by increasing RR, suggesting that the compensatory ventilator response to the increased resistance is not lost during sleep.[194] In addition to the variations in airway resistance and irritant sensitivity, there can be circadian influence on asthma in a number of ways including circadian variations in circulating catecholamines, circulating eosinophils and neutrophils, mast cell mediator response, adrenergic tone, cortisol level, cholinergic function.[114]

5.3. Epilepsy

Seizures can be associated with profound respiratory dysfunction, including apnea, with associated hypoxemia and hypercapnia.[195–199] This can contribute to SUDEP. SUDEP is the leading cause of death in patients with epilepsy.[200] There is still a great deal to be learned about the risk factors and pathophysiology of SUDEP; however, it is thought to ensue from a primarily respiratory, cardiac, autonomic, or electrocerebral etiology.[201–203] SUDEP tends to occur at night and it has been proposed that there may a circadian- and/or sleep-state dependence to SUDEP.[204,205] One could postulate that death could ensue if a seizure occurred in a susceptible individual at a time in the circadian cycle or sleep cycle in which the respiratory rhythm was already unstable or when the (P_{CO_2}) was already elevated.

5.4. Sudden infant death syndrome

SIDS is defined as the sudden death of an infant under a year of age that remains unexplained after autopsy, death scene investigation, and review of the clinical history.[206] In the triple risk model of SIDS, a susceptible baby experiences an environmental challenge during a critical developmental time period. The preponderance of data suggests that susceptible babies are ones that have abnormalities of their brainstem serotonergic system.[207–211] The environmental challenge may be a profound elevation in inspired CO_2 encountered via rebreathing expired CO_2 while lying face down. As mentioned above, 5-HT neurons are chemosensitive and as such mediate the HCVR[28] as well as the arousal response to hypercapnia.[130] Therefore, an abnormality in the brainstem 5-HT system could lead to an inability to respond to a CO_2 challenge with an increase in ventilation, an arousal, or both. The association between prone sleeping and SIDS was the impetus for the back-to-sleep campaign in the latter part of the twentieth century that led to a reduction in overall SIDS incidence.[212] The usual scenario in a SIDS case is that an otherwise normal-appearing infant is found dead in his/her crib following a sleep period, most commonly at night. Circadian variation and/or sleep-state variation in sensitivity to respiratory stimuli may contribute to why SIDS tends to occur during nighttime sleep.

6. CONCLUSIONS

Respiratory function is subject to both circadian-dependent and vigilance-state-dependent regulation. The specific nature of this regulation can vary considerably depending on the species studied and methodologies employed. Information regarding respiratory function can similarly influence both circadian timing and vigilance state. Again, there can be species- and methodology-dependent variation in these associations. Relatively little is known about the specific mechanisms underlying the regulation of breathing in a circadian- and state-dependent manner, or mechanisms by which respiration and respiratory stimuli can regulate circadian phase and vigilance state. Continuing to improve our understanding of the relationships between respiratory function, circadian timing, and vigilance-state control will improve our understanding of diseases such as sleep apnea, asthma, SUDEP, and SIDS, and thus help us devise strategies to reduce morbidity and mortality from these diseases.

REFERENCES

1. Duffin J. Functional organization of respiratory neurones: a brief review of current questions and speculations. *Exp Physiol.* 2004;89:517–529.
2. Gandevia SC, Allen GM, Butler JE, Gorman RB, McKenzie DK. Human respiratory muscles: sensations, reflexes and fatiguability. *Clin Exp Pharmacol Physiol.* 1998;25:757–763.
3. Kumar P. Sensing hypoxia in the carotid body: from stimulus to response. *Essays Biochem.* 2007;43:43–60.

4. Dean JB, Nattie EE. Central CO2 chemoreception in cardio-respiratory control. *J Appl Physiol.* 2010;108:976–978.
5. Baruscotti M, Barbuti A, Bucchi A. The cardiac pacemaker current. *J Mol Cell Cardiol.* 2010;48:55–64.
6. Feldman JL, Del Negro CA. Looking for inspiration: new perspectives on respiratory rhythm. *Nat Rev Neurosci.* 2006;7:232–242.
7. Hilaire G, Monteau R, Gauthier P, Rega P, Morin D. Functional significance of the dorsal respiratory group in adult and newborn rats: in vivo and in vitro studies. *Neurosci Lett.* 1990;111:133–138.
8. Stornetta RL, Sevigny CP, Guyenet PG. Inspiratory augmenting bulbospinal neurons express both glutamatergic and enkephalinergic phenotypes. *J Comp Neurol.* 2003;455:113–124.
9. Shen L, Duffin J. Caudal expiratory neurones in the rat. *Pflugers Arch.* 2002;444:405–410.
10. Ezure K, Tanaka I, Saito Y. Brainstem and spinal projections of augmenting expiratory neurons in the rat. *Neurosci Res.* 2003;45:41–51.
11. Smith JC, Ellenberger HH, Ballanyi K, Richter DW, Feldman JL. Pre-Botzinger complex: a brainstem region that may generate respiratory rhythm in mammals. *Science.* 1991;254:726–729.
12. Onimaru H, Homma I. A novel functional neuron group for respiratory rhythm generation in the ventral medulla. *J Neurosci.* 2003;23:1478–1486.
13. Onimaru H, Arata A, Homma I. Primary respiratory rhythm generator in the medulla of brainstem-spinal cord preparation from newborn rat. *Brain Res.* 1988;445:314–324.
14. Guyenet PG, Mulkey DK. Retrotrapezoid nucleus and parafacial respiratory group. *Respir Physiol Neurobiol.* 2010;173:244–255.
15. Hucksteppp RT, Dale N. Redefining the components of central CO2 chemosensitivity—towards a better understanding of mechanism. *J Physiol.* 2011;589:5561–5579.
16. Kubin L, Alheid GF, Zuperku EJ, McCrimmon DR. Central pathways of pulmonary and lower airway vagal afferents. *J Appl Physiol.* 2006;101:618–627.
17. Buckler KJ, Vaughan-Jones RD. Effects of hypercapnia on membrane potential and intracellular calcium in rat carotid body type I cells. *J Physiol.* 1994;478(Pt. 1):157–171.
18. Ramirez JM, Quellmalz UJ, Wilken B, Richter DW. The hypoxic response of neurones within the in vitro mammalian respiratory network. *J Physiol.* 1998;507 (Pt. 2):571–582.
19. Solomon IC, Edelman NH, Neubauer JA. Pre-Botzinger complex functions as a central hypoxia chemosensor for respiration *in vivo*. *J Neurophysiol.* 2000;83:2854–2868.
20. Mazza EJ, Edelman NH, Neubauer JA. Hypoxic excitation in neurons cultured from the rostral ventrolateral medulla of the neonatal rat. *J Appl Physiol.* 2000;88:2319–2329.
21. Dillon GH, Waldrop TG. In vitro responses of caudal hypothalamic neurons to hypoxia and hypercapnia. *Neuroscience.* 1992;51:941–950.
22. Honda Y. Respiratory and circulatory activities in carotid body-resected humans. *J Appl Physiol.* 1992;73:1–8.
23. Richerson GB, Wang W, Hodges MR, Dohle CI, Diez-Sampedro A. Homing in on the specific phenotype(s) of central respiratory chemoreceptors. *Exp Physiol.* 2005;90:259–266.
24. Guyenet PG, Stornetta RL, Bayliss DA, Mulkey DK. Retrotrapezoid nucleus: a litmus test for the identification of central chemoreceptors. *Exp Physiol.* 2005;90:247–253.
25. Nattie EE, Li A. Central chemoreception is a complex system function that involves multiple brainstem sites. *J Appl Physiol.* 2009;106:1464–1466.
26. Corcoran AE, Hodges MR, Wu Y, et al. Medullary serotonin neurons and central CO2 chemoreception. *Respir Physiol Neurobiol.* 2009;168:49–58.

27. Richerson GB. Response to CO_2 of neurons in the rostral ventral medulla *in vitro*. *J Neurophysiol.* 1995;73:933–944.
28. Richerson GB. Serotonergic neurons as carbon dioxide sensors that maintain pH homeostasis. *Nat Rev Neurosci.* 2004;5:449–461.
29. Dean JB, Bayliss DA, Erickson JT, Lawing WL, Millhorn DE. Depolarization and stimulation of neurons in nucleus tractus solitarii by carbon dioxide does not require chemical synaptic input. *Neuroscience.* 1990;36:207–216.
30. Pineda J, Aghajanian GK. Carbon dioxide regulates the tonic activity of locus coeruleus neurons by modulating a proton- and polyamine-sensitive inward rectifier potassium current. *Neuroscience.* 1997;77:723–743.
31. Oyamada Y, Ballantyne D, Muckenhoff K, Scheid P. Respiration-modulated membrane potential and chemosensitivity of locus coeruleus neurones in the in vitro brainstem-spinal cord of the neonatal rat. *J Physiol.* 1998;513(Pt. 2):381–398.
32. Dean JB, Kinkade EA, Putnam RW. Cell-cell coupling in CO(2)/H(+)-excited neurons in brainstem slices. *Respir Physiol.* 2001;129:83–100.
33. Nattie EE, Gdovin M, Li A. Retrotrapezoid nucleus glutamate receptors: control of CO2-sensitive phrenic and sympathetic output. *J Appl Physiol.* 1993;74:2958–2968.
34. Li A, Nattie EE. Focal central chemoreceptor sensitivity in the RTN studied with a CO2 diffusion pipette in vivo. *J Appl Physiol.* 1997;83:420–428.
35. Mulkey DK, Stornetta RL, Weston MC, et al. Respiratory control by ventral surface chemoreceptor neurons in rats. *Nat Neurosci.* 2004;7:1360–1369.
36. Williams RH, Jensen LT, Verkhratsky A, Fugger L, Burdakov D. Control of hypothalamic orexin neurons by acid and CO2. *Proc Natl Acad Sci USA.* 2007; 104: 10685–10690.
37. Li N, Li A, Nattie E. Focal microdialysis of CO(2) in the perifornical-hypothalamic area increases ventilation during wakefulness but not NREM sleep. *Respir Physiol Neurobiol.* 2013;185:349–355.
38. Nattie E, Li A. Respiration and autonomic regulation and orexin. *Prog Brain Res.* 2012;198:25–46.
39. Dias MB, Li A, Nattie E. The orexin receptor 1 (OX1R) in the rostral medullary raphe contributes to the hypercapnic chemoreflex in wakefulness, during the active period of the diurnal cycle. *Respir Physiol Neurobiol.* 2010;170:96–102.
40. Martino PF, Davis S, Opansky C, et al. The cerebellar fastigial nucleus contributes to CO2-H+ ventilatory sensitivity in awake goats. *Respir Physiol Neurobiol.* 2007;157:242–251.
41. Mohan R, Duffin J. The effect of hypoxia on the ventilatory response to carbon dioxide in man. *Respir Physiol.* 1997;108:101–115.
42. Smith CA, Rodman JR, Chenuel BJ, Henderson KS, Dempsey JA. Response time and sensitivity of the ventilatory response to CO_2 in unanesthetized intact dogs: central vs. peripheral chemoreceptors. *J Appl Physiol.* 2006;100:13–19.
43. Duffin J, McAvoy GV. The peripheral-chemoreceptor threshold to carbon dioxide in man. *J Physiol.* 1988;406:15–26.
44. Wang W, Pizzonia JH, Richerson GB. Chemosensitivity of rat medullary raphe neurones in primary tissue culture. *J Physiol.* 1998;511(Pt. 2):433–450.
45. Wang W, Tiwari JK, Bradley SR, Zaykin RV, Richerson GB. Acidosis-stimulated neurons of the medullary raphe are serotonergic. *J Neurophysiol.* 2001;85:2224–2235.
46. Bradley SR, Pieribone VA, Wang W, Severson CA, Jacobs RA, Richerson GB. Chemosensitive serotonergic neurons are closely associated with large medullary arteries. *Nat Neurosci.* 2002;5:401–402.
47. Wang W, Bradley SR, Richerson GB. Quantification of the response of rat medullary raphe neurones to independent changes in pH(o) and P(CO2). *J Physiol.* 2002;540:951–970.

48. Bouyer P, Bradley SR, Zhao J, Wang W, Richerson GB, Boron WF. Effect of extracellular acid-base disturbances on the intracellular pH of neurones cultured from rat medullary raphe or hippocampus. *J Physiol.* 2004;559:85–101.
49. Veasey SC, Fornal CA, Metzler CW, Jacobs BL. Response of serotonergic caudal raphe neurons in relation to specific motor activities in freely moving cats. *J Neurosci.* 1995;15:5346–5359.
50. Larnicol N, Wallois F, Berquin P, Gros F, Rose D. C-fos-like immunoreactivity in the cat's neuraxis following moderate hypoxia or hypercapnia. *J Physiol Paris.* 1994;88:81–88.
51. Pete G, Mack SO, Haxhiu MA, Walbaum S, Gauda EB. CO(2)-induced c-Fos expression in brainstem preprotachykinin mRNA containing neurons. *Respir Physiol Neurobiol.* 2002;130:265–274.
52. Haxhiu MA, Tolentino-Silva F, Pete G, Kc P, Mack SO. Monoaminergic neurons, chemosensation and arousal. *Respir Physiol.* 2001;129:191–209.
53. Johnson PL, Hollis JH, Moratalla R, Lightman SL, Lowry CA. Acute hypercarbic gas exposure reveals functionally distinct subpopulations of serotonergic neurons in rats. *J Psychopharmacol.* 2005;19:327–341.
54. Kanamaru M, Homma I. Compensatory airway dilation and additive ventilatory augmentation mediated by dorsomedial medullary 5-hydroxytryptamine 2 receptor activity and hypercapnia. *Am J Physiol Regul Integr Comp Physiol.* 2007;293:R854–R860.
55. Bernard DG, Li A, Nattie EE. Evidence for central chemoreception in the midline raphe. *J Appl Physiol.* 1996;80:108–115.
56. Nattie EE, Li A. CO_2 dialysis in the medullary raphe of the rat increases ventilation in sleep. *J Appl Physiol.* 2001;90:1247–1257.
57. Hodges MR, Martino P, Davis S, Opansky C, Pan LG, Forster HV. Effects on breathing of focal acidosis at multiple medullary raphe sites in awake goats. *J Appl Physiol.* 2004;97:2303–2309.
58. Mueller RA, Towle AC, Breese GR. Supersensitivity to the respiratory stimulatory effect of TRH in 5,7-dihydroxytryptamine-treated rats. *Brain Res.* 1984;298:370–373.
59. Messier ML, Li A, Nattie EE. Inhibition of medullary raphe serotonergic neurons has age-dependent effects on the CO_2 response in newborn piglets. *J Appl Physiol.* 2004;96:1909–1919.
60. Dreshaj IA, Haxhiu MA, Martin RJ. Role of the medullary raphe nuclei in the respiratory response to CO_2. *Respir Physiol.* 1998;111:15–23.
61. Messier ML, Li A, Nattie EE. Muscimol inhibition of medullary raphe neurons decreases the CO_2 response and alters sleep in newborn piglets. *Respir Physiol Neurobiol.* 2002;133:197–214.
62. Nattie EE, Li A, Richerson G, Lappi DA. Medullary serotonergic neurones and adjacent neurones that express neurokinin-1 receptors are both involved in chemoreception *in vivo*. *J Physiol.* 2004;556:235–253.
63. Hendricks TJ, Fyodorov DV, Wegman LJ, et al. Pet-1 ETS gene plays a critical role in 5-HT neuron development and is required for normal anxiety-like and aggressive behavior. *Neuron.* 2003;37:233–247.
64. Zhao ZQ, Scott M, Chiechio S, et al. Lmx1b is required for maintenance of central serotonergic neurons and mice lacking central serotonergic system exhibit normal locomotor activity. *J Neurosci.* 2006;26:12781–12788.
65. Hodges MR, Best S, Richerson GB. Altered ventilatory and thermoregulatory control in male and female adult Pet-1 null mice. *Respir Physiol Neurobiol.* 2011;177:133–140.
66. Hodges MR, Tattersall GJ, Harris MB, et al. Defects in breathing and thermoregulation in mice with near-complete absence of central serotonin neurons. *J Neurosci.* 2008;28:2495–2505.

67. Haxhiu MA, Yung K, Erokwu B, Cherniack NS. CO2-induced c-fos expression in the CNS catecholaminergic neurons. *Respir Physiol.* 1996;105:35–45.
68. Filosa JA, Dean JB, Putnam RW. Role of intracellular and extracellular pH in the chemosensitive response of rat locus coeruleus neurones. *J Physiol.* 2002;541:493–509.
69. Johnson SM, Haxhiu MA, Richerson GB. GFP expressing locus coeruleus neurons from Prp57 transgenic mice exhibit CO2/H+ responses in primary cell culture. *J Appl Physiol.* 2008;105:1301–1311.
70. Miles R. Does low pH stimulate central chemoreceptors located near the ventral medullary surface? *Brain Res.* 1983;271:349–353.
71. Guyenet PG, Stornetta RL, Bayliss DA. Retrotrapezoid nucleus and central chemoreception. *J Physiol.* 2008;586:2043–2048.
72. Mulkey DK, Rosin DL, West G, et al. Serotonergic neurons activate chemosensitive retrotrapezoid nucleus neurons by a pH-independent mechanism. *J Neurosci.* 2007;27:14128–14138.
73. Hodges MR, Richerson GB. Contributions of 5-HT neurons to respiratory control: neuromodulatory and trophic effects. *Respir Physiol Neurobiol.* 2008;164:222–232.
74. Guyenet PG. The 2008 Carl Ludwig Lecture: retrotrapezoid nucleus, CO2 homeostasis and breathing automaticity. *J Appl Physiol.* 2008;105:404–416.
75. Cross BA, Silver IA. Unit activity in the hypothalamus and the sympathetic response to hypoxia and hypercapnia. *Exp Neurol.* 1963;7:375–393.
76. Dillon GH, Waldrop TG. Responses of feline caudal hypothalamic cardiorespiratory neurons to hypoxia and hypercapnia. *Exp Brain Res.* 1993;96:260–272.
77. Nakamura A, Zhang W, Yanagisawa M, Fukuda Y, Kuwaki T. Vigilance state-dependent attenuation of hypercapnic chemoreflex and exaggerated sleep apnea in orexin knockout mice. *J Appl Physiol.* 2007;102:241–248.
78. Kuwaki T, Zhang W, Nakamura A, Deng BS. Emotional and state-dependent modification of cardiorespiratory function: role of orexinergic neurons. *Auton Neurosci.* 2008;142:11–16.
79. Feldman JL, Mitchell GS, Nattie EE. Breathing: rhythmicity, plasticity, chemosensitivity. *Annu Rev Neurosci.* 2003;26:239–266.
80. Kuwaki T, Li A, Nattie E. State-dependent central chemoreception: a role of orexin. *Respir Physiol Neurobiol.* 2010;173:223–229.
81. Campbell EJ, Freedman S, Smith PS, Taylor ME. The ability of man to detect added elastic loads to breathing. *Clin Sci.* 1961;20:223–231.
82. O'Donnell DE, Hamilton AL, Webb KA. Sensory-mechanical relationships during high-intensity, constant-work-rate exercise in COPD. *J Appl Physiol.* 2006;101:1025–1035.
83. Takahashi JS, Hong HK, Ko CH, McDearmon EL. The genetics of mammalian circadian order and disorder: implications for physiology and disease. *Nat Rev Genet.* 2008;9:764–775.
84. Moore RY, Eichler VB. Loss of a circadian adrenal corticosterone rhythm following suprachiasmatic lesions in the rat. *Brain Res.* 1972;42:201–206.
85. Stephan FK, Zucker I. Circadian rhythms in drinking behavior and locomotor activity of rats are eliminated by hypothalamic lesions. *Proc Natl Acad Sci USA.* 1972;69:1583–1586.
86. Mohawk JA, Green CB, Takahashi JS. Central and peripheral circadian clocks in mammals. *Annu Rev Neurosci.* 2012;35:445–462.
87. Mortola JP. Breathing around the clock: an overview of the circadian pattern of respiration. *Eur J Appl Physiol.* 2004;91:119–129.
88. Spengler CM, Shea SA. Endogenous circadian rhythm of pulmonary function in healthy humans. *Am J Respir Crit Care Med.* 2000;162:1038–1046.

89. Iizuka H, Sasaki K, Odagiri N, Obo M, Imaizumi M, Atai H. Measurement of respiratory function using whole-body plethysmography in unanesthetized and unrestrained nonhuman primates. *J Toxicol Sci.* 2010;35:863–870.
90. Piccione G, Giudice E, Fazio F, Mortola JP. The daily rhythm of body temperature, heart and respiratory rate in newborn dogs. *J Comp Physiol B.* 2010;180:895–904.
91. Peever JH, Stephenson R. Day-night differences in the respiratory response to hypercapnia in awake adult rats. *Respir Physiol.* 1997;109:241–248.
92. Iwase M, Ohshima Y, Izumizaki M, Homma I. Circadian changes in respiratory responses to acute hypoxia and histamine H1 receptors in mice. *New Front Resp Cont.* 2010;669:235–238.
93. Woodin M, Stephenson R. Circadian rhythms in diving behavior and ventilatory response to asphyxia in canvasback ducks. *Am J Physiol.* 1998;274:R686–R693.
94. Hicks JW, Riedesel ML. Diurnal ventilatory patterns in the Garter snake, Thaminophis elegans. *J Comp Physl A.* 2013;149:503–510.
95. Piccione G, Caola G, Mortola JP. Day/night pattern of arterial blood gases in the cow. *Respir Physiol Neurobiol.* 2004;140:33–41.
96. Reyes C, Milsom WK. Daily and seasonal rhythms in the respiratory sensitivity of red-eared sliders (Trachemys scripta elegans). *J Exp Biol.* 2009;212:3339–3348.
97. Teixeira LV, Waterhouse JM, Marques MD. Respiratory rhythms in stingless bee workers: circadian and ultradian components throughout adult development. *J Comp Physiol A Neuroethol Sens Neural Behav Physiol.* 2011;197:361–372.
98. Mortola JP, Seifert EL. Circadian patterns of breathing. *Respir Physiol Neurobiol.* 2002;131:91–100.
99. Adamczyk W, Tafil-Klawe M, Siekierka M, Zlomanczuk P, Weber P, Klawe JJ. Daily pattern of breathing in healthy young men. *J Physiol Pharmacol.* 2008;59(Suppl. 6): 115–122.
100. Borbely AA, Achermann P. Concepts and models of sleep regulation: an overview. *J Sleep Res.* 1992;2:63–79.
101. Czeisler CA, Weitzman E, Moore-Ede MC, Zimmerman JC, Knauer RS. Human sleep: its duration and organization depend on its circadian phase. *Science.* 1980;210:1264–1267.
102. Mills JN, Minors DS, Waterhouse JM. Adaptation to abrupt time shifts of the oscillator(s) controlling human circadian rhythms. *J Physiol.* 1978;285:455–470.
103. Spengler CM, Czeisler CA, Shea SA. An endogenous circadian rhythm of respiratory control in humans. *J Physiol.* 2000;526(Pt. 3):683–694.
104. Spengler CM, Shea SA. Explained and unexplained variability of CO2-sensitivity in humans. *Adv Exp Med Biol.* 2001;499:483–488.
105. Stupfel M, Pletan Y. Respiratory ultradian rhythms of mean and low frequencies: a comparative physiological approach. *Chronobiologia.* 1983;10:283–292.
106. Drorbaugh JE, Fenn WO. A barometric method for measuring ventilation in newborn infants. *Pediatrics.* 1955;16:81–87.
107. Mortola JP. Correlations between the circadian patterns of body temperature, metabolism and breathing in rats. *Respir Physiol Neurobiol.* 2007;155:137–146.
108. Guberan E, Williams MK, Walford J, Smith MM. Circadian variation of F.E.V. in shift workers. *Br J Ind Med.* 1969;26:121–125.
109. Hetzel MR. Circadian rhythms in respiration in health and disease with special reference to nocturnal asthma. *Bull Eur Physiopathol Respir.* 1987;23:536.
110. Hetzel MR, Clark TJ. Comparison of normal and asthmatic circadian rhythms in peak expiratory flow rate. *Thorax.* 1980;35:732–738.
111. Kerr HD. Diurnal variation of respiratory function independent of air quality: experience with an environmentally controlled exposure chamber for human subjects. *Arch Environ Health.* 1973;26:144–152.

112. Gervais P, Reinberg A, Gervais C, Smolensky M, DeFrance O. Twenty-four-hour rhythm in the bronchial hyperreactivity to house dust in asthmatics. *J Allergy Clin Immunol.* 1977;59:207–213.
113. de Vries K, Goei JT, Booy-Noord H, Orie NG. Changes during 24 hours in the lung function and histamine hyperreactivity of the bronchial tree in asthmatic and bronchitic patients. *Int Arch Allergy Appl Immunol.* 1962;20:93–101.
114. Barnes PJ. Circadian variation in airway function. *Am J Med.* 1985;79:5–9.
115. Barclay JL, Tsang AH, Oster H. Interaction of central and peripheral clocks in physiological regulation. *Prog Brain Res.* 2012;199:163–181.
116. Bando H, Nishio T, van der Horst GT, Masubuchi S, Hisa Y, Okamura H. Vagal regulation of respiratory clocks in mice. *J Neurosci.* 2007;27:4359–4365.
117. Stephenson R. Circadian rhythms and sleep-related breathing disorders. *Sleep Med.* 2007;8:681–687.
118. Stephenson R, Mohan RM, Duffin J, Jarsky TM. Circadian rhythms in the chemoreflex control of breathing. *Am J Physiol Regul Integr Comp Physiol.* 2000;278: R282–R286.
119. Seifert EL, Mortola JP. Circadian pattern of ventilation during acute and chronic hypercapnia in conscious adult rats. *Am J Physiol Regul Integr Comp Physiol.* 2002;282:R244–R251.
120. Saiki C, Mortola JP. Effect of CO2 on the metabolic and ventilatory responses to ambient temperature in conscious adult and newborn rats. *J Physiol.* 1996;491 (Pt. 1):261–269.
121. Ashkenazi IE, Ribak J, Avgar DM, Klepfish A. Altitude and hypoxia as phase shift inducers. *Aviat Space Environ Med.* 1982;53:342–346.
122. Samel A, Vejvoda M, Wittiber K, Wenzel J. Joint NASA-ESA-DARA Study. Part three: circadian rhythms and activity-rest cycle under different CO2 concentrations. *Aviat Space Environ Med.* 1998;69:501–505.
123. Jarsky TM, Stephenson R. Effects of hypoxia and hypercapnia on circadian rhythms in the golden hamster (Mesocricetus auratus). *J Appl Physiol.* 2000;89:2130–2138.
124. Mortola JP, Seifert EL. Hypoxic depression of circadian rhythms in adult rats. *J Appl Physiol.* 2000;88:365–368.
125. Berthon-Jones M, Sullivan CE. Ventilation and arousal responses to hypercapnia in normal sleeping humans. *J Appl Physiol.* 1984;57:59–67.
126. Berthon-Jones M, Sullivan CE. Ventilatory and arousal responses to hypoxia in sleeping humans. *Am Rev Respir Dis.* 1982;125:632–639.
127. Horner RL, Rivera MP, Kozar LF, Phillipson EA. The ventilatory response to arousal from sleep is not fully explained by differences in CO(2) levels between sleep and wakefulness. *J Physiol.* 2001;534:881–890.
128. Phillipson EA, Sullivan CE. Arousal: the forgotten response to respiratory stimuli. *Am Rev Respir Dis.* 1978;118:807–809.
129. Phillipson EA, Kozar LF, Rebuck AS, Murphy E. Ventilatory and waking responses to CO_2 in sleeping dogs. *Am Rev Resp Dis.* 1977;115:251–259.
130. Buchanan GF, Richerson GB. Central serotonin neurons are required for arousal to CO2. *Proc Natl Acad Sci USA.* 2010;107:16354–16359.
131. Buchanan GF, Hodges MR, Richerson GB. Contribution of chemosensitive serotonergic neurons to interactions between the sleep-wake cycle and respiratory control. In: Monti JM, Pandi-Perumal SR, Jacobs BL, Nutt DJ, eds. *Serotonin and Sleep: Molecular, Functional and Clinical aspects.* Switzerland: Birkhauser Verlag; 2008: 529–554.
132. Darnall RA. The carotid body and arousal in the fetus and neonate. *Respir Physiol Neurobiol.* 2013;185:132–143.

133. Darnall RA, Schneider RW, Tobia CM, Zemel BM. Arousal from sleep in response to intermittent hypoxia in rat pups is modulated by medullary raphe GABAergic mechanisms. *Am J Physiol Regul Integr Comp Physiol.* 2012;302:R551–R560.
134. Medanic M, Gillette MU. Suprachiasmatic circadian pacemaker of rat shows two windows of sensitivity to neuropeptide Y in vitro. *Brain Res.* 1993;620:281–286.
135. Chen D, Buchanan GF, Ding JM, Hannibal J, Gillette MU. Pituitary adenylyl cyclase-activating peptide: a pivotal modulator of glutamatergic regulation of the suprachiasmatic circadian clock. *Proc Natl Acad Sci USA.* 1999;96:13468–13473.
136. Piggins HD, Marchant EG, Goguen D, Rusak B. Phase-shifting effects of pituitary adenylate cyclase activating polypeptide on hamster wheel-running rhythms. *Neurosci Lett.* 2001;305:25–28.
137. Medanic M, Gillette MU. Serotonin regulates the phase of the rat suprachiasmatic circadian pacemaker in vitro only during the subjective day. *J Physiol.* 1992;450:629–642.
138. Mintz EM, Gillespie CF, Marvel CL, Huhman KL, Albers HE. Serotonergic regulation of circadian rhythms in Syrian hamsters. *Neuroscience.* 1997;79:563–569.
139. Richerson GB, Wang W, Tiwari J, Bradley SR. Chemosensitivity of serotonergic neurons in the rostral ventral medulla. *Respir Physiol.* 2001;129:175–189.
140. Severson CA, Wang W, Pieribone VA, Dohle CI, Richerson GB. Midbrain serotonergic neurons are central pH chemoreceptors. *Nat Neurosci.* 2003;6:1139–1140.
141. Wang W, Richerson GB. Development of chemosensitivity of rat medullary raphe neurons. *Neuroscience.* 1999;90:1001–1011.
142. Wu Y, Hodges MR, Richerson GB. Stimulation by hypercapnic acidosis in mouse 5-HT neurons is enhanced by age and increased temperature. *Soc Neurosci Abstr.* 2008;34:383–389.
143. Dias MB, Li A, Nattie E. Focal CO2 dialysis in raphe obscurus does not stimulate ventilation but enhances the response to focal CO2 dialysis in the retrotrapezoid nucleus. *J Appl Physiol.* 2008;105:83–90.
144. Penatti EM, Berniker AV, Kereshi B, et al. Ventilatory response to hypercapnia and hypoxia after extensive lesion of medullary sertonergic neurons in newborn conscious piglets. *J Appl Physiol.* 2006;101:1177–1188.
145. Veasey SC, Fornal CA, Metzler CW, Jacobs BL. Single-unit responses of serotonergic dorsal raphe neurons to specific motor challenges in freely moving cats. *Neuroscience.* 1997;79:161–169.
146. Seifert EL, Mortola JP. Circadian pattern of ventilation during prolonged hypoxia in conscious rats. *Respir Physiol Neurobiol.* 2002;133:23–34.
147. Fenelon K, Seifert EL, Mortola JP. Hypoxic depression of circadian oscillations in sino-aortic denervated rats. *Respir Physiol.* 2000;122:61–69.
148. Nolan PC, Dillon GH, Waldrop TG. Central hypoxic chemoreceptors in the ventrolateral medulla and caudal hypothalamus. *Adv Exp Med Biol.* 1995;393:261–266.
149. Horn EM, Dillon GH, Fan YP, Waldrop TG. Developmental aspects and mechanisms of rat caudal hypothalamic neuronal responses to hypoxia. *J Neurophysiol.* 1999;81: 1949–1959.
150. Stephenson R, Liao KS, Hamrahi H, Horner RL. Circadian rhythms and sleep have additive effects on respiration in the rat. *J Physiol.* 2001;536:225–235.
151. Snyder F, Hobson JA, Morrison DF, Goldfrank F. Changes in respiration, heart rate, and systolic blood pressure in human sleep. *J Appl Physiol.* 1964;19:417–422.
152. Trinder J, Whitworth F, Kay A, Wilkin P. Respiratory instability at sleep onset. *J Appl Physiol.* 1992;73:2462–2469.
153. Webb P. Periodic breathing during sleep. *J Appl Physiol.* 1974;37:899–903.
154. Nattie E. CO_2, brainstem chemoreceptors and breathing. *Prog Neurobiol.* 1999; 59:299–331.

155. Kuna ST, Smickley JS, Insalaco G. Posterior cricoarytenoid muscle activity during wakefulness and sleep in normal adults. *J Appl Physiol.* 1990;68:1746–1754.
156. Tangel DJ, Mezzanotte WS, White DP. Influence of sleep on tenor palatine EMG and upper airway resistance in normal men. *J Appl Physiol.* 1991;70:2574–2581.
157. Wiegand DA, Latz B, Zwillich CW, Wiegand L. Upper airway resistance and geniohyoid muscle activity in normal men during wakefulness and sleep. *J Appl Physiol.* 1990;69:1252–1261.
158. Tabachnik E, Muller NL, Bryan AC, Levison H. Changes in ventilation and chest wall mechanics during sleep in normal adolescents. *J Appl Physiol.* 1981;51:557–561.
159. Lopes LM, Trabachnik E, Muller NL, Levison H, Bryan AC. Total airway resistance and respiratory muscle activity during sleep. *J Appl Physiol.* 1983;54:773–777.
160. Orem J. Neural mechanisms of respiration in REM sleep. *Sleep.* 1980;3:251–252.
161. Aserinsky E. Periodic respiratory pattern occurring in conjunction with eye movements during sleep. *Science.* 1965;150:763–766.
162. Spreng LF, Johnson CL, Lubin A. Autonomic correlates of eye movement bursts during stage REM sleep. *Psychophysiology.* 1968;4:311–323.
163. Douglas NJ, Flenley DC. Breathing during sleep in patients with obstructive lung disease. *Am Rev Respir Dis.* 1990;141:1055–1070.
164. Lo YL, Jordan AS, Malhotra A, et al. Influence of wakefulness on pharyngeal airway muscle activity. *Thorax.* 2007;62:799–805.
165. Shea SA, Edwards JK, White DP. Effect of wake-sleep transitions and rapid eye movement sleep on pharyngeal muscle response to negative pressure in humans. *J Physiol.* 1999;520(Pt. 3):897–908.
166. Horner RL. Neuromodulation of hypoglossal motoneurons during sleep. *Respir Physiol Neurobiol.* 2008;164:179–196.
167. Horner RL, Innes JA, Morrell MJ, Shea SA, Guz A. The effect of sleep on reflex genioglossus muscle activation by stimuli of negative airway pressure in humans. *J Physiol.* 1994;476:141–151.
168. Cherniack NS. Respiratory dysrhythmias during sleep. *N Engl J Med.* 1981;305:325–330.
169. Kujirai T, Caramia MD, Rothwell JC, et al. Corticocortical inhibition in human motor cortex. *J Physiol.* 1993;471:501–519.
170. Chan E, Steenland HW, Liu H, Horner RL. Endogenous excitatory drive modulating respiratory muscle activity across sleep-wake states. *Am J Respir Crit Care Med.* 2006;174:1264–1273.
171. Bulow K. Respiration and wakefulness in man. *Acta Physiol Scand Suppl.* 1963;59:1–110.
172. Krieger J. Respiratory physiology: breathing in normal subjects. In: Kryger MH, Roth T, Dement WC, eds. *Principles and Practice of Sleep Medicine.* New York: Saunders; 2000:229–241.
173. Douglas NJ. Control of ventilation during sleep. *Clin Chest Med.* 1985;6:563–575.
174. Douglas NJ, White DP, Weil JV, et al. Hypoxic ventilatory response decreases during sleep in normal men. *Am Rev Respir Dis.* 1982;125:286–289.
175. White DP, Douglas NJ, Pickett CK, Weil JV, Zwillich CW. Hypoxic ventilatory response during sleep in normal premenopausal women. *Am Rev Respir Dis.* 1982;126:530–533.
176. Gothe B, Goldman MD, Cherniack NS, Mantey P. Effect of progressive hypoxia on breathing during sleep. *Am Rev Respir Dis.* 1982;126:97–102.
177. da Silva GS, Li A, Nattie E. High CO(2)/H(+) dialysis in the caudal ventrolateral medulla (Loeschcke's area) increases ventilation in wakefulness. *Respir Physiol Neurobiol.* 2010;171:46–53.

178. Nattie EE, Li A. CO2 dialysis in nucleus tractus solitarius region of rat increases ventilation in sleep and wakefulness. *J Appl Physiol.* 2002;92:2119–2130.
179. Nattie E, Li A. Central chemoreception in wakefulness and sleep: evidence for a distributed network and a role for orexin. *J Appl Physiol.* 2010;108:1417–1424.
180. Johnston RV, Grant DA, Wilkinson MH, Walker AM. The effects of repeated exposure to hypercapnia on arousal and cardiorespiratory responses during sleep in lambs. *J Physiol.* 2007;582:368–378.
181. Kinney HC, Richerson GB, Dymecki SM, Darnall RA, Nattie EE. The brainstem and serotonin in the sudden infant death syndrome. *Annu Rev Pathol.* 2009;4:517–550.
182. McGinty DJ, Harper RM. Dorsal raphe neurons: depression of firing during sleep in cats. *Brain Res.* 1976;101:569–575.
183. Trulson ME, Jacobs BL. Raphe unit activity in freely moving cats: correlation with level of behavioral arousal. *Brain Res.* 1979;163:135–150.
184. Saper CB, Fuller PM, Pedersen NP, Lu J, Scammell TE. Sleep state switching. *Neuron.* 2010;68:1023–1042.
185. Smith HR, Richerson GB, Buchanan GF. Increased seizure susceptibility and post-ictal respiratory suppression in mice with genetic deletion of 5-HT neurons. *Soc Neurosci Abstr.* 2012;799(7):799–807 [abstract].
186. Smith HR, Richerson GB, Buchanan GF. Direct stimulation of rostral 5-HT neurons with acidosis induces arousal from sleep in mice. *APSS Ann Mtg.* 2013;27(83) [abstract].
187. Brown JW, Sirlin EA, Benoit AM, Hoffman JM, Darnall RA. Activation of 5-HT1A receptors in medullary raphe disrupts sleep and decreases shivering during cooling in the conscious piglet. *Am J Physiol Regul Integr Comp Physiol.* 2008;294:R884–R894.
188. Hoffman JM, Brown JW, Sirlin EA, et al. Activation of 5-HT1A receptors in the paragigantocellularis lateralis decreases shivering during cooling in the conscious piglet. *Am J Physiol Regul Integr Comp Physiol.* 2007;293:R518–R527.
189. Darnall RA, Harris MB, Gill WH, Hoffman JM, Brown JW, Niblock MM. Inhibition of serotonergic neurons in the nucleus paragigantocellularis lateralis fragments sleep and decreases rapid eye movement sleep in the piglet: implications for sudden infant death syndrome. *J Neurosci.* 2005;25:8322–8332.
190. Darnall RA, Curran AK, Filiano JJ, Li A, Nattie EE. The effects of a GABA(A) agonist in the rostral ventral medulla on sleep and breathing in newborn piglets. *Sleep.* 2001;24:514–527.
191. Curran AK, Darnall RA, Filiano JJ, Li A, Nattie EE. Muscimol dialysis in the rostral ventral medulla reduced the CO(2) response in awake and sleeping piglets. *J Appl Physiol.* 2001;90:971–980.
192. Curran AK, Chen G, Darnall RA, Filiano JJ, Li A, Nattie EE. Lesion or muscimol in the rostral ventral medulla reduces ventilatory output and the CO(2) response in decerebrate piglets. *Respir Physiol.* 2000;123:23–37.
193. Stephenson R, Horner RL. The effect of time of day on apnoea index in the sleeping rat. *Respir Physiol Neurobiol.* 2006;154:351–355.
194. Ballard RD, Saathoff MC, Patel DK, Kelly PL, Martin RJ. Effect of sleep on nocturnal bronchoconstriction and ventilatory patterns in asthmatics. *J Appl Physiol.* 1989;67:243–249.
195. Nashef L, Walker F, Allen P, Sander JW, Shorvon SD, Fish DR. Apnoea and bradycardia during epileptic seizures: relation to sudden death in epilepsy. *J Neurol Neurosurg Psychiatry.* 1996;60:297–300.
196. Seyal M, Bateman LM, Albertson TE, Lin TC, Li CS. Respiratory changes with seizures in localization-related epilepsy: analysis of periictal hypercapnia and airflow patterns. *Epilepsia.* 2010;51:1359–1364.

197. Bateman LM, Li CS, Lin TC, Seyal M. Serotonin reuptake inhibitors are associated with reduced severity of ictal hypoxemia in medically refractory partial epilepsy. *Epilepsia.* 2010;51:2211–2214.
198. Seyal M, Bateman LM. Ictal apnea linked to contralateral spread of temporal lobe seizures: intracranial EEG recordings in refractory temporal lobe epilepsy. *Epilepsia.* 2009;50:2557–2562.
199. Bateman LM, Li CS, Seyal M. Ictal hypoxemia in localization-related epilepsy: analysis of incidence, severity and risk factors. *Brain.* 2008;131:3239–3245.
200. Shorvon S, Tomson T. Sudden unexpected death in epilepsy. *Lancet.* 2011;378:2028–2038.
201. Surges R, Thijs RD, Tan HL, Sander JW. Sudden unexpected death in epilepsy: risk factors and potential pathomechanisms. *Nat Rev Neurol.* 2009;5:492–504.
202. Tomson T, Nashef L, Ryvlin P. Sudden unexpected death in epilepsy: current knowledge and future directions. *Lancet Neurol.* 2008;7:1021–1031.
203. Richerson GB, Buchanan GF. The serotonin axis: shared mechanisms in seizures, depression, and SUDEP. *Epilepsia.* 2011;52(Suppl 1):28–38.
204. Nobili L, Proserpio P, Rubboli G, Montano N, Didato G, Tassinari CA. Sudden unexpected death in epilepsy (SUDEP) and sleep. *Sleep Med Rev.* 2011;15:237–246.
205. Lamberts RJ, Thijs RD, Laffan A, Langan Y, Sander JW. Sudden unexpected death in epilepsy: people with nocturnal seizures may be at highest risk. *Epilepsia.* 2012;53:253–257.
206. Willinger M, James LS, Catz C. Defining the sudden infant death syndrome (SIDS): deliberations of an expert panel convened by the National Institute of Child Health and Human Development. *Pediatr Pathol.* 1991;11:677–684.
207. Paterson DS, Trachtenberg FL, Thompson EG, et al. Multiple serotonergic brainstem abnormalities in sudden infant death syndrome. *JAMA.* 2006;296:2124–2132.
208. Kinney HC, Myers MM, Belliveau RA, et al. Subtle autonomic and respiratory dysfunction in sudden infant death syndrome associated with serotonergic brainstem abnormalities: a case report. *J Neuropathol Exp Neurol.* 2005;64:689–694.
209. Kinney HC, Randall LL, Sleeper LA, et al. Serotonergic brainstem abnormalities in northern plains Indians with the sudden infant death syndrome. *J Neuropathol Exp Neurol.* 2003;62:1178–1191.
210. Kinney HC, Filiano JJ, White WF. Medullary serotonergic network deficiency in the sudden infant death syndrome: review of a 15-year study of a single dataset. *J Neuropathol Exp Neurol.* 2001;60:228–247.
211. Panigraphy A, Filiano J, Sleeper LA, et al. Decreased serotonergic receptor binding in rhombic lip-derived regions of the medulla oblongata in the sudden infant death syndrome. *J Neuropathol Exp Neurol.* 2000;59:377–384.
212. Ponsonby AL, Dwyer T, Cochrane J. Population trends in sudden infant death syndrome. *Semin Perinatol.* 2002;26:296–305.

CHAPTER NINE

The Circadian Clock in Cancer Development and Therapy

Loning Fu[*,†,‡], **Nicole M. Kettner**[*,†]
[*]Department of Pediatrics/U.S. Department of Agriculture/Agricultural Research Service/Children's Nutrition Research Center, Baylor College of Medicine, Houston, Texas, USA
[†]Department of Molecular and Cellular Biology, Baylor College of Medicine, Houston, Texas, USA
[‡]Dan L. Duncan Cancer Center, Baylor College of Medicine, Houston, Texas, USA

Contents

Abstract

Most aspects of mammalian function display circadian rhythms driven by an endogenous clock. The circadian clock is operated by genes and comprises a central clock in the brain that responds to environmental cues and controls subordinate clocks in peripheral tissues via circadian output pathways. The central and peripheral clocks coordinately generate rhythmic gene expression in a tissue-specific manner *in vivo* to couple diverse physiological and behavioral processes to periodic changes in the environment.

Progress in Molecular Biology and Translational Science, Volume 119
ISSN 1877-1173
http://dx.doi.org/10.1016/B978-0-12-396971-2.00009-9

However, with the industrialization of the world, activities that disrupt endogenous homeostasis with external circadian cues have increased. This change in lifestyle has been linked to an increased risk of diseases in all aspects of human health, including cancer. Studies in humans and animal models have revealed that cancer development *in vivo* is closely associated with the loss of circadian homeostasis in energy balance, immune function, and aging, which are supported by cellular functions important for tumor suppression including cell proliferation, senescence, metabolism, and DNA damage response. The clock controls these cellular functions both locally in cells of peripheral tissues and at the organismal level via extracellular signaling. Thus, the hierarchical mammalian circadian clock provides a unique system to study carcinogenesis as a deregulated physiological process *in vivo*. The asynchrony between host and malignant tissues in cell proliferation and metabolism also provides new and exciting options for novel anticancer therapies.

1. INTRODUCTION

Circadian rhythms in physiological and behavioral processes in plants and animals have been known since the fourth century BC. These rhythms were originally attributed to a passive response of organisms to diurnal changes in external light cues but were later discovered to be generated by an endogenous clock in all species studied.[1] In mammals, circadian rhythms are generated by a central clock in the suprachiasmatic nucleus (SCN) located in the hypothalamus that constantly synchronizes with environmental cues via circadian input pathways and controls the peripheral clocks through circadian output pathways.[2,3]

Both central and peripheral clocks are operated by the same set of circadian genes expressed in all tissues studied. The molecular clockwork in mammals has been described in detail in several recent reviews.[4–6] Briefly, it is based on autoregulatory transcriptional feedback loops driven by the heterodimer of bHLH–PAS transcription factors BMAL1/CLOCK or BMAL1/NPAS2 that activate their downstream transcriptional repressor targets *Cryptochrome* (*Cry1,2*) and *Period* (*Per1–3*) at the beginning of a circadian day. The accumulation of PER and CRY proteins in the cytoplasm at the end of a circadian day, controlled by the Skp1–Cullin–F-box protein E3 ubiquitin ligase complexes, casein kinase 1ε/δ (CK1ε/δ) and adenosine monophosphate-activated protein kinase (AMPK), leads to the formation of a PER/CRY repressor complex that translocates into the nucleus at the beginning of a circadian night to inhibit the activity of BMAL1/CLOCK or BMAL1/NPAS2 heterodimers and recruit the transcriptional termination complex to the *Per* and *Cry* genes.[7] In addition, the transcription of *Bmal1* is alternatively regulated by its own transcription targets, the nuclear receptors *Rev-erbα/β* (the repressors) and *Rorα* (the activator).[8–10] The multiple interlocked autoregulatory feedback loops result in a robust circadian variation in the expression and activity of *Bmal1* over a 24-h period, providing a driving force for circadian oscillation of the molecular clockwork.

The circadian regulators also target clock-controlled genes to generate circadian rhythms in all major cellular processes in both SCN neurons and peripheral organs, resulting in a rhythmic expression of 3–10% of all mRNAs expressed in a given tissue due to time-dependent interactions between the circadian regulators with specific gene promoter sequences, transcription factors, or transcriptional initiation, elongation, and termination complexes, as well as the key factors controlling chromatin remodeling.[7,11–15] The clock-controlled genes usually do not share overlapping expression patterns between tissues, suggesting a key role for the circadian clock in controlling tissue-specific function *in vivo*. Clock-controlled genes expressed in all the tissues studied include the key regulators of cell proliferation, metabolism, senescence, and DNA damage response.[16–23]

The molecular clock in SCN neurons and peripheral tissues can be entrained or phase-shifted by cellular signaling. The most potent circadian time cue for the SCN clock is light, which is received by a subset of melanopsin-expressing retinal ganglion cells and transmitted directly to the SCN neurons via the retinohypothalamic tract (RHT). Upon activation, the RHT produces neurotransmitters that activate a cascade of signal transduction events leading to circadian phase resetting.[24,25] Although the SCN clock is capable of generating autonomic circadian outputs on its own, the constant coupling of the central clock with environmental cues provides a survival advantage by synchronizing daily physiology and behavior with local time cues.[25,26] A shift in environmental cues, such as traveling across several time zones on an aircraft, induces a phase-shift in the central clock and the subsequent SCN-controlled phase shift in peripheral tissues via circadian output pathways to reestablish the endogenous circadian homeostasis to the new local time. The number of days needed to fully adjust to the new time zone is dependent on the number of time zones crossed during the trip. Constant back-and-forth phase shifts of environmental light cues resulting from rotating work schedules or chronic jet lag disrupt endogenous circadian homeostasis by uncoupling the central and peripheral clock coordination.[27–29]

The best-studied circadian output pathways include the autonomic nervous system (ANS) and the neuroendocrine system (NES) that control all aspects of mammalian physiology as well as the peripheral clocks via cellular signaling. The rhythmic activities of these systems provide a mechanism for the central clock to control peripheral tissues directly and indirectly via peripheral clocks.[24,29–33]

Dramatic changes in lifestyles since the industrial revolution due to increased use of artificial lighting, night-shift working schedules, or rapid long-distance transmeridian traveling have led to frequent disruptions of endogenous circadian homeostasis in modern societies. These changes in lifestyle are coupled with a significant increase in the risk of diseases in all aspects of human health, including cancer.

2. CIRCADIAN DYSFUNCTION PROMOTES CANCER DEVELOPMENT IN HUMANS

2.1. Circadian disruption is an independent cancer risk factor for humans

Recent epidemiology studies have linked circadian disruption to increased susceptibility to cancer development in all key organ systems in humans. The cancers observed from these studies included breast, ovarian, lung, pancreatic, prostate, colorectal and endometrial cancers, non-Hodgkin's lymphoma (NHL), osteosarcoma, acute myeloid leukemia (AML), head and neck squamous cell carcinoma, and hepatocellular carcinoma.[34–49] Circadian dysfunction-induced cancer risk increases with the number of years, the frequency of rotating work schedules, and the number of hours per week working at night among human night-shift workers.[45,46,50–52] Together, these findings suggest that loss of circadian homeostasis could be an independent cancer risk factor for humans.[53] Due to the prevalence of night-shift work schedules in modern societies, the World Health Organization's International Agency for Research on Cancer (IARC) listed "shift work that involves circadian disruption" as a probable carcinogen in 2007.

2.2. Circadian disruption is associated with poor prognosis and early mortality of cancer patients

Loss of circadian homeostasis not only promotes cancer development but is also associated with poor performance with regard to anticancer treatments and early mortality among cancer patients. After adjusting for other factors that might affect survival, circadian rhythm in salivary and serum cortisol levels as well as daily rest/activity patterns are used as independent prognosis factors for survival and therapeutic response of patients with metastatic breast, lung, and colorectal cancers.[54–61]

2.3. Disruption of the molecular clockworks in human cancers

Ample evidence has linked dysfunction of the molecular clock with pathogenesis of human cancers (Table 9.1). The mechanisms of dysregulation of the core circadian genes in human cancers discovered to date include epigenetic silencing by promoter methylation, deregulation at the transcriptional and posttranscriptional levels, and structural variations of clock proteins due to circadian gene polymorphisms.

In comparison to normal host tissues, decreased expression and polymorphism of the core circadian genes *Per1*, *Per2*, and *Per3* are frequently found in human breast, endometrial, prostate, pancreatic, colorectal, and nonsmall cell lung cancers (NSCLC), as well as hepatocellular carcinoma, neck squamous cell carcinoma, glioma, AML, and chronic myeloid lymphoma (CML).[62,63,80,85,86,88,90,91,98,99,102,103] In CML, breast, endometrial, and NSCLC, this deregulation is often linked to hypermethylation of CpG islands or aberrant acetylation in the promoters of *Per* genes, which leads to gene silencing.[64,78,84,91,100] Other core circadian genes are also frequently deregulated or silenced in human cancers. For example, the epigenetic inactivation of *Bmal1* is often linked to hematologic malignancies including NHL, diffuse large B-cell lymphoma, acute lymphocytic leukemia (ALL), and AML, whereas polymorphisms in *Clock*, *Cry1*, *Cry2*, and *Npas2* gene are often found to be associated with an increased risk or recurrence of NHL, AML, endometrial ovarian, colorectal, and breast cancers.[65,75,81,84,94,104] In most studies examining the role of the molecular clock in human cancers, deregulation or polymorphism of multiple or all core circadian genes is observed. For example, deregulation or polymorphism of *Per1*, *Per2*, and *Per3*, *Clock*, *Bmal1*, *Cry1*, *Cry2*, *Clock*, *Npas2*, and/or *CK1ε* is often found in human CML, prostate, pancreatic and epithelial ovarian cancers, leukemia, pleural mesothelioma, hepatocellular carcinoma, glioma, and neck squamous cell carcinoma.[76,78,87,89,92,93,95,96,99,101,104] Based on these discoveries, a combined deregulation of *Cry1* and *Bmal1*, or *Cry1* and *Per2*, has been suggested as a negative prognostic marker for epithelial ovarian cancer and CML, respectively.[76,77,96]

Table 9.1 Deregulation of the core clock genes in human cancers

Cancer type	Circadian genes deregulated	Deregulated targets	Cellular functions affected	References
Breast cancer	*Bmal1*, *Clock*, *Cry1*, *Cry2*, *Per1*, *Per2*, *Per3*, and *Npas2*	BCCIP, BCL2, BRAC1, ERα, estrogen, EXO1, c-AMP, CDKN1A, cortisol, Cyclin D1, c-ERBB2, GADD45A, HERC5, Melatonin, MCM5, MSH2, $p21^{WAF1/CPI1}$, p38, p53, PARP, PKA, PPP1R15A, SIRT1, SUMO1, TERT, TIP60, and UBA1	Apoptosis, cell cycle control, chromatin remodeling, DNA damage repair, and telomere length	38, 39, 42, 44, 45, 48–50, 53, 55, 62–74
Acute lymphocytic leukemia (ALL)	*Bmal1* and *Clock*	Catalase, c-MYC, and p300	Cell cycle control and chromatin remodeling	75
Acute myeloid leukemia (AML)	*Bmal1*, *Per1*, *Per2*, and *Per3*	Catalase, c-MYC, and p300	Cell cycle control and chromatin remodeling	75
Chronic lymphocytic leukemia (CLL)	*Cry1* and *Per2*	ZAP70	Cell cycle control, chromatin remodeling, and DNA damage repair	76, 77
Chronic myeloid leukemia (CML)	*Bmal1*, *Cry1*, *Cry2*, *Per1*, *Per2*, and *Per3*	c-MYC, Cyclin B1, and p53	Apoptosis, cell cycle control, and chromatin remodeling	78, 79

Colorectal cancer	*Clock*, *Per1*, *Per2*, and *Per3*	Cortisol, ATM, EGFR, ER-β, EXO1, IL-6, MSH2, p53, PARP, TGFα, and TNFα	Chromatin remodeling and DNA damage repair	34, 43, 54, 58, 70, 80–83
Endometrial cancer	*Cry1*, *Per1*, *Per2*, and *Per3*	Melatonin	Chromatin remodeling	36, 42, 84
Glioma	*Cry1*, *Cry2*, *Per1*, *Per2*, and *Per3*	N/A	Apoptosis and cell cycle control	85–87
Head and neck squamous cell carcinoma	*Bmal1*, *CKIε*, *Cry1*, *Cry2*, *Per1*, *Per2*, *Per3*, and *Tim*	TIP60	N/A	52, 88, 89, 71
Hepatocellular carcinoma	*Cry2*, *Per1*, *Per2*, *Per3*, and *Tim*	CDC2, Cyclin B1, EZH2, GR, IGF-1, and WEE1	Cell cycle control, chromatin remodeling	47, 90
Lung cancer	*Clock*, *Per1*, *Per2*, and *Per3*	Cortisol and TIP60	Cell cycle control, chromatin remodeling, DNA damage repair	43, 56, 57, 71, 82, 91
Malignant pleural mesothelioma	*Bmal1*, *Cry2*, *Per1*, *Per3*, *Npas2*, *Rev-erbα*, *Rev-erbβ*, and *Tim*	CASP3, Cyclin B, Cyclin E, $p21^{WAF1/CIP1}$, and WEE1	Apoptosis, cell cycle control, chromatin remodeling, and deregulated chemotherapy drug response	92, 93

Continued

Table 9.1 Deregulation of the core clock genes in human cancers—cont'd

Cancer type	Circadian genes deregulated	Deregulated targets	Cellular functions affected	References
Non-Hodgkin's lymphoma (NHL)	*Npas2*	DMC1, EXO1, and MSH2	Cell cycle control, DNA damage repair, and immune deficiency	37, 70, 94
Diffuse large B-cell lymphoma	*Bmal1* and *Clock*	Catalase, c-MYC, and p300	Chromatin remodeling	75
Osteosarcoma	*CK1ε* and *Per2*	CASP3, Cyclin B, and Cyclin A2	Apoptosis and cell cycle control	40, 95
Ovarian cancer	*Bmal1*, *CK1ε*, *Clock*, *Cry1*, *Cry2*, *Per1*, *Per2*, and *Per3*	Cortisol	Apoptosis, cell cycle control, and deregulated chemotherapy drug response	39, 96, 97
Pancreatic cancer	*Bmal1*, *CK1ε*, *Clock*, *Cry1*, *Cry2*, *DEC1*, *Per1*, *Per2*, *Per3*, *Tim*, and *Tipin*	BCL-XL, CDC2, Cyclin B1, TNF-α, and USP30	Apoptosis, cell cycle control, and chromatin remodeling	98–101
Prostate cancer	*Bmal1*, *CK1ε*, *Clock*, *Cry1*, *Cry2*, *Npas2*, *Per1*, *Per2*, and *Per3*	Melatonin, SIRT1, and TIP60	Apoptosis, cell cycle control, DNA damage repair, and transactivation of AR	35, 43, 45, 71, 102, 103, 104, 102–105

Deregulation of the core circadian genes in human cancers is closely associated with a constitutive activation of intracellular inflammatory and oncogenic signaling pathways including the constitutive activation of p38, c-MYC, NF-κB, BCL-XL, and protein kinase A (PKA)[66,75,79,98,106]; aberrant chromatin remodeling; deregulation of inflammatory cytokines, catalase, TIP60, telomerase, PARP [poly (ADP-ribose) polymerase], SIRT1 and p300[66,67,75,105–107]; overexpression of ERα, G1, and S-phase cyclins; and suppression of tumor suppressors, ATM, p53, $p21^{WAF1/CPI1}$, and WEE1.[68,69,79,82,92] Deregulation of the molecular clock is correlated with the loss of control in cell proliferation, metabolism, DNA replication and repair, senescence, apoptosis, and DNA damage response, and increased drug resistance in all types of human cancer cells studied (Table 9.1).[67,68,70–75,79,82,83,92,95,97,98]

2.4. Central clock dysfunction increases cancer risk in humans

In the hierarchical organization of the mammalian circadian clock, the peripheral clock can only sense changes in environmental light cues via central clock-controlled circadian output pathways. Thus, central clock dysfunction induced by frequent back-and-forth phase shifts of environmental cues may play a key role in promoting cancer development among human night-shift workers by disrupting the homeostasis of neuroendocrine function.[107–110] This hypothesis is supported by the facts that visually impaired people who are insensitive to changes in environmental circadian light cues and largely or completely depend on a free-running endogenous clock to organize their daily physiology display a lower cancer risk compared to the general population.[111–113]

In summary, ample evidence obtained from human studies suggests that the mammalian circadian clock plays a key role in tumor suppression. Therefore, disruption of circadian homeostasis of mammalian physiology is a novel risk factor for cancer (Table 9.1).

3. CIRCADIAN DISRUPTION PROMOTES CANCER DEVELOPMENT IN ANIMAL MODELS

3.1. The central clock suppresses tumor initiation and progression in animal models

Pioneering studies to investigate the role of circadian disruption in cancer development using experimental animal models started in the late 1960s. These studies demonstrate that disruption of circadian endocrine rhythms by either constant light exposure or pinealectomy increases spontaneous and carcinogen-induced mammary gland and hepatocellular carcinogenesis in rodents.[114–116] Similar experiments conducted in recent years have also shown that disruption of circadian homeostasis by a short period of back-and-forth or consecutive phase-advance shifts of environmental light cues, or by constant light exposure significantly accelerates tumor growth in animals.[117–121] Compared to sham-operated animals, mice lacking a central clock due to surgical ablation of the SCN were unable to maintain circadian rhythmicity in locomotor activity, body temperature, and immune function. This loss of circadian homeostasis in SCN-lesioned mice is coupled with a significant decrease in survival time due to the increased rate of tumor growth compared to control tumor-bearing mice with an intact SCN.[122] Together, these studies agree with the findings from human studies in that circadian dysfunction increases the risk of cancer by demonstrating that disruption of the central circadian clock promotes cancer development and progression in rodents.

3.2. Variation in cancer phenotypes reported for circadian gene-mutant mouse models

The role of mammalian circadian genes in cancer genetics was first reported in 2002 in a study showing that mice expressing a mutant PER2 (*Per2*$^{m/m}$) defective in PER2-mediated protein/protein interactions due to an 85-amino acid in-frame deletion in the PAS domain of *Per2* gene display multiple tumor-prone phenotypes including increased spontaneous and γ-radiation-induced lymphoma, hyperplastic growth in salivary and preputial glands, resistance to radiation-induced apoptosis in thymocytes, and deregulation of key tumor suppressors, cyclins, and proto-oncogenes, such as *p53*, *Gadd45α*, *Cyclin D1*, *Cyclin A*, *c-Myc*, and *Mdm2*.[123,124] The same study also shows that *Per2*-null (*Per2*$^{-/-}$) mice display similar cancer-prone phenotypes as *Per2*$^{m/m}$ mice (Supplemental data).[124]

In contrast, other reports indicate that mice deficient in other core circadian genes either lack neoplastic phenotypes or are tumor resistant. For example, mice lacking *Bmal1* (*Bmal1*$^{-/-}$) show a significantly reduced life span and premature aging, but not spontaneous tumor development.[125] Hepatocytes in mice lacking both *Cry1* and *Cry2* (*Cry1*$^{-/-}$;*Cry2*$^{-/-}$) proliferated slower than wild-type (Wt) hepatocytes in the first 72 h immediately after partial hepatectomy.[126] Ablation of both *Cry1* and *Cry2* reduced cancer risk for *p53*-null mice.[127] *Clock*-mutant (*Clock*$^{m/m}$) and *Cry1*$^{-/-}$;*Cry2*$^{-/-}$ mice did not show predisposition to cancer in response to a low dose of γ-irradiation.[128,129] Furthermore, MEFs isolated from *Cry1*$^{-/-}$;*Cry2*$^{-/-}$ mice do not show deficiencies in γ-radiation-induced cell cycle arrest, whereas *Clock*$^{m/m}$ MEFs show lower levels of DNA synthesis and cell proliferation than wild-type controls.[19,128] Together, the studies described above led to the conclusion that the cancer-prone phenotypes discovered in *Per2*$^{m/m}$ and *Per2*$^{-/-}$ mice are a result of the loss of a "non-clock function" of the *Per2* gene and not the function of the mammalian circadian clock.[6]

3.3. The molecular clock suppresses tumor development in mice

In contrast, we suggest that a more detailed analysis of the available information supports a direct role for the molecular clock in tumor suppression in mice.

First, the observation of a temporary slowdown in *Cry1*$^{-/-}$;*Cry2*$^{-/-}$ hepatocyte proliferation immediately after partial hepatectomy cannot predict whether *Cry1*$^{-/-}$;*Cry2*$^{-/-}$ mice are tumor resistant under normal physiological conditions. Surgical stress caused by partial hepatectomy can suppress the growth of hepatocellular carcinoma in mouse livers until the third postoperative day.[130] In fact, compared to wild-type controls, most mouse models prone to spontaneous hepatocellular carcinoma show an initial delay in hepatocyte proliferation after partial hepatectomy. For example, compared to wild-type controls, mice lacking the nuclear receptor FXR display a delay in hepatocyte proliferation after partial hepatectomy until the ninth postoperative day. However, *Fxr*$^{-/-}$ mice quickly regain the ability of rapid hepatocyte proliferation and develop malignant liver tumors after 12 months of age.[131,132] Since *Cry1*$^{-/-}$;*Cry2*$^{-/-}$ mice show significantly dampened *Bmal1* expression and deregulation of the mitogen-activated protein kinase/extracellular signal-regulated kinase (MAPK/ERK) pathway in the liver,[133–135] it would be important to examine whether the temporary delay in hepatocyte proliferation in *Cry1*$^{-/-}$;*Cry2*$^{-/-}$ mice after partial hepatectomy is caused by deficiencies in cellular signaling essential for G1 cell cycle initiation.[136,137]

Second, the conclusions that *Per*-mutants are cancer prone but *Cry*-mutants are tumor resistant are confounded by significant differences in the phenotypes of control wild-type and *p53*-null mice but not *Per*- and *Cry*-mutant mice in different studies.[29,124,127,128] For example, the conclusion that a single 4-Gy sublethal dose of γ irradiation led to a similar rate of decline in the survival of irradiated wild-type and *Cry1*$^{-/-}$;*Cry2*$^{-/-}$ mice was based on an unusual sensitivity of wild-type mice to γ-irradiation,[128] which was not observed in the earlier study indicating increased sensitivity of the *Per2*-mutant mice to the same 4-Gy sublethal γ-radiation.[124]

Similarly, discrepancies in the effects of *Per* or *Cry* gene ablation on the survival and tumor-developing rate of *p53*-null mice can also be attributed to a significant difference in the average and maximal life spans of *p53*-null mice,[29,127] which vary between 15–30 and 28–60 weeks of age, respectively, as reported in different studies.[138–145] In accordance with our studies demonstrating that chronic back-and-forth jet lag significantly accelerated tumor development and reduced survival of *p53*-null mice,[29] mice harboring both a mutant *p53* corresponding to the $p53^{R175H}$ hotspot mutation in humans and a mutant *Per2* allele ($Per2^{S662G}$),[146] which leads to a short and phase-advanced behavior rhythm among human patients suffering from familial advanced sleep phase syndrome due to a single serine to glycine mutation within the CKIε-binding region in PER2,[147] also display an increased tumor incidence rate and decreased survival compared to $p53^{R175H}$ mice.[148] These findings suggest strongly that circadian dysfunction cooperates with loss of *p53* to promote tumor development.

Different research teams have also independently reached the same conclusion that *Per*- and *Cry*-mutant mice display a similar neoplastic growth of osteoblasts in bone,[149,150] and that disruption of *Period* genes increases cancer risk in mice.[29,124,148,151–153] Therefore, it would be important to verify the role of *Per* and *Cry* genes in cancer risk since *Per* and *Cry* genes are both indispensable for operating the same negative loop in the molecular clock and display the same deregulated behavioral phenotypes.[123,154,155] In addition, despite of a high *Per2* mRNA expression, PER2 protein is not detected in peripheral tissues of $Cry1^{-/-};Cry2^{-/-}$ mice.[156] Indeed, when *Per*- and *Cry*-mutant mice of the same mouse strain were studied under exactly the same conditions, the two mouse models displayed the same increased rate of tumor development in the skeletal, immune, reproductive, and digestive systems when kept in a steady-state 12-h light/12-h dark (24-h LD) condition, in response to a sublethal dose of γ-radiation, or treated with chronic jet lag after γ-radiation.[29] Together, the evidence obtained from studying mice lacking *Per* or *Cry* under the same conditions strongly argues that as found in human studies, the *Cryptochrome* genes also function in tumor suppression in rodents.

Third, $Clock^{m/m}$ mice show a significant decrease in survival compared to wild-type controls at 80 weeks of age after a sublethal dose of γ-radiation. No significant difference in tumor incidence or the rate of radiation-induced apoptosis between wild-type and $Clock^{m/m}$ splenocytes was reported. The decrease in the survival of irradiated $Clock^{m/m}$ mice was attributed to accelerated aging but not tumor development.[129] Since only the apoptotic response of $Clock^{m/m}$ splenocytes cultured *in vitro* to a lethal, and not a sublethal, dose of γ-radiation was studied, and the aging phenotypes displayed by irradiated *Clock*$^{m/m}$ mice are also commonly observed in other irradiated circadian gene-mutant mouse models,[29,124,128,129] a role for *Clock* in tumor suppression cannot be ruled out without examining the cancer risk of *Clock*$^{-/-}$ and *Clock*$^{-/-}$;*Npas2*$^{-/-}$ mice as well as irradiated *Clock*$^{m/m}$ mice after 80 weeks of age. This is because aging is considered a primary risk factor for cancer,[157] and the mammalian CLOCK may play a direct role in DNA damage repair after γ-radiation.[158] In addition, CLOCK and NPAS2 play overlapping roles in the molecular clock.[159,160] Unlike *Per2*$^{m/m}$ and *Per2*$^{-/-}$ mice that show a similar deregulated circadian phenotype,[123,124,155,161] *Clock*$^{m/m}$ and *Clock*$^{-/-}$ mice display different circadian behavioral phenotypes and patterns of deregulation of gene expression in somatic cells.[162–165]

Fourth, the observations of aggressive aging phenotypes and lack of tumor incidence among *Bmal1*-null mice are not sufficient to exclude a role for *Bmal1* in tumor suppression. When circadian gene-mutant mice are kept in stable 24-h LD cycles, most spontaneous tumors are identified after 50 weeks of age,[29] which is an age most *Bmal1*$^{-/-}$ mice would not live to.[125] After being treated with a sublethal dose of γ-radiation, although the maximal life span of *Bmal1*$^{-/-}$ mice was further decreased, about 6–7% of them developed malignant lymphomas before or at about 40 weeks of age. This rate of tumor development in irradiated *Bmal1*$^{-/-}$ mice is very similar to the reported rate of tumor development among irradiated *Per*- and *Cry*-mutant mice.[29,124,128] In addition, *Bmal1*$^{+/-}$ mice that have a similar life span as *Per*- and *Cry*-mutants display the same rate of spontaneous and radiation-induced tumor development in the same organ systems as *Per*- and *Cry*-mutant mice.[29]

Bmal1-null mice also display a delay in anagen progression and decreased cell proliferation in secondary hair germ cells, which is coupled with increased G1 cell cycle block as shown by decreased levels of RB phosphorylation, increased expression of cyclin-dependent kinase inhibitors *p21*$^{WAF1/CIP1}$ and *p16*^{Ink4A}, and accelerated epidermal aging.[166] In contrast, mice lacking *Bmal1* only in keratinocytes show constitutive elevation of cell proliferation and intracellular redox levels as well as deregulated UVB-induced DNA damage in the epidermis at linear growth age after wean.[22] However, in a different study, the same keratinocyte-specific *Bmal1*$^{-/-}$ mouse model was reported to display aging phenotypes in the skin starting from 10 months of age, an age most *Bmal1*-null mice could not survive to.[125] The decreased regeneration of keratinocyte-specific *Bmal1*$^{-/-}$ hair germ cells reported in this study cannot be rescued by overexpressing oncogenic SOS, a Ras activator,[167] suggesting an early onset of replicative or cellular senescence (explained later in this review). Together with the findings that *Bmal1*-null mice display normal skin regeneration and aggressive hyperplastic growth in bone at a young age as well as increased lymphoma development after γ-radiation, and that targeted silencing of *Bmal1* in tumors induces immune suppression and accelerated tumor growth in mice,[29,125,149,168] the studies described earlier suggest that cellular senescence resulting from hyperplastic growth, oncogenic activation, and accumulated free radical-induced DNA damage is intrinsic to $Bmal1^{-/-}$ somatic cells. However, in tumors and somatic tissues that can overcome the barriers of cellular senescence and reactive oxygen species (ROS)-induced apoptosis, loss of *Bmal1* only accelerates tumor initiation and growth.[22,29,166,168] $Bmal1^{-/-}$ mice are specially distinct from other circadian gene-mutant mouse models in that they lack circadian homeostasis even when kept in 24-h LD conditions.[169] This severe disruption of endogenous homeostasis may also contribute to increased senescence at the organismal level. However, if *Bmal1*-null mice can overcome aggressive aging, they are likely cancer prone.

3.4. Cellular-based studies using mouse primary cells lacking core circadian genes

The role of core circadian genes in controlling cell proliferation and DNA damage response has also been studied using various types of primary cells isolated from different circadian gene mouse models *in vitro*.[19,29,124,128,129,149,170] The results obtained from these studies should be explained with caution. For example, primary MEFs cultured *in vitro* behave very differently from somatic cells in tissues prone to tumor development. MEFs are known to be resistant to γ-radiation-induced aopotosis regardless of genotypes and, therefore, should not display a high rate of apoptotic death after a sublethal dose of γ-radiation in the absence of aberrant oncogenic activation.[128,171,172] In addition, the cell cycle clock is different from the molecular clock in that it does not free run (Fig. 9.1).[173,176] Therefore, the serum shock protocol for setting free-running status of the molecular clock in cultured MEFs is not suitable for studying the role of a core circadian gene in cell cycle control because this protocol requires confluent cell culture condition and only provides growth factor-containing serum for a few hours at the initial serum shocking stage, which leads to uncoupling of the cell cycle clock from the molecular clock due to growth arrest induced by contact inhibition and lack of proper extracellular signals to induce immediate early genes essential for G1 cell cycle progression after the first day of the experiment.[19,177,178]

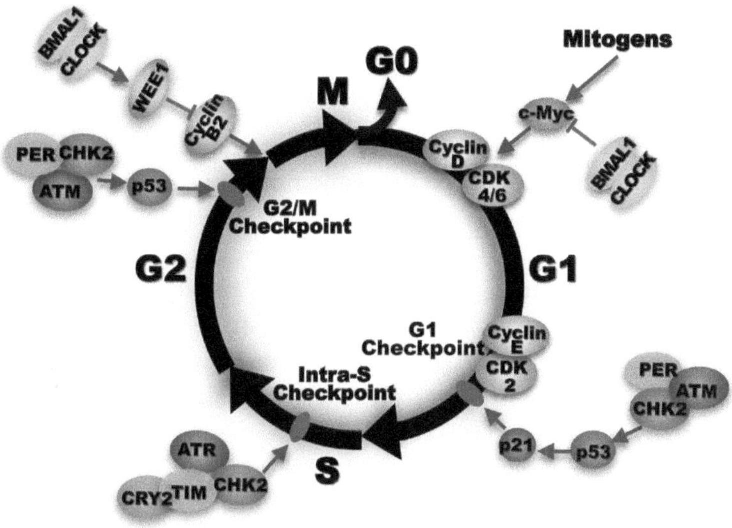

Figure 9.1 Control of cell proliferation by the molecular clock. Unlike the molecular clock, the cell cycle does not free-run before passing the G1/S-phase transition. The initiation of cell cycle progression is strictly controlled by extracellular mitogenic signals that transiently activate immediate early genes such as *c-Myc*, which then induces Cyclin D leading to activation of Cyclin D/CDK4/6 complex that in turn activates E2F-dependent Cyclin E expression by suppressing tumor suppressor RB (not shown). The interaction of Cyclin E with CDK2 allows G1- to S-phase transition. G1 is the longest phase in the cell cycle during which most biosynthesis essential for supporting cell cycle progression occurs. c-MYC or E2F oncogenic activation-induced elevation of G1 Cyclin expression or genomic DNA damage leads to activation of G1 checkpoint mediated by $p16^{Ink4A}$ and $p21^{WAF1/CIP1}$, controlled by RB and p53, respectively. $p16^{Ink4A}$ disrupts Cyclin D/CDK4/6 complex (not shown), whereas $p21^{WAF1/CIP1}$ disrupts Cyclin E/CDK2 interaction. The activation of G1 checkpoint leads cells to pause before entering the S phase to repair damaged DNA or exit the cell cycle to enter the G0 phase (nondividing status). Under certain conditions, excessive DNA damage or uncontrolled oncogenic signaling can activate RB and/or p53 tumor suppression pathways leading to cellular senescence. DNA damage induced by UV radiation leads to activation of ATR/CHK1-mediated intra-S checkpoint that couples DNA damage repair with replication whereas double-stranded DNA damage induced by γ-radiation activates ATM/CHK2-mediated G1/S and G2/M checkpoints to prevent damaged cells from entering the S or mitotic (M) phase. Prolonged G2/M checkpoint arrest is also often associated with p53-mediated apoptosis. G2/M transition is also regulated by WEE1, a kinase that phosphorylates and inactivates the Cyclin B1/cell division cycle 2 (CDC2) complex essential for G2/M transition. Upon the completion of mitosis, cells either enter the next cell cycle stimulated by extracellular mitogen, or withdraw from cell cycle to enter the G0 phase in the absence of mitogenic signals.[173,174] The molecular clock functions in all phases of the cell cycle to prevent neoplastic growth. At the early G1 phase, the BMAL1/CLOCK heterodimer downregulates *Myc* transcription to prevent its overexpression.[29,124,149] PER1 directly interacts with ATM and CHK2 to control G1 checkpoint in response to double-strand DNA damage.[82] In the S phase, CRY2/TIM complex directly interacts with ATR/CHK1 to control intra-S checkpoint.[175] In the G2 phase, PER-mediated ATM/CHK2/p53 signaling in response to DNA double-strand breaks and BMAL1/CLOCK activated *Wee1* expression both lead to activation of G2/M checkpoint to prevent inappropriate M phase entry.[82,126] (See Color Insert.)

The choice of cell types and *in vitro* cell culture conditions used in a study may lead to different conclusions on the role of a gene in cell cycle control. For example, under confluent culture condition, serum-shocked differentiated skeletal muscle cells and hepatocytes from $Bmal1^{-/-}$ mice show high level expression of $p21^{WAF1/CIP1}$, leading to the conclusion that loss of *Bmal1* decreases the rate of cell proliferation.[179] However, when a human RNAi library targeting 8000 human genes was studied to identify modulators of p53 function using the BJ-TERT-tsLT cells under subconfluent condition, which were originally isolated from normal human diploid foreskin fibroblasts, *Bmal1* was identified as a novel positive regulator of the tumor suppressor p53. Inhibition of *Bmal1* expression in this system led to loss of p53-mediated G1 cell cycle arrest at least in part due to an inability to activate the p53 target $p21^{WAF1/CIP1}$.[69]

In summary, genetic studies using various circadian gene-mutant mouse models strongly suggest that as found in human studies, both positive and negative loops of the molecular clock function in tumor suppression in rodents (Table 9.2).

4. THE ROLE OF THE MAMMALIAN CIRCADIAN CLOCK IN TUMOR SUPPRESSION

Cancer is a multifactorial disease *in vivo*. Its initiation and progression need various manifestations of abnormal physiological conditions. As the master regulator of mammalian physiology, the circadian clock acts at the molecular, cellular, tissue/organ, and organismal levels to suppress tumor development by maintaining homeostasis of physiology.

4.1. The role of peripheral clock in tumor suppression

In cells of peripheral tissues, the clock orchestrates diverse cellular functions in a diurnal oscillating pattern via generation of a network of gene expression at the transcriptional and posttranscriptional levels.[9,197–199] Disruption of the circadian profiles of this gene expression network leads to loss of homeostasis in cell/tissue function, a key mechanism in circadian dysfunction-induced diseases. Recent studies have revealed that in both human and experimental animal models, circadian disruption specially increases the risk of cancers in the immune, digestive, and reproductive systems that need daily cell proliferation to support their functions. These findings highlight the importance of circadian control of cell cycle progression in both homeostasis of tissue function and tumor suppression *in vivo*. In peripheral tissues, the molecular

Table 9.2 Phenotypes of circadian gene-mutant mouse models

Circadian genes	Mouse models	Circadian behavior phenotypes	Disease phenotype	Key targets affected	References
Bmal1	*Bmal1*-heterozygous	Normal	Premature aging and cancer prone	N/A	29
	Bmal1-null	Arrhythmic	Premature aging, metabolic syndrome, immune deficiency, caner prone, deregulated drug response	CDC2, Cyclin B1, Cyclin D1, Cyclin E, $p21^{WAF1/CIP1}$, $p16^{Ink4a}$, p53, RB, and WEE1	23, 29, 124, 125, 149, 166, 168, 169, 180–182
	Keratinocyte-specific *Bmal1*-null	Normal	Hyperproliferation and deregulated UVB DNA damage to hair cells at a young age	DAB2, DKK3, LEF1, $p16^{Ink4A}$, SMAD3, TGFBR2, and WNT10a	22
	Keratinocyte-specific *Bmal1*/oncogenic *Sos*	Normal	Resistance to cutaneous squamous tumors, senescence of hair cells at 10 months of age	$CD34^{+}$ and Integrin-α6	167
	Macrophage-specific *Bmal1*-null	Normal	Immune deficiency	IL-6	183
Clock	*Clock*-mutant (*Clock* Δ19)	Lengthened period and arrhythmic in constant darkness	Metabolic syndrome, premature aging, and deregulated drug response	AKT1, ATR1, Cdk2, Chk1, Chk2, Cyclin D3, Cyclin E1, EGF, ERα, JAK2, $p21^{WAF1/CPI1}$, $p27^{KIP1}$, PBEF, TGFβ, and WEE1	19, 129, 163, 166, 181, 184–187
	Clock-null	Shortened period and failure of phase delay/advance	Premature aging and immune deficiency	NF-κB and WEE1	162, 164

Continued

Table 9.2 Phenotypes of circadian gene-mutant mouse models—cont'd

Circadian genes	Mouse models	Circadian behavior phenotypes	Disease phenotype	Key targets affected	References
Cry	*Cry1*-null	Shortened period	Cancer prone	N/A	29, 154
	Cry2-null	Lengthened period	Cancer prone	N/A	29, 150, 154
	Cry1, *Cry2* double knockout	Arrhythmic	Cancer prone, immune deficiency, premature aging, metabolic syndrome, cell cycle control, neuroendocrine deficiency, and deregulated drug and DNA damage response	IL-6, MAPK/ERK, TNFα, WEE1, and XPA, GR, and Gsα	29, 106, 126, 134, 149, 154, 170, 188–191
			Tumor resistant		127, 128
Per	*Per1*-null	Unstable period length	Immune deficiency	IFNγ, Perforin, Granzyme B	155, 192
	Per2-null	Shortened period and arrhythmic in constant darkness	Cancer prone, premature aging, immune deficiency, and deregulated drug response	c-MYC, Cyclin D1, Cyclin A, GADD45α, IFNγ, IL-1β, LY49C, MDM2, and p53	29, 124, 161, 193
	Per2 mutant (*Per2*^{Brdm1})	Shortened period and arrhythmic in constant darkness	Cancer prone, vascular senescence, premature aging, immune and neuroendocrine deficiency, and metabolic syndrome	AKT, β-catenin, CAR, c-MYC, Cyclin D1, Cyclin A, GADD45α, IFN-γ, IL-1, IL-6, IL-10, MDM2, p53, and NKG2D	29, 123, 124, 149–153, 155, 183, 193–196

	Per2 mutant (*Per2*^{S662G})	Shortened period and phase-advance rhythm	Cancer prone	Cyclin D and p21$^{WAF1/CPI1}$	146, 148
	Per1, *Per2* double knockout	Arrhythmic	Premature aging, cancer prone, and metabolic syndrome	AP1, ATM, c-MYC, Cyclin D1, Cyclin E, MDM2, and p53	29, 149, 167
Rev-erb	*Rev-Erbα*-null	Shortened period	Metabolic syndrome and immune deficiency	CCL2, CXCL6, CXCL11, IL-6, IL-19, and IL-10	9, 184
	Rev-Erbα; *Rev-erbβ* double knockout	Arrhythmic	Metabolic syndrome	POR, PPARα, and SCO2	9
Rorα	*Rorα*-null	Shortened period	Metabolic syndrome and cerebellar ataxia	TUBα8 and RASD1	10

clock suppresses neoplastic growth by controlling cell proliferation, metabolism, senescence, and DNA damage response.

4.1.1 Control of cell proliferation by the molecular clock

Both cell cycle and molecular clocks are operated by interlocked feedback loops of genes that display periodic and sequential phases of activation and repression at the transcriptional, posttranscriptional, and posttranslational levels.[200] However, although both are considered as "intracellular clocks," a cell cycle clock is fundamentally different from the molecular clock. First, unlike the molecular clock, the cell cycle clock does not free-run in normal somatic cells. Therefore, the activities of the peripheral clock and cell cycle clock can be separated in peripheral tissues in the absence of proper extracellular mitogenic signals.[173,201–203] Second, the period of a cell cycle clock is not always fixed as 24 h throughout one's life span. It can vary from only a few hours in early embryogenesis to 24 h in rapidly renewing somatic tissues in adult life or indefinitely long due to cellular senescence.[204–207] Since tumor cells often display the properties of dedifferentiation and rapid self-renewal with a cell cycle period shorter than 24 h, loss of circadian coupling of cell cycle progression in adult life may play a key role in the initiation of neoplastic growth *in vivo*.

Circadian variation of mitotic activity in normal human tissues has been described since 1938.[208] The uncoupling of mitotic rhythm between normal host tissue and metastasizing cancer was first reported in 1940.[209] It is now well known that cell proliferation in all the rapidly renewing mammalian tissues studied follows a diurnal oscillating rhythm under normal physiological conditions but is altered in tumors.[75,83,210–216] The coupling of cell proliferation rhythm between host and tumor has been observed in slow-growing tumors that show considerably higher levels of DNA synthesis and mitotic indices than host tissues throughout a 24-h period.[216–218] On the other hand, an ultradian rhythm less than 24 h in cell proliferation is often found in metastasizing cancers.[218–221]

Genome-wide studies have identified a number of genes controlling the key steps of initiation, progression, and checkpoint functions of the cell cycle clock as clock-controlled genes.[7,11–14] These genes encode proto-oncogenes, tumor suppressors, caspases, cyclins, transcription factors, and ubiquitin-associated factors essential for regulating cell proliferation and death.[16–23] Clock-controlled cell cycle regulators are also expressed in all circadian gene-mutant mouse models studied except that they display significant changes in expression profiles and amplitudes over a 24-h period, which is coupled with loss of cell cycle control in adult tissues of mutant mice.[22,124,126,149,166,167] Thus, cell cycle clock can function independent of the molecular clock. However, the control of the expression of key cell cycle genes by an intact molecular clock is indispensable for coupling the rate of cell proliferation to diurnal changes in mammalian physiology *in vivo*.

Both positive and negative loops of the molecular clock are involved in circadian control of cell cycle gene expression in peripheral tissues. The best-studied clock-controlled cell cycle regulators include oncogenes *c-Myc*, *Mdm2*, and *β-catenin*, *Cyclins D1*, *B*, and *A*, and tumor suppressors *p53*, *Wee1*, and *p21*$^{WAF1/CIP1}$.[82,124,126,149,166,179] *c-Myc* is an immediate early responsive gene to diverse cell proliferation stimuli. It plays a key role in G1 cell cycle initiation as well as cell growth and death (Fig. 9.1).[222] The expression of *c-Myc* is tightly controlled in somatic cells in response to diverse stimuli at the transcriptional level via multiple *cis*-acting sequences in its proximal promoter region.[223] Deregulation of *c-Myc* in cooperation with loss of function in p53 promotes neoplastic transformation, and tumor initiation, maintenance, and metastasis.[224] In addition to clock-controlled chromatin remodeling, BMAL1/CLOCK and BMAL1/NPAS2 directly regulate *c-Myc* transcription via the two E-boxes in the *P1* promoter of *c-Myc*.[124,149,225] Disruption of circadian rhythm leads to deregulation of *c-Myc* which is coupled with increased neoplastic growth in mice, suggesting that the control of G1 cell cycle initiation is one of the key mechanisms for circadian control of tumor suppression.[29,124,149] Many key cell cycle regulators, such as *Cdk4*, *Itga6*, *Wnt3*, *LHx2*, *Tcf4*, *Sox 9*, *Smad7*, and *Wee1* are also directly regulated at the transcriptional level by BMAL1/CLOCK heterodimer via E-box-mediated interaction in the gene promoters.[126,167,225] p53 and MDM2 are controlled indirectly by the *Per* genes via the tumor suppressor ataxia-telangiectasia mutated (ATM) at the posttranscriptional level.[29,82] *Bmal1* has also been reported to regulate p53–*p21*$^{WAF1/CIP1}$ signaling, although the molecular mechanism of this regulation is still not clear.[69,92,179] The rhythmic expression of cell cycle genes and tumor suppressor p53 is synchronized with the oscillation patterns of the core circadian genes in normal somatic tissues in both humans and mouse models.[29,124,126,226–228]

The core circadian regulators also regulate the activity of β-Catenin via direct control of intracellular signaling pathways. Constitutive activation of the core circadian regulator CKIε mimics the effect of WNT signaling, resulting in cytoplasmic accumulation of β-Catenin and its subsequent nuclear localization.[229,230] In the nucleus, β-Catenin interacts with transcription factors of the T-cell-specific transcription factor/lymphoid enhancer factor-1 (TCF/LEF) family to regulate transcription and promote tumorigenesis.[231,232] Genes activated by β-Catenin/TCF/LEF include members of the AP1 transcription family, *c-Myc* and *Cyclin D1*.[233–238] Interestingly, the molecular clock itself also responds to β-Catenin nuclear localization to regulate the expression of genes via BMAL1/CLOCK-mediated transcriptional regulation.[167] Aberrant activation of β-Catenin also disrupts the molecular clock to promote neoplastic transformation by inducing β-transducin repeat-containing protein-mediated PER2 degradation.[151]

In the absence of WNT signaling, β-Catenin is destabilized by glycogen synthase kinase-3β (GSK3β), a functional homologue of the core circadian gene *Shaggy* in the fruit fly, which functions in regulating the period length of circadian cycles by indirectly controlling PER stability and nuclear entry.[239,240] Recent studies have revealed that deregulation of GSK3β promotes tumor cell survival, proliferation, invasion, and resistance to chemo- and radiation therapy in humans by inhibiting p53 and RB tumor suppressors, inducing intracellular NF-κB signaling, *Cyclin D1* overexpression, and local chronic inflammation.[241,242] The activity of GSK3β exhibits robust circadian rhythm in both SCN and peripheral tissues, suggesting that GSK3β may also indirectly target PER2 in the mammalian molecular clock.[243] Together, the evidence discussed above suggests that the molecular clock couples cell proliferation with mammalian daily physiology by rhythmically pacing the key cell proliferation and tumor suppression pathways at the cellular level. Since the molecular clockworks operate as interlocked feedback loops, disruption of either a positive or a negative loop would disrupt the stability of the molecular clock leading to loss of control in the circadian homeostasis of cell cycle progression.[29,135,156,244,245] Indeed, deregulation of both positive and negative loops of molecular clock frequently occurs in the same type of tumor in humans.[65,75,76,79,84,92,94–96,104,246,247]

4.1.2 Control of DNA damage response by the molecular clock

An average human body contains about 10^{14} cells, each receiving tens of thousands of DNA lesions every day. If not repaired, these lesions induce harmful mutations that could affect the survival of cells or even an organism.[248] DNA damage activates genes encoding key enzymes in DNA repair machinery and cell cycle checkpoints that pause cell cycle progression to give the cell time to repair damaged DNA before continuing to divide.[249] When DNA damage exceeds the capacity of the cell to repair, efficient elimination of damaged cells by apoptosis or necrosis plays a key role in suppressing tumor development.[250]

DNA damage response in mammals is controlled mainly by two master kinases, ATM and ataxia-telangiectasia and Rad3-related (ATR).[251,252] ATM/ATR targets the protein kinases CHK1 and CHK2, which together with ATM and ATR suppress cyclin-dependent kinase (CDK) activity and activate CKIs such as $p21^{WAF1/CIP1}$ in a p53-dependent manner.[174,224] Inhibition of CDKs and activation of CKIs lead to arrest of cell cycle progression at the G1/S, intra-S, or G2/M checkpoints to allow DNA damage repair.[253,254] ATM/ATR signaling also enhances DNA repair by transcriptionally and post-transcriptionally activating DNA repair genes and by recruiting repair factors to the sites of damage. Activation of p53 by ATM/ATR signaling in response to genomic DNA damage often leads to p53-mediated apoptosis (Fig. 9.1).[255]

A large number of key players in DNA replication, recombination, and repair have been identified as clock-controlled genes in mice.[7,11–14] Some of these genes, such as Xeroderma pigmentosum A (XPA) that plays an essential role in nucleotide excision repair, are deregulated in $Cry1^{-/-};Cry2^{-/-}$ mice, which correlates with a dampened circadian rhythm in nucleotide excision repair after UVB irradiation in $Cry1^{-/-};Cry2^{-/-}$ epidermis.[256,257] The evidence of direct involvement of *Cryptochromes* in DNA damage repair in mammals is still missing, although *Cryptochromes* are structurally related to evolutionarily conserved DNA photolyases that catalyze light-dependent DNA repair in plants.[258] However, mammalian CLOCK may play a direct role in DNA damage repair because it directly locates to the sites of γ-radiation-induced DNA double-strand breaks.[158] PER1 was found directly interacting with ATM and CHK2 in response to γ-radiation-induced DNA damage. Overexpression of PER1 in human cancer cells sensitizes cells to radiation-induced apoptosis by activation of *Myc*-mediated pro-apoptotic pathways.[82] The human CRY2 has been reported to interact with ATR and CHK1 to regulate intra-S checkpoint function in UV-induced DNA damage response via Timeless (TIM), a natural partner of PER in *Drosophila* and is necessary for maintaining the robustness of the mammalian central clock.[175,259] The role for BMAL1 in DNA damage response is shown by a recent study in which knock-down *Bmal1* expression abolishes γ-radiation-induced p53 activation as well as p53-dependent $p21^{WAF1/CIP1}$ induction in human colorectal carcinoma cells.[69] All core circadian genes studied are activated by exogenous DNA damage agents such as γ-radiation in peripheral tissues following a time-dependent profile over a 24-h period in mice. Wild-type mouse thymocytes display a time-dependent response to γ-radiation-induced apoptosis *in vivo*, while loss of function in *Per2* leads to resistance to radiation-induced apoptosis in thymocytes throughout a 24-h period and an increased risk of radiation-induced lymphoma in $Per2^{m/m}$ and $Per2^{-/-}$ mice.[29,124,260,261] Thus, the potentiation of the molecular clock to respond to DNA damage agents varies at different times during a day and plays a key role in determining the outcomes of DNA damage response.

4.1.3 Control of cell metabolism by the molecular clock

Cancer is classically considered as a disease originated from genetic and epigenetic abnormalities that lead to uncontrolled cell growth and division, and formation of metastasizing tumors. The prevalence of metabolic syndromes worldwide and its coherence to cancer has led to a renewed interest in the Warburg effect, which describes an essential role of deregulation of cell metabolism in cancer initiation. In 1956, Otto Warburg observed that normal quiescent somatic cells metabolize glucose to CO_2 and H_2O via a low rate of glycolysis followed by oxidation of pyruvate in TCA cycle in mitochondria. However, cancer cells predominantly use glucose to produce energy through a high rate of glycolysis followed by lactic acid fermentation in the cytosol even in the presence of abundance oxygen. Warburg predicted that this metabolic reprogram is a fundamental cause of cancer.[262] Studies in the past decade have revealed that cancer cells display an array of metabolic abnormalities and that both oncoproteins and tumor suppressors can influence the switch between aerobic glycolysis and the use of TCA cycle to generate ATP.[263–268] The predominant use of the aerobic glycolysis pathway in cancer cells leads to not only a high level of intracellular ROS, a major source of endogenous DNA damage agents promoting cancer and aging, but also the accumulation of intermediate products including purines, pyrimidines, nonessential amino acids, and free fatty acids that can be used for anabolic synthesis, cell growth, and division.[266,269–271]

The mammalian circadian clock is a master regulator of metabolic homeostasis both at the organismal and peripheral tissue levels.[6,272–275] In peripheral tissues, the molecular clock regulates nutrient uptake, metabolism, energy storage, mitochondria biosynthesis, and intracellular redox levels by targeting key metabolic genes including those controlling the Warburg effect, such as glucose-6-phosphatase, pyruvate kinase, and glucose transporter 2 (GLUT).[17,20,198,276,277] It also responds to food cues and nutrient sensors to shift metabolic balance independent of SCN clock function.[12,278–283] The peripheral clock also indirectly controls Warburg switch by regulating the expression of oncoproteins and tumor suppressors. For example, p53 inhibits aerobic glycolysis and decreases intracellular ROS levels by suppressing GLUT1–4 and fructose-2,6-bisphosphate, a critical substrate of aerobic glycolysis, via the tumor suppressor p53-induced glycolysis and apoptosis regulator,[284,285] and promotes the use of the TCA cycle by inhibiting glycolic enzyme phosphoglycerol mutatase.[286] Whereas c-MYC stimulates biosynthesis to support cell growth and proliferation via upregulation of lipogenetic, glycolytic, and mitochondrial genes,[287–290] and increases glutamine uptake to compensate for the progressive loss of glucose as a mitochondrial substrate in cancer cells due to the Warburg effect,[291] Myc also upregulates mitochondrial biosynthesis by regulating genes including nuclear respiratory factor 1 (NRF1) and the transcription factor A, mitochondrial as well as *cytochrome C*, to stimulate the production of ROS.[292–296] Although elevated ROS leads to *Myc*-mediated apoptosis in normal somatic cells,[297] loss of function or deregulation of p53 in combination with *Myc* oncogenic activation results in the metabolic switch to support cell proliferation, apoptotic resistance, and neoplastic transformation.[271,298–302]

The molecular clock not only controls metabolic homeostasis by regulating nuclear receptors and their coactivators and suppressors as well as chromatin remodeling in metabolically active peripheral tissues[6,272–275,303] but is also regulated by nuclear receptors REV-ERBα/β, RORα, and PPARγ coactivator PGC-1α, which play an active role in energy metabolism, adipogenesis, and lipid storage in peripheral tissues, and by nutrient sensors such as AMPK.[8–10,282,304] Therefore, disruption of the molecular clockworks would inevitably shift the homeostasis of metabolism and energy balance in peripheral tissues to provide an intracellular environment that favors tumor initiation and progression.

4.1.4 Control of cell senescence by the molecular clock

Aging is frequently associated with disruption of circadian rhythmicity in humans and animal models.[305–307] Premature aging is commonly observed among circadian gene-mutant mouse models. Among them, *Bmal1*$^{-/-}$ mice display the most aggressive aging phenotypes leading to a significantly reduced life span.[125] Other circadian gene-mutant mouse models including mice carrying a mutated *Clock* (*Clock*$^{m/m}$) or lacking *Per* or *Cry* genes (*Cry1*$^{-/-}$;*Cry2*$^{-/-}$, *Per1*$^{-/-}$;*Per2*$^{-/-}$, *Per2*$^{m/m}$, or *Per2*$^{-/-}$) also display premature aging phenotypes, which become more evident in response to DNA damage agents such as γ-radiation.[29,124,128,129] These premature aging phenotypes of circadian gene-mutant mouse models are closely related to deregulation of cell proliferation and DNA damage response *in vivo*.

Aging is a universal process for all multicell organisms on earth, which is measured chronologically by biological changes over time and is accompanied by a dramatic increase in the risk of various diseases at the mid-point of the life span.[308] Although aging is marked by progressive degeneration of tissue and cell function *in vivo*, it is also coupled with an increased risk of neoplastic growth in renewable tissues in mammals, leading to the conclusion that aging is a primary risk factor of cancer.[309,310] The mechanisms linking aging to cancer include immunosenescence and age-related endocrine dysfunction at the organismal level as well as telomere shortening, reproductive cell cycle, and accumulation of DNA damage over a lifetime.[311–316] These adverse age-related pathological changes have led to the conclusions that if humans live long enough, they would all eventually develop cancer.[317,318]

Throughout the life span, mammals need continuous cell proliferation to support their daily physiology. However, the inherent limitations in DNA repair mechanisms inevitably lead to accumulation of errors in genomic DNA, which often results in replicative or cellular senescence, a direct cause of aging.[319–323] Senescence refers to a state of permanent and irreversible withdrawal from the cell cycle resulting from accumulation of cellular damages including DNA lesions, oncogenic activation, and/or overexpression of tumor suppressors.[157] Since cancer cells do not have a limited replicative life span, cellular senescence is often used to enforce the idea that it suppresses cancer development.[324] However, mounting evidence suggests that cellular senescence also promotes cancer initiation and progression. Senescent cells are still metabolically active and often show changes in chromatin organization and gene expression, leading to secretion of proinflammatory cytokines, proteases, and growth factors. The paracrine activities of senescent cells have been found to stimulate proliferation and migration of neighboring cells and promote the development of metastatic tumors.[325–328] The activation of *Myc*, Ras/MAPK oncogenic signaling and/or p53/$p21^{WAF1/CIP1}$, and pRB/$p16^{INK4a}$ tumor suppressing pathways are established molecular mechanisms for cellular senescence.[329–332] Under certain conditions, suppression or even a subtle change in the expression of these tumor suppressors could lead to senescent cells to rapidly regain the ability to proliferate.[333]

Increased oxidative stress in circadian gene-mutant mice may lead to a higher level of accumulation of DNA damage that induces early cellular senescence to promote aging.[125,306,334] The modulators of the molecular clock, such as NAD-dependent deacetylase sirtuin-1 (SIRT1), may play a role in bridging the aging and cancer-prone phenotypes found in circadian gene-mutant mice.[6] SIRT1 is a class III histone deacetylase that promotes cell survival by inhibiting apoptosis and cellular senescence in mammals.[335] SIRT1 expression follows a robust circadian expression *in vivo*. It directly interacts with CLOCK and deacetylates BMAL1 to regulate the activity of the molecular clock,[12,13] and also plays a role in regulating cell proliferation and apoptosis by deacetylating key tumor suppressors and oncoproteins including p53, β-Catenin, and DNA repair protein KU70.[336–338] Both *Bmal1* and *Per2* are found to inhibit cellular senescence *in vivo* possibly by distinct mechanisms. *Per2* mutation leads to AKT-dependent vascular senescence that impairs endothelial progenitor cell function, while loss of *Bmal1* promotes senescence *in vivo* via a p53-independent mechanism.[23,194] The increased cellular senescence found in circadian gene-mutant mice may be explained by their inherent high risk of deregulation of oncogenic and tumor suppression pathways, high intracellular levels of redox and its associated accumulation of genomic DNA damage, and abnormal internal physiological environments that promote oncogenic extracellular signaling. These abnormalities together with a lack of proper control of gene expression could increase paracrine activities and loss of senescence surveillance, which leads not only to increased local tissue damage and inflammation that stimulate cell regeneration but also to the possibility of reentering the cell cycle of senescent cells. Since cancer is a clonogenic disease *in vivo*, one or a few premalignant cells that successfully escape senescence surveillance would be sufficient enough to promote cancer development *in vivo*.[224,339]

4.2. Tumor suppression *in vivo* is a clock-controlled physiological function

Although an intact molecular clock can provide self-sustained circadian oscillations in peripheral tissues, peripheral organs rely on daily entrainment signals from the central pacemaker to maintain the synchrony of the internal physiology. Disruption of central clock function leads to phase desynchrony among peripheral tissues.[201] Such desynchronization of the internal circadian homeostasis is closely related to increased risk of tumor development and accelerates tumor progression in both humans and experimental animal models.[34–49,54–59,117–122] The key entrainment signals from the central clock include extracellular signaling controlled by the ANS and NES.[24] Loss of homeostasis of ANS and NES disrupts circadian homeostasis of cancer immune surveillance and energy balance as well as G1 cell cycle progression in renewable peripheral tissues, which synergistically promote tumor development.

The mechanisms of SCN control of peripheral tissues have been discussed in detail in several recent reviews.[24,30–33,275,340] Briefly, via direct and indirect targeting, the SCN clock controls brain centers, especially those in the hypothalamus including the paraventricular nucleus (PVN), arcuate nucleus (ARC), dorsomedial hypothalamus (DMH), lateral hypothalamus (LHA), and endocrine neurons producing corticotropin-releasing hormone

(CRH), thyrotropin-releasing hormone (TRH), and gonadotropin-releasing hormone (GnRH). The CRH, TRH, and GnRH control the activity of NES via the hypothalamic–pituitary–adrenal (HPA) and hypothalamic–pituitary–gonadal (HPG) axes, while ARC, DMH, LHA, and PVN control energy expenditure and food intake in response to both central and peripheral signals. The autonomic paraventricular neurons (aPVN) directly project to the preganglionic parasympathetic and sympathetic neurons in the brainstem nuclei, dorsal motor nucleus of the vagus (DMV), and intermediolateral cell columns (IML) of the spinal cord to control parasympathetic and sympathetic nervous systems. The SCN control of aPVN neurons leads to a robust circadian oscillation in the function of the ANS *in vivo*. The example of SCN control of NES is demonstrated by the rhythmic activity of the HPA axis. The SCN pacemaker indirectly generates circadian oscillation of adrenocorticotropic hormone-controlled corticosterone production from the adrenal gland into the blood via control of the hypothalamic CRH endocrine neurons.[340] *In vivo*, the ANS innervates all peripheral tissues except skeletal muscle through G-protein-coupled transmembrane receptor (GPCR) in a tissue and/or cell type-specific manner.[341–343] Hormones produced by the pineal gland, and the HPA and HPG axes, such as melatonin, glucocorticoids, and estrogen, target a wide range of peripheral tissues, especially the immune, metabolically active, and reproductive organs.[344–351] The rhythmic intracellular signaling generated by neurotransmitters and hormones plays an essential role in maintaining homeostasis of tissue microenvironment (Fig. 9.2).

4.2.1 Control of G1 cell cycle progression in peripheral tissues by the central pacemaker

In the central pacemaker, light stimuli activate a cascade of intracellular signal transduction pathways in the SCN neurons to phase-shift the center pacemaker including the MAPK, ERK, protein kinase C alpha (PKCα), calcium/calmodulin-dependent protein kinases II (CaM kinases II), c-Jun N-terminal kinase (JNK), c-AMP–PKA, and nitric oxide/c-GMP pathways that differentially regulate the expression of the immediate early genes c-*Fos* and *JunB* and the core circadian genes *Per1* and *Per2* in a time-dependent manner.[352–361] The peripheral clocks do not directly respond to light stimuli, but are instead synchronized by cyclic changes in the levels of neurotransmitters, growth factors, cytokines, and hormones in the tissue microenvironment.[340,362,363] Despite the sensitivity of the peripheral clock to non-SCN cues such as food cues, metabolites, or fluctuation in body

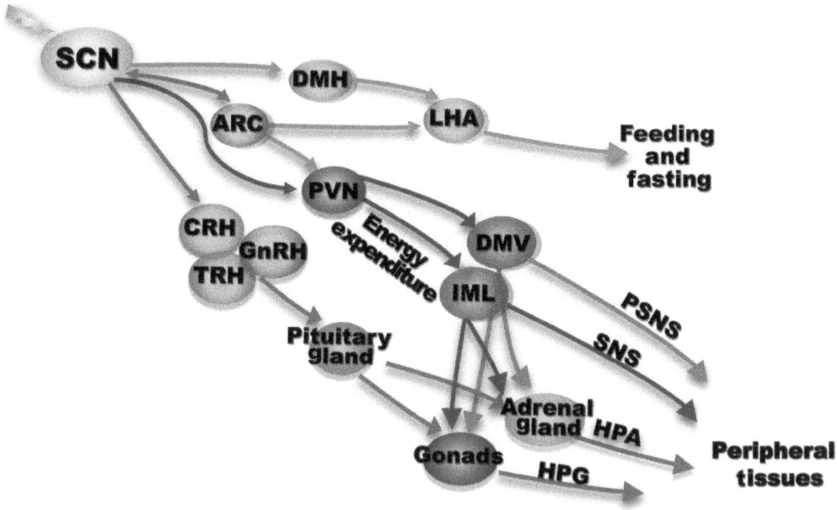

Figure 9.2 Peripheral control by the SCN pacemaker. The SCN clock targets a variety of brain centers within the hypothalamus to control homeostasis of endogenous physiology. It controls nutrient intake and energy expenditure by targeting the brain energy homeostasis center arcuate nucleus (ARC) and catabolic center paraventricular nucleus (PVN) directly, and feeding and satiety center LHA indirectly via ARC and dorsomedial hypothalamus (DMH). It also controls the NES by directly targeting the corticotropin-releasing hormone (CRH), thyrotropin-releasing hormone (TRH), and gonadotropin-releasing hormone (GnRH) neurons that control the adrenal and gonadal glands via the pituitary glands. The SCN pacemaker directly targets the autonomic paraventricular (aPVN) neurons that project to the preganglionic parasympathetic and sympathetic neurons in the dorsal motor nucleus of the vagus (DMV) and intermediolateral cell columns (IML) of the spinal cord to control parasympathetic and sympathetic nervous systems (PSNS and SNS). Both PSNS and SNS also cross-talk with the HPA and HPG axes by directly innervating the adrenal and gonadal glands. The NES and ANS innervate all peripheral tissues *in vivo* to generate circadian rhythm of internal physiology by controlling extracellular signaling and peripheral clock activity.[24,30–33,275,340] (See Color Insert.)

temperature, the central pacemaker plays a dominant role in peripheral clock control to maintain the integrity of the internal physiology.[24]

The best understood intracellular signaling pathways for entraining the peripheral clock by ANS and NES include glucocorticoid and beta-adrenergic receptor (ADRβ)-mediated activation of the molecular clock. The interaction of glucocorticoid with glucocorticoid receptor (GR) in the cytoplasm stimulates GR nuclear localization and activation of GR-mediated transcription via glucocorticoid-responsive elements (GREs) in gene promoters.[364,365] GREs are found in the regulatory regions of *Per1*, *Per2*, *Bmal1*, and *Cry1* genes,[366–368] and both CRY1 and 2 directly interact with GR in a ligand-dependent fashion to modulate the transcriptional activity of GR.[188] Administration of dexamethasone, a glucocorticoid analog, phase-shifts the molecular clock in cultured rat-1 fibroblasts as well as in mouse livers. Although unable to phase-shift the SCN clock, dexamethasone can resynchronize about 60% of the circadian transcriptome in the livers of SCN-lesioned mice via at least in part directly phase-shifting the molecular clock.[31,366] Adrenalectomy results in deregulation of the core clock genes, and desynchronization and dampening of the molecular clock in multiple peripheral tissues.[369]

The SNS directly innervates all peripheral tissues by releasing norepinephrine (NE) to target adrenergic receptors (ADRs) on cell membranes. It also controls the production and secretion of epinephrine (EPI) from the adrenal medulla, which then targets all cells expressing ADRs in the body via blood circulation.[370] ADRβ2 is the best-studied ADR that responds to ligand binding by activating c-AMP response element-binding protein (CREB), which then interacts with the c-AMP response element (CRE) in promoters to regulate gene transcription. Both *Per1* and *Per2* contain CREs in the promoters and are among the immediate early genes activated by CREB in cultured primary osteoblasts, NIH3T3 cells, and mouse liver slides in response to administration of isoproterenol, a synthetic agonist of ADRβ2, EPI, or NE *in vitro* and in mouse livers after intraperitoneal injection of adrenalines *in vivo*.[29,149,371] Loss of function in *Per1* and *Per2* or sympathectomy abolishes SNS-induced peripheral clock activation in affected tissues in rodents.[29,149,372–375]

The activation of ADRβ is followed by β-Arrestin-mediated receptor desensitization. As multifunctional scaffold proteins, the interaction of β-Arrestins with ADRβ leads to activation of other signal transduction pathways including the MAPK, Ras/ERK, JNK, nuclear factor kappa-light-chain-enhancer of activated B cells (NF-κB), β-Catenin, CaM kinases II, phosphatidylinositide 3-kinases/protein kinase B/mammalian target of rapamycin (PI3K/AKT/mTOR), Janus kinase 3/signal transducer and activator of transcription (JAK3/STAT), insulin-like growth factor 1 (IGF-1), and MDM2-p53 pathways.[134,376–380] Glucocorticoid signaling also cross-talks with pathways controlled by NF-κB, β-Catenin, PI3K, and epidermal growth factor receptor.[381–385] The pathways stimulated by glucocorticoid and ADRβ signaling modulate not only the molecular clock at transcriptional and posttranslational levels but also cell proliferation, apoptosis, and metabolism in a tissue and cell type-specific manner, leading to coupling of peripheral clock activity with tissue-specific functions *in vivo* (Fig. 9.3).[263,269,271,333,395–398]

The initiation of G1 cell cycle progression *in vivo* is strictly controlled by extracellular signals that activate proto-oncogenes *c-Myc* and/or *E2f* via intracellular signaling including c-AMP–PKA, MAPK, Ras/ERK, JNK, β-Catenin, and/or PI3K pathways in a cell type-specific manner.[173] Loss of homeostasis in HPA axis and ANS signaling is frequently associated with increased risk of cancer.[399,400] The role of the central clock in controlling cell cycle progression in peripheral tissues can be explained by a model obtained from studying circadian control of *Myc* transcriptional activation. In cultured primary osteoblasts, isoproterenol-mediated ADRβ signaling stimulates cell cycle progression via a coupled activation of the peripheral clock, the cell cycle clock, and p53 controlled by immediate early genes including *Per1* and *Per2*, *Ap1* and *Myc*, and ATM, respectively. Activation of *Myc* leads to cell cycle progression, while activation of peripheral clock prevents *Myc* overexpression and also stimulates ATM-mediated p53 induction. Loss of function in the *Per* genes, or elevated concentration of isoproterenol, prevents the activation of peripheral clock and ATM but not AP1-controlled *Myc* activation, leading to suppression of p53 induction and uncontrolled osteoblast proliferation due to *Myc* overexpression. The PER proteins may be directly involved in SNS-stimulated ATM activation since PER1 has been found to interact with ATM in response to γ-radiation,[82] whereas CRY proteins may prevent uncontrolled c-AMP-PKA-CREB-AP1-*c-Myc* signaling in response to ADRβ activation by directly inhibiting the G_s alpha subunit (Gsα) essential for activating adenylate cyclase.[189] *In vivo*, *Per*- and *Cry*-mutant mice both show uncontrolled SNS signaling and display a phenotype of neoplastic growth of osteoblasts in bone.[29,149,150,189,190] Disruption of homeostasis of SNS signaling by chronic jet lag is associated with the disruption of peripheral clock function, suppression of p53 and *Myc* oncogenic activation, which is coupled with increased tumor development in all mouse models studied.[29] Together with the previous reports on increased tumor development and progression in rodent models treated with constant light exposure, pinealectomy, chronic jet lag, or SCN lesion, these findings provide an explanation on how circadian dysfunction induces tumor development in the absence of gene mutations,[117–122] which is especially relevant for developing novel strategies for cancer prevention in the modern world in which frequent disruption of endogenous circadian homeostasis due to

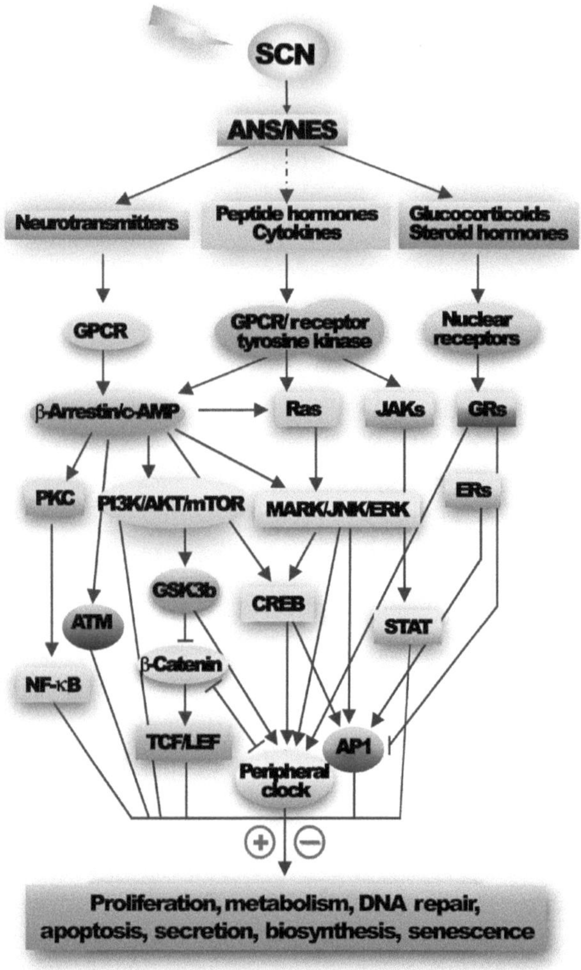

Figure 9.3 Circadian control of intracellular signaling. The central pacemaker controlled autonomic nervous and neuroendocrine systems (ANS and NES) rhythmically signal to all of their target tissues. The resulting circadian rhythm in peripheral tissue function also generates local and/or circulating signaling molecules that rhythmically act on their targets. Together, these extracellular signals including neurotransmitters, steroid hormones, peptide hormones, chemokines, growth factors, and cytokines activate intracellular signaling mediated by G-protein-coupled receptors (GPCRs), tyrosine kinase receptors, integrins (not shown), and nuclear receptors in a tissue and cell type-specific manner. These same intracellular signaling pathways also activate the peripheral clock. The coordinated activities of the central and peripheral clocks orchestrate the complicated extracellular and intracellular signaling to maintain tissue homeostasis by controlling a network of gene expression. Disruption of the central clock-controlled extracellular signaling or mutations in core circadian genes both abolish peripheral clock activity leading to loss of circadian homeostasis in peripheral tissues. The

lifestyle change is associated with a dramatic increase in the risk of sporadic cancers (Fig. 9.4).[401]

4.2.2 Circadian control of cancer immune surveillance

The current concept of cancer immuno-editing is based on the evidence of sequential steps of elimination of transformed cells *in vivo* by the immune system.[402] When transformed cells accumulate above a threshold, they are recognized by lymphocytes including Nature Killing T, Nature Killing (NK), and gamma delta T cells that are stimulated by transformed cells to produce interferon γ (IFN-γ). This triggers a cascade of innate immunity including the induction of chemokines CXCL9, 10, and 11 to block neovascularization in the tumor and the recruitment of NK cells, dendritic cells, macrophages, and other immune effector cells to the tumor site. The antiproliferative effects of IFN-γ on transformed cells and the cytocidal activities of macrophages and NK cells result in the death of tumor cells, which are ingested by dendritic cells and trafficked to the draining lymph node, where the tumor-specific $CD4^+$ and $CD8^+$ T lymphocytes are developed. These tumor-specific T lymphocytes are then directed to the tumor site along a chemokine gradient, where they act together with NK cells and activated macrophages to recognize and destroy tumor cells.[402,403] Mice deficient in cancer immuno-editing display a significantly higher risk of spontaneous tumor development in the immune, digestive, respiration, and reproductive organs.[404–408] Cancer immuno-editing is usually abolished by cancer-induced immunosuppression in human cancer patients.[402,403]

The mechanisms of cancer immunosuppression include deregulation or loss of expression of cancer cell surface markers leading to the lack of recognition of transformed cells by cytotoxic T lymphocytes, resistance to cell death induced by cytotoxic T lymphocytes due to deregulation of apoptotic factors and death receptors, production of immunosuppressive factors

representative intracellular signaling pathways directly or indirectly controlled by the central clock shown in the figure include the c-AMP/PKA/CREB/AP1, Ras/MARK/JNK/ERK, and PI3K/AKT/β-catenin/TCF/LEF pathways essential for *c-Myc* activation and cell cycle progression,[386–388] the PI3K/AKT/mTOR signaling controlling biosynthesis and drug resistance,[389,390] the GPCR/ATM signaling for p53 activation,[29] the GPCR/PKC/NF-κB pathway that regulates stress and immune response,[391] the JAK/STAT pathway controlling apoptotic response,[392] and the GR and ERα signaling pathways cross-talking with the AP1 signaling.[393,394] These signaling pathways also control the expression and function of circadian genes leading to a coupled activation of the molecular clock with tissue-specific function *in vivo* including cell proliferation, metabolism, apoptosis, DNA repair, biosynthesis, secretion, and senescence.[395–398] (See Color Insert.)

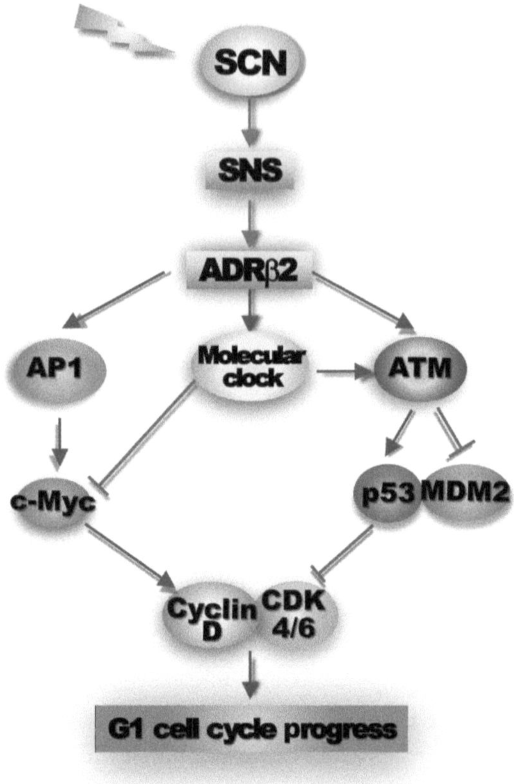

Figure 9.4 Control of G1 cell cycle progression by the peripheral clock and the SNS. The activation of the β-adrenergic receptor 2 (ADRβ2) by SNS signaling leads a coupled induction of *Ap1* and *Period* genes via CREB-mediated transcriptional regulation, which in turn activates AP1-controlled *Myc* induction and *Myc*-dependent G1 cell cycle progression as well as peripheral clock that prevents *Myc* overexpression via BMAL1/CLOCK-mediated transcriptional regulation. The activation of ADRβ2 intracellular signaling and the peripheral clock also synergistically activate ATM, which induces p53 by blocking p53–MDM2 interaction to provide an additional mechanism for preventing MYC oncogenic activation. Disruption of the central clock-SNS-peripheral clock axis in mice by chronic jet lag or ablation of *Per* genes suppresses peripheral clock activation in response to ADRβ2 activation and abolishes ATM-mediated p53 induction but has no effect on *Ap1-Myc* signaling. Together, these events lead to uncontrolled G1 cell cycle progression and neoplastic growth of osteoblasts both *in vitro* and *in vivo*.[29,149] (See Color Insert.)

including free radicals, cytokines, and growth factors that negatively affect cancer immuno-editing by impeding the proliferation and/or function of $CD4^+$ and $CD8^+$ T cells. The immunosuppressive microenvironment in tumors also stimulates the generation and/or promotion of immunosuppressive cells such as type 2 macrophages, myeloid-derived suppressor cells, immature dendritic cells, and regulatory T lymphocytes.[402,403,409,410]

Both primary and secondary lymphoid organs including thymus, spleen, and lymph nodes are intensively innervated by ANS and NES.[411–416] Under normal physiological conditions, the production of cytokines and cytolytic factors, proliferation of leukocytes, activities of NK cells, and redistribution of T and B lymphocytes, dendritic cells, leukocytes, and macrophages to lymphoid organs, all follow a robust circadian rhythm *in vivo*.[195,417–424] Disruption of circadian homeostasis is closely related to immune suppression.[425] Ablation or deregulation of the core circadian genes *Per1*, *Per2*, *Bmal1*, *Rev-erbα*, or *Clock* in mice induces an array of abnormalities in the immune system, including the deregulation of proinflammatory cytokines, cytotoxic receptors, immunoregulatory genes, and NK and mast cell activities, and inhibition of B lymphocyte differentiation.[164,180,183,184,192,193,196,418,426] Mice lacking both *Cry1* and *Cry2* display constitutive elevation of proinflammatory cytokines and are prone to chronic inflammation, a common underlying mechanism for cancer.[106] Importantly, the immune function is not controlled at the cell autonomous level *in vivo*. Consecutive phase-advance shifts of environmental light cues disrupt the molecular clock and circadian homeostasis of NK cell function in rats.[427] Ablation of sympathetic innervation abolishes the circadian oscillation of cytokines and cytolytic factors in splenocytes and NK cells,[374] hematopoietic stem cell trafficking, and the expression of chemokine CXCL12.[375] Adoptive transfer of *Bmal1*$^{-/-}$ bone marrow deficient in B lymphocyte differentiation to lethally irradiated BALB/c*Rag2*$^{-/-}$ recipients that are unable to generate mature B lymphocytes due to lack of V(D)J recombination activating gene 2 (*Rag2*) resulted in normal T and B lymphocyte differentiation from *Bmal1*$^{-/-}$ bone marrow. However, reciprocal transfer of BALB/c *Rag2*$^{-/-}$ bone marrow to lethally irradiated *Bmal1*$^{-/-}$ mice did not lead to normal B lymphocyte development, suggesting that the SCN control of tissue microenvironment in bone marrow plays a dominant role in lymphocyte precursor proliferation and differentiation *in vivo*.[180]

The central pacemaker controls the function of the immune system, but it can also be modulated by immune products such as proinflammatory cytokines interleukin 1 and 6 (IL-1 and IL-6), tumor necrosis factor-α (TNF-α) and anti-inflammatory drugs, which alter intracellular expression of *Bmal1*, *Npas2*, *Cry1*, and/or *Per2*[428–430] and the activity of the SCN clock and HPA axis at the organismal level.[431,432] Such modulation provides a circadian-paced immune-regulatory feedback loop that acts as an internal cue to the central clock to control the homeostasis of internal physiology.

4.2.3 Circadian homeostasis of energy balance suppresses tumor development

Human night-shift workers show a coupled increase in the risk of metabolic syndromes and cancers.[29,46,433–437] Disruption of circadian rhythm in mouse models also induces metabolic syndromes and whole body, local and/or organ-specific adiposity,[6,181,185,186,272–275] which is coupled with a significantly higher risk of tumor development in the digestive system.[29] As discovered from cancer epidemiological studies, deregulation, epigenetic silencing, and/or polymorphisms of core circadian genes including *Per2*, *Bmal1*, *Cry2*, and *Clock* are often associated with an increased risk of obesity and metabolic syndromes in humans.[438–443] Together, these findings strongly argue that disruption of circadian homeostasis of energy balance is an important cancer risk factor.

Energy homeostasis *in vivo* is maintained by the reciprocal interaction of peripheral adiposity signals and the central nervous system,[444,445] which are both under close surveillance of the circadian clock.[275,446] Although beyond the scope of this review, the evidence of direct and indirect interactions between the SCN pacemaker and the brain energy homeostasis center ARC, feeding and satiety center LHA and catabolic center PVN suggest that although the central circadian clock does not directly respond to food cues,[447] it plays an active role in the brain circuits of energy balance.[448–453] Together with its role in modulating the daily activity of NES and ANS, which controls the production and circulating levels of most, if not all, peripheral adiposity signals as well as metabolic balance in peripheral tissues,[33,454–460] the central pacemaker acts as a master regulator of energy homeostasis *in vivo* (Fig. 9.2).

Although the biological mechanisms linking obesity and cancer are still not well understood, obesity-induced deregulation of extracellular signaling, angiogenesis, and chronic inflammation are considered as the major pathways promoting cancer development.[389,461,462] Obesity leads to increased production of metabolic hormones from the fat and livers of obese subjects, such as Leptin and bioavailable IGF-1, which stimulate biosynthesis, cell growth, proliferation, and survival via activation of intracellular signaling pathways including the PI3K/AKT/mTOR, MAPK, and JAK3/STAT pathways that are often aberrantly activated in tumors.[463–465] Obesity also promotes chronic inflammation by increasing circulating free fatty acids, proinflammatory cytokines IL-1β, IL-6, TNF-α, and monocyte chemoattractant protein-1, stimulating type 2 macrophage polarization, and activating NF-κB-mediated proinflammatory and proproliferation pathways.[462,466–468] Such changes in tissue microenvironment lead to tissue damage due to increased cell necrosis induced by constant stress of increasing in biosynthesis, oxidative stress, and DNA damage, which in turn stimulates tissue regeneration that needs active cell proliferation to support. However, the deregulation of multiple oncogenic and tumor suppression pathways as well as immunosuppression due to circadian disruption would result in an increased risk of neoplastic growth that would accelerate cancer development in obese subjects.[469–471]

5. ANTICANCER CHRONOTHERAPY

Chemotherapy using one or more cytotoxic drugs in conjunction with radiation therapy or surgery is one of the most common procedures for anticancer therapy. Since chemotherapeutic drugs indiscriminately target both cancerous cells and normal host cells in renewable tissues, these drugs often generate intolerable side effects that impair the treatment.[472] The "targeted" anticancer therapy developed in recent years is aimed at blocking the growth of cancer cells via monoclonal antibodies recognizing tumor-specific antigens on the tumor cell surface or small molecules that block intracellular tyrosine kinase signaling including MARK/ERK, JAK, PI3K, CDKs, estrogen receptor (ER), epithelial growth factor receptor (EGFR), and vascular endothelial growth factor receptor-controlled pathways.[473] However, since these signaling pathways are also important for mediating neuroendocrine functions in host tissues, these small inhibitors also potentially increase the risk of fatal side effects among cancer patients.[474] Thus, to maximally increase tumor targeting efficiency and protect normal host tissues are the biggest challenges for successful anticancer treatments.

Tumors are derived from host tissues after regaining the properties of rapid self-renewing and dedifferentiation phenotypes. Most biological processes supporting tumor growth are also essential for normal physiological functions of the host except that the biological processes in normal host tissues are properly controlled and integrated with the daily physiology. Thus, anticancer treatments would inevitably have an adverse effect on patients. Biological processes determining the efficiency of anticancer drugs including drug absorption, distribution, intracellular metabolism, and elimination follow circadian rhythms in the host.[475] In addition, once inside the tumor cells, the function of a cytotoxic drug is largely determined by the circadian phase of cell proliferation in the tumors, which, even in advanced stage tumors, is not temporally disorganized.[211,216–220,476] Therefore, to apply anticancer treatment at a selected time during the day based on circadian variation in host internal physiology and the asynchronies in cell proliferation and drug metabolic rhythms between normal and malignant tissues could maximize drug toxicity to tumors and increase the efficiency of anticancer treatment.[477]

Studies using mouse models have revealed that both host tolerability and drug efficacy are affected by circadian timing, and that the best therapeutic index is achieved by coupling the time for drug delivery with host endogenous circadian rhythms.[478] These results have been extrapolated to randomized clinical trials of patients undergoing treatment for advanced stage cancers including metastatic ovarian, lung, colorectal, and breast cancers using conventional chemotherapeutic drugs.[58,479,480] The results of these studies have revealed that current procedures of anticancer chronotherapy indeed leads to better therapeutic outcomes and is especially more beneficial to patients who still maintain the endogenous circadian rhythmicity. The improved therapeutic index is shown by reduced drug toxicity, improved tumor response rate and the duration of the response, and decreased frequency of tumor metastasis. However, although chronotherapy has been shown to significantly increase the survival time for children with ALL,[481] it does not increase the long-term survival of patients with metastasizing cancers,[187] suggesting that anticancer chronotherapy is still at its initial stage of practice and needs further improvement. The facts that the molecular clock directly responds to genotoxic insults and that mice deficient in the *Clock*, *Bmal1*, and *Cry* genes respond differently to anticancer drugs compared to wild-type controls suggest that the response to anticancer treatment is a complicated clock-controlled physiological function *in vivo*.[124,182,260,261] Further investigation of the mechanisms of cancer chronotherapy, especially the mechanisms controlling the response of the circadian clock to anticancer treatment, the consequence of such a response for host physiology and tumor biology, the effect of tumor development on host circadian homeostasis, and the ability of the clock to respond to anticancer treatment would contribute greatly to improve the current anticancer chronotherapy.

6. CONCLUSIONS

The rate of cancer continues to rise as more people live to an old age and changes in lifestyle affect more developing countries.[482] Cancer is usually not detected at the early stage due to the absence of any obvious symptoms.[191,483] However, the fact that only 5–10% of all cancers diagnosed are caused by inherited genetic mutations suggests that an unhealthy lifestyle is the major risk factor for cancer. Therefore, cancer is preventable.[401] Mounting evidence obtained from recent studies suggests that disruption of endogenous circadian homeostasis is a novel and independent risk factor for cancer. In addition, as a master regulator of mammalian physiology, disruption of the clock likely manifests cancer development and progression induced by previously identified exogenous and endogenous cancer risk factors including diet choices, tobacco and alcohol usage, viral infection, air pollutions, aging, endocrine dysfunction, metabolic syndromes, and immune deficiencies.[484,485] Although still at the initial stage, recent advancements strongly suggest that both tumor suppression and response to anticancer treatments are clock-controlled physiological functions *in vivo*. Therefore, the mammalian circadian clock provides an invaluable and exciting system to study the mechanism of cancer and anticancer chronotherapy at the molecular, cellular, tissue, organ, and organismal levels *in vivo*. Such studies will have a significant impact on human health in the future by improving both cancer prevention and treatment.

ACKNOWLEDGMENTS

The authors would like to thank D. D. Moore for input in preparing the manuscript. This work is supported by the grants from USDA/ARS (6250-51000-055) and NIH/NCI (R01 CA137019-01A) to L. F. and the training grant award from NIH/NIDDK to N. M. K. (5T32DK007696-20).

REFERENCES

1. Young MW, Kay SA. Time zones: a comparative genetics of circadian clocks. *Nat Rev Genet.* 2001;2(9):702–715.
2. Reppert SM, Weaver DR. Coordination of circadian timing in mammals. *Nature.* 2002;418(6901):935–941.
3. Buijs RM, Scheer FA, Kreier F, et al. Organization of circadian functions: interaction with the body. *Prog Brain Res.* 2006;153:341–360.
4. Guo H, Brewer JM, Lehman MN, Bittman EL. Suprachiasmatic regulation of circadian rhythms of gene expression in hamster peripheral organs: effects of transplanting the pacemaker. *J Neurosci.* 2006;26(24):6406–6412.

5. Takahashi JS, Hong HK, Ko CH, McDearmon EL. The genetics of mammalian circadian order and disorder: implications for physiology and disease. *Nat Rev Genet.* Oct 2008;9(10):764–775.
6. Sahar S, Sassone-Corsi P. Metabolism and cancer: the circadian clock connection. *Nat Rev Cancer.* 2009;9(12):886–896.
7. Padmanabhan K, Robles MS, Westerling T, Weitz CJ. Feedback regulation of transcriptional termination by the mammalian circadian clock PERIOD complex. *Science.* 2012;337(6094):599–602.
8. Preitner N, Damiola F, Lopez-Molina L, et al. The orphan nuclear receptor REV-ERBalpha controls circadian transcription within the positive limb of the mammalian circadian oscillator. *Cell.* 2002;110(2):251–260.
9. Cho H, Zhao X, Hatori M, et al. Regulation of circadian behaviour and metabolism by REV-ERB-alpha and REV-ERB-beta. *Nature.* 2012;485(7396):123–127.
10. Sato TK, Panda S, Miraglia LJ, et al. A functional genomics strategy reveals Rora as a component of the mammalian circadian clock. *Neuron.* 2004;43(4):527–537.
11. Duong HA, Robles MS, Knutti D, Weitz CJ. A molecular mechanism for circadian clock negative feedback. *Science.* 2011;332(6036):1436–1439.
12. Nakahata Y, Kaluzova M, Grimaldi B, et al. The NAD+-dependent deacetylase SIRT1 modulates CLOCK-mediated chromatin remodeling and circadian control. *Cell.* 2008;134(2):329–340.
13. Asher G, Gatfield D, Stratmann M, et al. SIRT1 regulates circadian clock gene expression through PER2 deacetylation. *Cell.* 2008;134(2):317–328.
14. Feng D, Liu T, Sun Z, et al. A circadian rhythm orchestrated by histone deacetylase 3 controls hepatic lipid metabolism. *Science.* 2011;331(6022):1315–1319.
15. Stratmann M, Suter DM, Molina N, Naef F, Schibler U. Circadian Dbp transcription relies on highly dynamic BMAL1-CLOCK interaction with E boxes and requires the proteasome. *Mol Cell.* 2012;48(2):277–287.
16. Akhtar RA, Reddy AB, Maywood ES, et al. Circadian cycling of the mouse liver transcriptome, as revealed by cDNA microarray, is driven by the suprachiasmatic nucleus. *Curr Biol.* 2002;12(7):540–550.
17. Duffield GE, Best JD, Meurers BH, Bittner A, Loros JJ, Dunlap JC. Circadian programs of transcriptional activation, signaling, and protein turnover revealed by microarray analysis of mammalian cells. *Curr Biol.* 2002;12(7):551–557.
18. Hughes ME, DiTacchio L, Hayes KR, et al. Harmonics of circadian gene transcription in mammals. *PLoS Genet.* Apr 2009;5(4):e1000442.
19. Miller BH, McDearmon EL, Panda S, et al. Circadian and CLOCK-controlled regulation of the mouse transcriptome and cell proliferation. *Proc Natl Acad Sci USA.* 2007;104(9):3342–3347.
20. Panda S, Antoch MP, Miller BH, et al. Coordinated transcription of key pathways in the mouse by the circadian clock. *Cell.* 2002;109(3):307–320.
21. Storch KF, Lipan O, Leykin I, et al. Extensive and divergent circadian gene expression in liver and heart. *Nature.* 2002;417(6884):78–83.
22. Geyfman M, Kumar V, Liu Q, et al. Brain and muscle Arnt-like protein-1 (BMAL1) controls circadian cell proliferation and susceptibility to UVB-induced DNA damage in the epidermis. *Proc Natl Acad Sci USA.* 2012;109(29):11758–11763.
23. Khapre RV, Kondratova AA, Susova O, Kondratov RV. Circadian clock protein BMAL1 regulates cellular senescence in vivo. *Cell Cycle.* 2011;10(23): 4162–4169.
24. Mohawk JA, Green CB, Takahashi JS. Central and peripheral circadian clocks in mammals. *Annu Rev Neurosci.* 2012;35:445–462.
25. Golombek DA, Rosenstein RE. Physiology of circadian entrainment. *Physiol Rev.* 2010;90(3):1063–1102.
26. Do MT, Yau KW. Intrinsically photosensitive retinal ganglion cells. *Physiol Rev.* 2010;90(4):1547–1581.

27. Davidson AJ, Castanon-Cervantes O, Leise TL, Molyneux PC, Harrington ME. Visualizing jet lag in the mouse suprachiasmatic nucleus and peripheral circadian timing system. *Eur J Neurosci*. 2009;29(1):171–180.
28. Kennaway DJ, Owens JA, Voultsios A, Varcoe TJ. Functional central rhythmicity and light entrainment, but not liver and muscle rhythmicity, are Clock independent. *Am J Physiol Regul Integr Comp Physiol*. 2006;291(4):R1172–R1180.
29. Lee S, Donehower LA, Herron AJ, Moore DD, Fu L. Disrupting circadian homeostasis of sympathetic signaling promotes tumor development in mice. *PLoS One*. 2010;5(6): e10995.
30. Barclay JL, Tsang AH, Oster H. Interaction of central and peripheral clocks in physiological regulation. *Prog Brain Res*. 2012;199:163–181.
31. Balsalobre A, Brown SA, Marcacci L, et al. Resetting of circadian time in peripheral tissues by glucocorticoid signaling. *Science*. 2000;289(5488):2344–2347.
32. Richter HG, Torres-Farfan C, Rojas-Garcia PP, Campino C, Torrealba F, Seron-Ferre M. The circadian timing system: making sense of day/night gene expression. *Biol Res*. 2004;37(1):11–28.
33. Kalsbeek A, Bruinstroop E, Yi CX, Klieverik LP, La Fleur SE, Fliers E. Hypothalamic control of energy metabolism via the autonomic nervous system. *Ann N Y Acad Sci*. 2010;1212:114–129.
34. Schernhammer ES, Laden F, Speizer FE, et al. Night-shift work and risk of colorectal cancer in the nurses' health study. *J Natl Cancer Inst*. 2003;95(11):825–828.
35. Kubo T, Ozasa K, Mikami K, et al. Prospective cohort study of the risk of prostate cancer among rotating-shift workers: findings from the Japan collaborative cohort study. *Am J Epidemiol*. 2006;164(6):549–555.
36. Viswanathan AN, Hankinson SE, Schernhammer ES. Night shift work and the risk of endometrial cancer. *Cancer Res*. 2007;67(21):10618–10622.
37. Lahti TA, Partonen T, Kyyronen P, Kauppinen T, Pukkala E. Night-time work predisposes to non-Hodgkin lymphoma. *Int J Cancer*. 2008;123(9):2148–2151.
38. Stevens RG. Working against our endogenous circadian clock: breast cancer and electric lighting in the modern world. *Mutat Res*. 2009;680(1–2):106–108.
39. Touitou Y, Bogdan A, Levi F, Benavides M, Auzeby A. Disruption of the circadian patterns of serum cortisol in breast and ovarian cancer patients: relationships with tumour marker antigens. *Br J Cancer*. 1996;74(8):1248–1252.
40. Panzer A. Melatonin in osteosarcoma: an effective drug? *Med Hypotheses*. 1997;48 (6):523–525.
41. Skibola CF, Holly EA, Forrest MS, et al. Body mass index, leptin and leptin receptor polymorphisms, and non-Hodgkin lymphoma. *Cancer Epidemiol Biomarkers Prev*. 2004;13(5):779–786.
42. Viswanathan AN, Schernhammer ES. Circulating melatonin and the risk of breast and endometrial cancer in women. *Cancer Lett*. 2009;281(1):1–7.
43. Kloog I, Haim A, Stevens RG, Portnov BA. Global co-distribution of light at night (LAN) and cancers of prostate, colon, and lung in men. *Chronobiol Int*. 2009;26 (1):108–125.
44. Knutsson A, Alfredsson L, Karlsson B, et al. Breast cancer among shift workers: results of the WOLF longitudinal cohort study. *Scand J Work Environ Health*. 2013; 39(2):170–177.
45. Bhatti P, Mirick DK, Davis S. Invited commentary: shift work and cancer. *Am J Epidemiol*. 2012;176(9):760–763, discussion 764–765.
46. Haus EL, Smolensky MH. Shift work and cancer risk: potential mechanistic roles of circadian disruption, light at night, and sleep deprivation. *Sleep Med Rev*. 2013; 17(4):273–284.
47. Buzzelli G, Dattolo P, Pinzani M, Brocchi A, Romano S, Gentilini P. Circulating growth hormone and insulin-like growth factor-I in nonalcoholic liver cirrhosis with

or without superimposed hepatocarcinoma: evidence of an altered circadian rhythm. *Am J Gastroenterol.* 1993;88(10):1744–1748.
48. Davis S, Mirick DK, Stevens RG. Night shift work, light at night, and risk of breast cancer. *J Natl Cancer Inst.* 2001;93(20):1557–1562.
49. Rafnsson V, Tulinius H, Jonasson JG, Hrafnkelsson J. Risk of breast cancer in female flight attendants: a population-based study (Iceland). *Cancer Causes Control.* 2001;12 (2):95–101.
50. Schernhammer ES, Laden F, Speizer FE, et al. Rotating night shifts and risk of breast cancer in women participating in the nurses' health study. *J Natl Cancer Inst.* 2001;93 (20):1563–1568.
51. Stevens RG, Hansen J, Costa G, et al. Considerations of circadian impact for defining 'shift work' in cancer studies: IARC Working Group Report. *Occup Environ Med.* 2011;68(2):154–162.
52. Pukkala E, Aspholm R, Auvinen A, et al. Cancer incidence among 10,211 airline pilots: a Nordic study. *Aviat Space Environ Med.* 2003;74(7):699–706.
53. Keith LG, Oleszczuk JJ, Laguens M. Circadian rhythm chaos: a new breast cancer marker. *Int J Fertil Womens Med.* 2001;46(5):238–247.
54. Rich T, Innominato PF, Boerner J, et al. Elevated serum cytokines correlated with altered behavior, serum cortisol rhythm, and dampened 24-hour rest-activity patterns in patients with metastatic colorectal cancer. *Clin Cancer Res.* 2005;11 (5):1757–1764.
55. Sephton SE, Sapolsky RM, Kraemer HC, Spiegel D. Diurnal cortisol rhythm as a predictor of breast cancer survival. *J Natl Cancer Inst.* 2000;92(12):994–1000.
56. Kim KS, Kim YC, Oh IJ, Kim SS, Choi JY, Ahn RS. Association of worse prognosis with an aberrant diurnal cortisol rhythm in patients with advanced lung cancer. *Chronobiol Int.* 2012;29(8):1109–1120.
57. Sephton SE, Lush E, Dedert EA, et al. Diurnal cortisol rhythm as a predictor of lung cancer survival. *Brain Behav Immun.* 2012;30:S163–S170.
58. Innominato PF, Giacchetti S, Bjarnason GA, et al. Prediction of overall survival through circadian rest-activity monitoring during chemotherapy for metastatic colorectal cancer. *Int J Cancer.* 2012;131(11):2684–2692.
59. Wu HS, Davis JE, Natavio T. Fatigue and disrupted sleep-wake patterns in patients with cancer: a shared mechanism. *Clin J Oncol Nurs.* 2012;16(2):E56–E68.
60. Levi F, Filipski E, Iurisci I, Li XM, Innominato P. Cross-talks between circadian timing system and cell division cycle determine cancer biology and therapeutics. *Cold Spring Harb Symp Quant Biol.* 2007;72:465–475.
61. Mormont MC, Waterhouse J, Bleuzen P, et al. Marked 24-h rest/activity rhythms are associated with better quality of life, better response, and longer survival in patients with metastatic colorectal cancer and good performance status. *Clin Cancer Res.* 2000;6 (8):3038–3045.
62. Winter SL, Bosnoyan-Collins L, Pinnaduwage D, Andrulis IL. Expression of the circadian clock genes Per1 and Per2 in sporadic and familial breast tumors. *Neoplasia.* Oct 2007;9(10):797–800.
63. Chen ST, Choo KB, Hou MF, Yeh KT, Kuo SJ, Chang JG. Deregulated expression of the PER1, PER2 and PER3 genes in breast cancers. *Carcinogenesis.* 2005;26 (7):1241–1246.
64. Kuo SJ, Chen ST, Yeh KT, et al. Disturbance of circadian gene expression in breast cancer. *Virchows Arch.* Apr 2009;454(4):467–474.
65. Dai H, Zhang L, Cao M, et al. The role of polymorphisms in circadian pathway genes in breast tumorigenesis. *Breast Cancer Res Treat.* 2011;127(2):531–540.
66. Blask DE, Hill SM, Dauchy RT, et al. Circadian regulation of molecular, dietary, and metabolic signaling mechanisms of human breast cancer growth by the nocturnal

melatonin signal and the consequences of its disruption by light at night. *J Pineal Res.* 2011;51(3):259–269.
67. Xiang S, Coffelt SB, Mao L, Yuan L, Cheng Q, Hill SM. Period-2: a tumor suppressor gene in breast cancer. *J Circadian Rhythms.* 2008;6:4.
68. Gery S, Virk RK, Chumakov K, Yu A, Koeffler HP. The clock gene Per2 links the circadian system to the estrogen receptor. *Oncogene.* 2007;26(57):7916–7920.
69. Mullenders J, Fabius AW, Madiredjo M, Bernards R, Beijersbergen RL. A large scale shRNA barcode screen identifies the circadian clock component ARNTL as putative regulator of the p53 tumor suppressor pathway. *PLoS One.* 2009;4(3):e4798.
70. Hoffman AE, Zheng T, Ba Y, Zhu Y. The circadian gene NPAS2, a putative tumor suppressor, is involved in DNA damage response. *Mol Cancer Res.* 2008;6 (9):1461–1468.
71. Miyamoto N, Izumi H, Noguchi T, et al. Tip60 is regulated by circadian transcription factor clock and is involved in cisplatin resistance. *J Biol Chem.* 2008;283 (26):18218–18226.
72. Hoffman AE, Zheng T, Ba Y, et al. Phenotypic effects of the circadian gene Cryptochrome 2 on cancer-related pathways. *BMC Cancer.* 2010;10:110.
73. Yang X, Wood PA, Ansell C, Hrushesky WJ. Circadian time-dependent tumor suppressor function of period genes. *Integr Cancer Ther.* 2009;8(4):309–316.
74. Yang X, Wood PA, Ansell CM, et al. The circadian clock gene Per1 suppresses cancer cell proliferation and tumor growth at specific times of day. *Chronobiol Int.* 2009;26 (7):1323–1339.
75. Taniguchi H, Fernandez AF, Setien F, et al. Epigenetic inactivation of the circadian clock gene BMAL1 in hematologic malignancies. *Cancer Res.* 2009;69(21): 8447–8454.
76. Lewintre EJ, Martin CR, Ballesteros CG, et al. Cryptochrome-1 expression: a new prognostic marker in B-cell chronic lymphocytic leukemia. *Haematologica.* 2009;94 (2):280–284.
77. Eisele L, Prinz R, Klein-Hitpass L, et al. Combined PER2 and CRY1 expression predicts outcome in chronic lymphocytic leukemia. *Eur J Haematol.* 2009;83 (4):320–327.
78. Yang MY, Chang JG, Lin PM, et al. Downregulation of circadian clock genes in chronic myeloid leukemia: alternative methylation pattern of hPER3. *Cancer Sci.* Dec 2006;97(12):1298–1307.
79. Sun CM, Huang SF, Zeng JM, et al. Per2 inhibits k562 leukemia cell growth in vitro and in vivo through cell cycle arrest and apoptosis induction. *Pathol Oncol Res.* 2010;16 (3):403–411.
80. Mostafaie N, Kallay E, Sauerzapf E, et al. Correlated downregulation of estrogen receptor beta and the circadian clock gene Per1 in human colorectal cancer. *Mol Carcinog.* 2009;48(7):642–647.
81. Zhou F, He X, Liu H, et al. Functional polymorphisms of circadian positive feedback regulation genes and clinical outcome of Chinese patients with resected colorectal cancer. *Cancer.* 2012;118(4):937–946.
82. Gery S, Komatsu N, Baldjyan L, Yu A, Koo D, Koeffler HP. The circadian gene per1 plays an important role in cell growth and DNA damage control in human cancer cells. *Mol Cell.* 2006;22(3):375–382.
83. Brandi G, Calabrese C, Pantaleo MA, et al. Circadian variations of rectal cell proliferation in patients affected by advanced colorectal cancer. *Cancer Lett.* 2004;208 (2):193–196.
84. Shih MC, Yeh KT, Tang KP, Chen JC, Chang JG. Promoter methylation in circadian genes of endometrial cancers detected by methylation-specific PCR. *Mol Carcinog.* 2006;45(10):732–740.

85. Xia HC, Niu ZF, Ma H, et al. Deregulated expression of the Per1 and Per2 in human gliomas. *Can J Neurol Sci*. 2010;37(3):365–370.
86. Fujioka A, Takashima N, Shigeyoshi Y. Circadian rhythm generation in a glioma cell line. *Biochem Biophys Res Commun*. 2006;346(1):169–174.
87. Luo Y, Wang F, Chen LA, et al. Deregulated expression of cry1 and cry2 in human gliomas. *Asian Pac J Cancer Prev*. 2012;13(11):5725–5728.
88. Hsu CM, Lin SF, Lu CT, Lin PM, Yang MY. Altered expression of circadian clock genes in head and neck squamous cell carcinoma. *Tumour Biol*. 2012;33(1): 149–155.
89. Hsu CM, Lin PM, Wang YM, Chen ZJ, Lin SF, Yang MY. Circulating miRNA is a novel marker for head and neck squamous cell carcinoma. *Tumour Biol*. Dec 2012;33 (6):1933–1942.
90. Lin YM, Chang JH, Yeh KT, et al. Disturbance of circadian gene expression in hepatocellular carcinoma. *Mol Carcinog*. Dec 2008;47(12):925–933.
91. Gery S, Komatsu N, Kawamata N, et al. Epigenetic silencing of the candidate tumor suppressor gene Per1 in non-small cell lung cancer. *Clin Cancer Res*. 2007;13 (5):1399–1404.
92. Elshazley M, Sato M, Hase T, et al. The circadian clock gene BMAL1 is a novel therapeutic target for malignant pleural mesothelioma. *Int J Cancer*. 2012;131 (12):2820–2831.
93. Roe OD, Anderssen E, Helge E, et al. Genome-wide profile of pleural mesothelioma versus parietal and visceral pleura: the emerging gene portrait of the mesothelioma phenotype. *PLoS One*. 2009;4(8):e6554.
94. Zhu Y, Leaderer D, Guss C, et al. Ala394Thr polymorphism in the clock gene NPAS2: a circadian modifier for the risk of non-Hodgkin's lymphoma. *Int J Cancer*. 2007;120 (2):432–435.
95. Yang WS, Stockwell BR. Inhibition of casein kinase 1-epsilon induces cancer-cell-selective, PERIOD2-dependent growth arrest. *Genome Biol*. 2008;9(6):R92.
96. Tokunaga H, Takebayashi Y, Utsunomiya H, et al. Clinicopathological significance of circadian rhythm-related gene expression levels in patients with epithelial ovarian cancer. *Acta Obstet Gynecol Scand*. 2008;87(10):1060–1070.
97. Klevecz RR, Shymko RM, Blumenfeld D, Braly PS. Circadian gating of S phase in human ovarian cancer. *Cancer Res*. 1987;47(23):6267–6271.
98. Oda A, Katayose Y, Yabuuchi S, et al. Clock gene mouse period2 overexpression inhibits growth of human pancreatic cancer cells and has synergistic effect with cisplatin. *Anticancer Res*. 2009;29(4):1201–1209.
99. Pogue-Geile KL, Lyons-Weiler J, Whitcomb DC. Molecular overlap of fly circadian rhythms and human pancreatic cancer. *Cancer Lett*. 2006;243(1):55–57.
100. Suzuki T, Sato F, Kondo J, et al. Period is involved in the proliferation of human pancreatic MIA-PaCa2 cancer cells by TNF-alpha. *Biomed Res*. 2008;29(2):99–103.
101. Relles D, Sendecki J, Chipitsyna G, Hyslop T, Yeo CJ, Arafat HA. Circadian gene expression and clinicopathologic correlates in pancreatic cancer. *J Gastrointest Surg*. 2012;17:443–450.
102. Jung-Hynes B, Huang W, Reiter RJ, Ahmad N. Melatonin resynchronizes dysregulated circadian rhythm circuitry in human prostate cancer cells. *J Pineal Res*. 2010;49(1):60–68.
103. Cao Q, Gery S, Dashti A, et al. A role for the clock gene per1 in prostate cancer. *Cancer Res*. 2009;69(19):7619–7625.
104. Zhu Y, Stevens RG, Hoffman AE, et al. Testing the circadian gene hypothesis in prostate cancer: a population-based case–control study. *Cancer Res*. 2009;69(24):9315–9322.
105. Jung-Hynes B, Schmit TL, Reagan-Shaw SR, Siddiqui IA, Mukhtar H, Ahmad N. Melatonin, a novel Sirt1 inhibitor, imparts antiproliferative effects against prostate

cancer in vitro in culture and in vivo in TRAMP model. *J Pineal Res.* Mar 2011;50 (2):140–149.
106. Narasimamurthy R, Hatori M, Nayak SK, Liu F, Panda S, Verma IM. Circadian clock protein cryptochrome regulates the expression of proinflammatory cytokines. *Proc Natl Acad Sci USA.* 2012;109(31):12662–12667.
107. Blask DE. Melatonin, sleep disturbance and cancer risk. *Sleep Med Rev.* 2009;13 (4):257–264.
108. Morris CJ, Aeschbach D, Scheer FA. Circadian system, sleep and endocrinology. *Mol Cell Endocrinol.* 2012;349(1):91–104.
109. Boersma GJ, Salton SR, Spritzer PM, Steele CT, Carbone DL. Models and mechanisms of metabolic regulation: genes, stress, and the HPA and HPG axes. *Horm Metab Res.* 2012;44(8):598–606.
110. Wirth M, Burch J, Violanti J, et al. Shiftwork duration and the awakening cortisol response among police officers. *Chronobiol Int.* 2011;28(5):446–457.
111. Feychting M, Osterlund B, Ahlbom A. Reduced cancer incidence among the blind. *Epidemiology.* 1998;9(5):490–494.
112. Kliukiene J, Tynes T, Andersen A. Risk of breast cancer among Norwegian women with visual impairment. *Br J Cancer.* 2001;84(3):397–399.
113. Pukkala E, Verkasalo P, Ojamo M, Rudanko SL. Response to the letter by Maria Feychting and Anders Ahlbom (Cancer Causes and Control 10: 637, 1999.) Why is the cancer pattern so different among visually impaired persons in Finland and Sweden? *Cancer Causes Control.* 2000;11(1):99–100.
114. Hamilton T, Sneddon A. Endocrine differences between rats bearing simple and malignant mammary tumours induced by 9,10-dimethyl-1,2-benzanthracene. *Br J Cancer.* 1968;22(2):324–329.
115. van den Heiligenberg S, Depres-Brummer P, Barbason H, Claustrat B, Reynes M, Levi F. The tumor promoting effect of constant light exposure on diethylnitrosamine-induced hepatocarcinogenesis in rats. *Life Sci.* 1999;64(26): 2523–2534.
116. Shah PN, Mhatre MC, Kothari LS. Effect of melatonin on mammary carcinogenesis in intact and pinealectomized rats in varying photoperiods. *Cancer Res.* Aug 1984;44 (8):3403–3407.
117. Li JC, Xu F. Influences of light–dark shifting on the immune system, tumor growth and life span of rats, mice and fruit flies as well as on the counteraction of melatonin. *Biol Signals.* 1997;6(2):77–89.
118. Wu M, Zeng J, Chen Y, et al. Experimental chronic jet lag promotes growth and lung metastasis of Lewis lung carcinoma in C57BL/6 mice. *Oncol Rep.* 2012;27 (5):1417–1428.
119. Filipski E, Subramanian P, Carriere J, Guettier C, Barbason H, Levi F. Circadian disruption accelerates liver carcinogenesis in mice. *Mutat Res.* 2009;680(1–2):95–105.
120. Logan RW, Zhang C, Murugan S, et al. Chronic shift-lag alters the circadian clock of NK cells and promotes lung cancer growth in rats. *J Immunol.* 2012;188 (6):2583–2591.
121. Yasuniwa Y, Izumi H, Wang KY, et al. Circadian disruption accelerates tumor growth and angio/stromagenesis through a Wnt signaling pathway. *PLoS One.* 2010;5(12): e15330.
122. Filipski E, King VM, Li X, et al. Host circadian clock as a control point in tumor progression. *J Natl Cancer Inst.* 2002;94(9):690–697.
123. Zheng B, Larkin DW, Albrecht U, et al. The mPer2 gene encodes a functional component of the mammalian circadian clock. *Nature.* 1999;400(6740):169–173.
124. Fu L, Pelicano H, Liu J, Huang P, Lee C. The circadian gene Period2 plays an important role in tumor suppression and DNA damage response in vivo. *Cell.* 2002;111 (1):41–50.

125. Kondratov RV, Kondratova AA, Gorbacheva VY, Vykhovanets OV, Antoch MP. Early aging and age-related pathologies in mice deficient in BMAL1, the core component of the circadian clock. *Genes Dev.* 2006;20(14):1868–1873.
126. Matsuo T, Yamaguchi S, Mitsui S, Emi A, Shimoda F, Okamura H. Control mechanism of the circadian clock for timing of cell division in vivo. *Science.* 2003;302 (5643):255–259.
127. Ozturk N, Lee JH, Gaddameedhi S, Sancar A. Loss of cryptochrome reduces cancer risk in p53 mutant mice. *Proc Natl Acad Sci USA.* 2009;106(8):2841–2846.
128. Gauger MA, Sancar A. Cryptochrome, circadian cycle, cell cycle checkpoints, and cancer. *Cancer Res.* 2005;65(15):6828–6834.
129. Antoch MP, Gorbacheva VY, Vykhovanets O, et al. Disruption of the circadian clock due to the Clock mutation has discrete effects on aging and carcinogenesis. *Cell Cycle.* 2008;7(9):1197–1204.
130. Barbeito CG, Garcia MN, Flamini MA, Andrini LB, Badran AF. Effect of partial and sham hepatectomy on the growth of a hepatocellular carcinoma. *J Exp Clin Cancer Res.* 2001;20(1):153–158.
131. Huang W, Ma K, Zhang J, et al. Nuclear receptor-dependent bile acid signaling is required for normal liver regeneration. *Science.* 2006;312(5771):233–236.
132. Yang F, Huang X, Yi T, Yen Y, Moore DD, Huang W. Spontaneous development of liver tumors in the absence of the bile acid receptor farnesoid X receptor. *Cancer Res.* 2007;67(3):863–867.
133. Doi M, Hirayama J, Sassone-Corsi P. Circadian regulator CLOCK is a histone acetyltransferase. *Cell.* 2006;125(3):497–508.
134. Tsuchiya Y, Minami I, Kadotani H, Todo T, Nishida E. Circadian clock-controlled diurnal oscillation of Ras/ERK signaling in mouse liver. *Proc Jpn Acad Ser B Phys Biol Sci.* 2013;89(1):59–65.
135. Oster H, van der Horst GT, Albrecht U. Daily variation of clock output gene activation in behaviorally arrhythmic mPer/mCry triple mutant mice. *Chronobiol Int.* 2003;20 (4):683–695.
136. Hoshino R, Chatani Y, Yamori T, et al. Constitutive activation of the 41-/43-kDa mitogen-activated protein kinase signaling pathway in human tumors. *Oncogene.* 1999;18(3):813–822.
137. Zou Y, Bao Q, Kumar S, Hu M, Wang GY, Dai G. Four waves of hepatocyte proliferation linked with three waves of hepatic fat accumulation during partial hepatectomy-induced liver regeneration. *PLoS One.* 2012;7(2):e30675.
138. Lawler J, Miao WM, Duquette M, Bouck N, Bronson RT, Hynes RO. Thrombospondin-1 gene expression affects survival and tumor spectrum of p53-deficient mice. *Am J Pathol.* 2001;159(5):1949–1956.
139. Yamamoto M, Tsukamoto T, Sakai H, et al. p53 knockout mice (−/−) are more susceptible than (+/−) or (+/+) mice to N-methyl-N-nitrosourea stomach carcinogenesis. *Carcinogenesis.* 2000;21(10):1891–1897.
140. Tyner SD, Venkatachalam S, Choi J, et al. p53 mutant mice that display early ageing-associated phenotypes. *Nature.* 2002;415(6867):45–53.
141. Oda H, Zhang S, Tsurutani N, et al. Loss of p53 is an early event in induction of brain tumors in mice by transplacental carcinogen exposure. *Cancer Res.* 1997;57 (4):646–650.
142. Hursting SD, Perkins SN, Phang JM. Calorie restriction delays spontaneous tumorigenesis in p53-knockout transgenic mice. *Proc Natl Acad Sci USA.* 1994;91(15):7036–7040.
143. Nozawa H, Oda E, Nakao K, et al. Loss of transcription factor IRF-1 affects tumor susceptibility in mice carrying the Ha-ras transgene or nullizygosity for p53. *Genes Dev.* 1999;13(10):1240–1245.

144. Liu G, McDaniel TK, Falkow S, Karlin S. Sequence anomalies in the Cag7 gene of the Helicobacter pylori pathogenicity island. *Proc Natl Acad Sci USA*. 1999;96 (12):7011–7016.
145. Donehower LA, Harvey M, Slagle BL, et al. Mice deficient for p53 are developmentally normal but susceptible to spontaneous tumours. *Nature*. 1992;356(6366):215–221.
146. Xu Y, Toh KL, Jones CR, Shin JY, Fu YH, Ptacek LJ. Modeling of a human circadian mutation yields insights into clock regulation by PER2. *Cell*. 2007;128(1):59–70.
147. Toh KL, Jones CR, He Y, et al. An hPer2 phosphorylation site mutation in familial advanced sleep phase syndrome. *Science*. 2001;291(5506):1040–1043.
148. Gu X, Xing L, Shi G, et al. The circadian mutation PER2(S662G) is linked to cell cycle progression and tumorigenesis. *Cell Death Differ*. 2012;19(3):397–405.
149. Fu L, Patel MS, Bradley A, Wagner EF, Karsenty G. The molecular clock mediates leptin-regulated bone formation. *Cell*. 2005;122(5):803–815.
150. Maronde E, Schilling AF, Seitz S, et al. The clock genes Period 2 and Cryptochrome 2 differentially balance bone formation. *PLoS One*. 2010;5(7):e11527.
151. Yang X, Wood PA, Ansell CM, et al. Beta-catenin induces beta-TrCP-mediated PER2 degradation altering circadian clock gene expression in intestinal mucosa of ApcMin/+ mice. *J Biochem*. 2009;145(3):289–297.
152. Wood PA, Yang X, Taber A, et al. Period 2 mutation accelerates ApcMin/+ tumorigenesis. *Mol Cancer Res*. 2008;6(11):1786–1793.
153. Miyazaki K, Wakabayashi M, Hara Y, Ishida N. Tumor growth suppression in vivo by overexpression of the circadian component, PER2. *Genes Cells*. 2010;15(4):351–358.
154. van der Horst GT, Muijtjens M, Kobayashi K, et al. Mammalian Cry1 and Cry2 are essential for maintenance of circadian rhythms. *Nature*. 1999;398(6728):627–630.
155. Zheng B, Albrecht U, Kaasik K, et al. Nonredundant roles of the mPer1 and mPer2 genes in the mammalian circadian clock. *Cell*. 2001;105(5):683–694.
156. Tanioka M, Yamada H, Doi M, et al. Molecular clocks in mouse skin. *J Invest Dermatol*. 2009;129(5):1225–1231.
157. Campisi J. Aging, cellular senescence, and cancer. *Annu Rev Physiol*. 2012;75:685–705.
158. Cotta-Ramusino C, McDonald 3rd ER, Hurov K, Sowa ME, Harper JW, Elledge SJ. A DNA damage response screen identifies RHINO, a 9-1-1 and TopBP1 interacting protein required for ATR signaling. *Science*. 2011;332(6035):1313–1317.
159. DeBruyne JP, Weaver DR, Reppert SM. CLOCK and NPAS2 have overlapping roles in the suprachiasmatic circadian clock. *Nat Neurosci*. 2007;10(5):543–545.
160. Bertolucci C, Cavallari N, Colognesi I, et al. Evidence for an overlapping role of CLOCK and NPAS2 transcription factors in liver circadian oscillators. *Mol Cell Biol*. 2008;28(9):3070–3075.
161. Bae K, Jin X, Maywood ES, Hastings MH, Reppert SM, Weaver DR. Differential functions of mPer1, mPer2, and mPer3 in the SCN circadian clock. *Neuron*. 2001;30(2):525–536.
162. Debruyne JP, Noton E, Lambert CM, Maywood ES, Weaver DR, Reppert SM. A clock shock: mouse CLOCK is not required for circadian oscillator function. *Neuron*. 2006;50(3):465–477.
163. Asher G, Schibler U. A CLOCK-less clock. *Trends Cell Biol*. 2006;16(11):547–549.
164. Spengler ML, Kuropatwinski KK, Comas M, et al. Core circadian protein CLOCK is a positive regulator of NF-kappaB-mediated transcription. *Proc Natl Acad Sci USA*. 2012;109(37):E2457–E2465.
165. Antoch MP, Song EJ, Chang AM, et al. Functional identification of the mouse circadian Clock gene by transgenic BAC rescue. *Cell*. 1997;89(4):655–667.
166. Lin KK, Kumar V, Geyfman M, et al. Circadian clock genes contribute to the regulation of hair follicle cycling. *PLoS Genet*. 2009;5(7):e1000573.

167. Janich P, Pascual G, Merlos-Suarez A, et al. The circadian molecular clock creates epidermal stem cell heterogeneity. *Nature*. 2011;480(7376):209–214.
168. Zeng ZL, Wu MW, Sun J, et al. Effects of the biological clock gene Bmal1 on tumour growth and anti-cancer drug activity. *J Biochem*. 2010;148(3):319–326.
169. Bunger MK, Wilsbacher LD, Moran SM, et al. Mop3 is an essential component of the master circadian pacemaker in mammals. *Cell*. 2000;103(7):1009–1017.
170. Gaddameedhi S, Reardon JT, Ye R, Ozturk N, Sancar A. Effect of circadian clock mutations on DNA damage response in mammalian cells. *Cell Cycle*. 2012;11(18):3481–3491.
171. Maclean KH, Keller UB, Rodriguez-Galindo C, Nilsson JA, Cleveland JL. c-Myc augments gamma irradiation-induced apoptosis by suppressing Bcl-XL. *Mol Cell Biol*. 2003;23(20):7256–7270.
172. Yuan ZM, Huang Y, Ishiko T, Kharbanda S, Weichselbaum R, Kufe D. Regulation of DNA damage-induced apoptosis by the c-Abl tyrosine kinase. *Proc Natl Acad Sci USA*. 1997;94(4):1437–1440.
173. Massague J. G1 cell-cycle control and cancer. *Nature*. 2004;432(7015):298–306.
174. Jackson SP, Bartek J. The DNA-damage response in human biology and disease. *Nature*. 2009;461(7267):1071–1078.
175. Unsal-Kacmaz K, Mullen TE, Kaufmann WK, Sancar A. Coupling of human circadian and cell cycles by the timeless protein. *Mol Cell Biol*. 2005;25(8):3109–3116.
176. Marks PA, Rifkind RA, Gambari R, Epner E, Chen ZX, Banks J. Commitment to terminal differentiation and the cell cycle. *Curr Top Cell Regul*. 1982;21:189–203.
177. Balsalobre A, Damiola F, Schibler U. A serum shock induces circadian gene expression in mammalian tissue culture cells. *Cell*. 1998;93(6):929–937.
178. Skehan P. Control models of cell cycle transit, exit, and arrest. *Biochem Cell Biol*. 1988;66(6):467–477.
179. Grechez-Cassiau A, Rayet B, Guillaumond F, Teboul M, Delaunay F. The circadian clock component BMAL1 is a critical regulator of p21WAF1/CIP1 expression and hepatocyte proliferation. *J Biol Chem*. 2008;283(8):4535–4542.
180. Sun Y, Yang Z, Niu Z, et al. MOP3, a component of the molecular clock, regulates the development of B cells. *Immunology*. 2006;119(4):451–460.
181. Marcheva B, Ramsey KM, Buhr ED, et al. Disruption of the clock components CLOCK and BMAL1 leads to hypoinsulinaemia and diabetes. *Nature*. 2010;466 (7306):627–631.
182. Gorbacheva VY, Kondratov RV, Zhang R, et al. Circadian sensitivity to the chemotherapeutic agent cyclophosphamide depends on the functional status of the CLOCK/ BMAL1 transactivation complex. *Proc Natl Acad Sci USA*. 2005;102(9):3407–3412.
183. Arjona A, Sarkar DK. The circadian gene mPer2 regulates the daily rhythm of IFN-gamma. *J Interferon Cytokine Res*. 2006;26(9):645–649.
184. Gibbs JE, Blaikley J, Beesley S, et al. The nuclear receptor REV-ERBalpha mediates circadian regulation of innate immunity through selective regulation of inflammatory cytokines. *Proc Natl Acad Sci USA*. 2012;109(2):582–587.
185. Arble DM, Bass J, Laposky AD, Vitaterna MH, Turek FW. Circadian timing of food intake contributes to weight gain. *Obesity*. Nov 2009;17(11):2100–2102.
186. Turek FW, Joshu C, Kohsaka A, et al. Obesity and metabolic syndrome in circadian Clock mutant mice. *Science*. 2005;308(5724):1043–1045.
187. Librodo P, Buckley M, Luk M, Bisso A. Chronotherapeutic drug delivery. *J Infus Nurs*. 2012;35(5):329–334.
188. Lamia KA, Papp SJ, Yu RT, et al. Cryptochromes mediate rhythmic repression of the glucocorticoid receptor. *Nature*. 2011;480(7378):552–556.
189. Zhang EE, Liu Y, Dentin R, et al. Cryptochrome mediates circadian regulation of cAMP signaling and hepatic gluconeogenesis. *Nat Med*. 2010;16(10):1152–1156.

190. Ikeda H, Yong Q, Kurose T, et al. Clock gene defect disrupts light-dependency of autonomic nerve activity. *Biochem Biophys Res Commun*. 2007;364(3):457–463.
191. Verma M, Srivastava S. New cancer biomarkers deriving from NCI early detection research. *Recent Results Cancer Res*. 2003;163:72–84, discussion 264–266.
192. Logan RW, Wynne O, Levitt D, Price D, Sarkar DK. Altered circadian expression of cytokines and cytolytic factors in splenic natural killer cells of Per1(−/−) mutant mice. *J Interferon Cytokine Res*. 2013;33(3):108–114.
193. Luo Y, Tian W, Cai L, et al. Expression profiling reveals a positive regulation by mPer2 on circadian rhythm of cytotoxicity receptors: Ly49C and Nkg2d. *Chronobiol Int*. 2009;26(8):1514–1544.
194. Wang CY, Wen MS, Wang HW, et al. Increased vascular senescence and impaired endothelial progenitor cell function mediated by mutation of circadian gene Per2. *Circulation*. 2008;118(21):2166–2173.
195. Arjona A, Sarkar DK. Circadian oscillations of clock genes, cytolytic factors, and cytokines in rat NK cells. *J Immunol*. 2005;174(12):7618–7624.
196. Liu J, Malkani G, Shi X, et al. The circadian clock Period 2 gene regulates gamma interferon production of NK cells in host response to lipopolysaccharide-induced endotoxic shock. *Infect Immun*. 2006;74(8):4750–4756.
197. Koike N, Yoo SH, Huang HC, et al. Transcriptional architecture and chromatin landscape of the core circadian clock in mammals. *Science*. 2012;338(6105):349–354.
198. Masri S, Zocchi L, Katada S, Mora E, Sassone-Corsi P. The circadian clock transcriptional complex: metabolic feedback intersects with epigenetic control. *Ann N Y Acad Sci*. 2012;1264(1):103–109.
199. Reddy AB, Karp NA, Maywood ES, et al. Circadian orchestration of the hepatic proteome. *Curr Biol*. 2006;16(11):1107–1115.
200. Hunt T, Sassone-Corsi P. Riding tandem: circadian clocks and the cell cycle. *Cell*. 2007;129(3):461–464.
201. Yoo SH, Yamazaki S, Lowrey PL, et al. PERIOD2::LUCIFERASE real-time reporting of circadian dynamics reveals persistent circadian oscillations in mouse peripheral tissues. *Proc Natl Acad Sci USA*. 2004;101(15):5339–5346.
202. Schibler U, Sassone-Corsi P. A web of circadian pacemakers. *Cell*. 2002;111 (7):919–922.
203. Nagoshi E, Saini C, Bauer C, Laroche T, Naef F, Schibler U. Circadian gene expression in individual fibroblasts: cell-autonomous and self-sustained oscillators pass time to daughter cells. *Cell*. 2004;119(5):693–705.
204. Florian MC, Geiger H. Concise review: polarity in stem cells, disease, and aging. *Stem Cells*. 2010;28(9):1623–1629.
205. de Boer BA, van den Berg G, Soufan AT, et al. Measurement and 3D-visualization of cell-cycle length using double labelling with two thymidine analogues applied in early heart development. *PLoS One*. 2012;7(10):e47719.
206. Rubin P, Williams JP, Devesa SS, Travis LB, Constine LS. Cancer genesis across the age spectrum: associations with tissue development, maintenance, and senescence. *Semin Radiat Oncol*. 2010;20(1):3–11.
207. Shackleton M. Normal stem cells and cancer stem cells: similar and different. *Semin Cancer Biol*. 2010;20(2):85–92.
208. Cooper ZK. Mitotic rhythm in human epidermis. *Proc Soc Exp Biol Med*. 1938;39:323–324.
209. Dublin WB, Gregg RO, Broders AC. Mitosis in specimens removed during day and night from carcinoma of large intestine. *Arch Pathol*. 1940;30:893–895.
210. Smaaland R, Laerum OD, Lote K, Sletvold O, Sothern RB, Bjerknes R. DNA synthesis in human bone marrow is circadian stage dependent. *Blood*. 1991;77(12):2603–2611.

211. Buchi KN, Moore JG, Hrushesky WJ, Sothern RB, Rubin NH. Circadian rhythm of cellular proliferation in the human rectal mucosa. *Gastroenterology*. 1991;101 (2):410–415.
212. Marra G, Anti M, Percesepe A, et al. Circadian variations of epithelial cell proliferation in human rectal crypts. *Gastroenterology*. 1994;106(4):982–987.
213. Sothern RB, Smaaland R, Moore JG. Circannual rhythm in DNA synthesis (S-phase) in healthy human bone marrow and rectal mucosa. *FASEB J*. 1995;9(5):397–403.
214. Rydell R, Wennerberg J, Willen R. Circadian variations in cell cycle phase distribution in a squamous cell carcinoma xenograft; effects of cisplatin and fluorouracil treatment. *In Vivo*. 1990;4(6):385–389.
215. Rotstein J, Sarma DS, Farber E. Sequential alterations in growth control and cell dynamics of rat hepatocytes in early precancerous steps in hepatocarcinogenesis. *Cancer Res*. 1986;46(5):2377–2385.
216. Izquierdo JN. Increased cell proliferation with persistence of circadian rhythms in hamster cheek pouch neoplasms. *Cell Tissue Kinet*. 1977;10(4):313–322.
217. Hrushesky WJ, Lannin D, Haus E. Evidence for an ontogenetic basis for circadian coordination of cancer cell proliferation. *J Natl Cancer Inst*. 1998;90(19):1480–1484.
218. Echave Llanos JM, Nash RE. Mitotic circadian rhythm in a fast-growing and a slow-growing hepatoma: mitotic rhythm in hepatomas. *J Natl Cancer Inst*. 1970;44(3):581–585.
219. Klevecz RR, Braly PS. Circadian and ultradian rhythms of proliferation in human ovarian cancer. *Chronobiol Int*. 1987;4(4):513–523.
220. Klevecz RR, Braly PS. Circadian and ultradian cytokinetic rhythms of spontaneous human cancer. *Ann N Y Acad Sci*. 1991;618:257–276.
221. Smaaland R, Lote K, Sothern RB, Laerum OD. DNA synthesis and ploidy in non-Hodgkin's lymphomas demonstrate intrapatient variation depending on circadian stage of cell sampling. *Cancer Res*. 1993;53(13):3129–3138.
222. Dang CV. MYC on the path to cancer. *Cell*. 2012;149(1):22–35.
223. Levens D. You don't muck with MYC. *Genes Cancer*. 2010;1(6):547–554.
224. Evan GI, Vousden KH. Proliferation, cell cycle and apoptosis in cancer. *Nature*. 2001;411(6835):342–348.
225. Repouskou A, Sourlingas TG, Sekeri-Pataryas KE, Prombona A. The circadian expression of c-MYC is modulated by the histone deacetylase inhibitor trichostatin A in synchronized murine neuroblastoma cells. *Chronobiol Int*. 2010;27(4):722–741.
226. Bjarnason GA, Jordan R. Circadian variation of cell proliferation and cell cycle protein expression in man: clinical implications. *Prog Cell Cycle Res*. 2000;4:193–206.
227. Bjarnason GA, Jordan RC, Sothern RB. Circadian variation in the expression of cell-cycle proteins in human oral epithelium. *Am J Pathol*. 1999;154(2):613–622.
228. Bjarnason GA, Jordan RC, Wood PA, et al. Circadian expression of clock genes in human oral mucosa and skin: association with specific cell-cycle phases. *Am J Pathol*. 2001;158(5):1793–1801.
229. Casagolda D, Del Valle-Perez B, Valls G, et al. A p120-catenin-CK1epsilon complex regulates Wnt signaling. *J Cell Sci*. 2010;123(Pt. 15):2621–2631.
230. Lee E, Salic A, Kirschner MW. Physiological regulation of [beta]-catenin stability by Tcf3 and CK1epsilon. *J Cell Biol*. 2001;154(5):983–993.
231. Anastas JN, Moon RT. WNT signalling pathways as therapeutic targets in cancer. *Nat Rev Cancer*. 2013;13(1):11–26.
232. Valenta T, Hausmann G, Basler K. The many faces and functions of beta-catenin. *EMBO J*. 2012;31(12):2714–2736.
233. Batlle E, Henderson JT, Beghtel H, et al. Beta-catenin and TCF mediate cell positioning in the intestinal epithelium by controlling the expression of EphB/ephrinB. *Cell*. 2002;111(2):251–263.

234. van de Wetering M, de Lau W, Clevers H. WNT signaling and lymphocyte development. *Cell*. 2002;109(Suppl.):S13–S19.
235. van de Wetering M, Sancho E, Verweij C, et al. The beta-catenin/TCF-4 complex imposes a crypt progenitor phenotype on colorectal cancer cells. *Cell*. 2002;111 (2):241–250.
236. Tetsu O, McCormick F. Beta-catenin regulates expression of cyclin D1 in colon carcinoma cells. *Nature*. 1999;398(6726):422–426.
237. Crawford HC, Fingleton B, Gustavson MD, et al. The PEA3 subfamily of Ets transcription factors synergizes with beta-catenin-LEF-1 to activate matrilysin transcription in intestinal tumors. *Mol Cell Biol*. 2001;21(4):1370–1383.
238. Mann B, Gelos M, Siedow A, et al. Target genes of beta-catenin-T cell-factor/lymphoid-enhancer-factor signaling in human colorectal carcinomas. *Proc Natl Acad Sci USA*. 1999;96(4):1603–1608.
239. Ahmed Y, Hayashi S, Levine A, Wieschaus E. Regulation of armadillo by a Drosophila APC inhibits neuronal apoptosis during retinal development. *Cell*. 1998;93(7):1171–1182.
240. Martinek S, Inonog S, Manoukian AS, Young MW. A role for the segment polarity gene shaggy/GSK-3 in the Drosophila circadian clock. *Cell*. 2001;105(6):769–779.
241. Shimasaki T, Kitano A, Motoo Y, Minamoto T. Aberrant glycogen synthase kinase 3beta in the development of pancreatic cancer. *J Carcinog*. 2012;11:15.
242. Shimura T. Acquired radioresistance of cancer and the AKT/GSK3beta/cyclin D1 overexpression cycle. *J Radiat Res*. 2011;52(5):539–544.
243. Iitaka C, Miyazaki K, Akaike T, Ishida N. A role for glycogen synthase kinase-3beta in the mammalian circadian clock. *J Biol Chem*. 2005;280(33):29397–29402.
244. Kondratov RV, Shamanna RK, Kondratova AA, Gorbacheva VY, Antoch MP. Dual role of the CLOCK/BMAL1 circadian complex in transcriptional regulation. *FASEB J*. 2006;20(3):530–532.
245. Doi M, Takahashi Y, Komatsu R, et al. Salt-sensitive hypertension in circadian clock-deficient Cry-null mice involves dysregulated adrenal Hsd3b6. *Nat Med*. 2010;16 (1):67–74.
246. Hua H, Wang Y, Wan C, et al. Circadian gene mPer2 overexpression induces cancer cell apoptosis. *Cancer Sci*. 2006;97(7):589–596.
247. Relles D, Sendecki J, Chipitsyna G, Hyslop T, Yeo CJ, Arafat HA. Circadian gene expression and clinicopathologic correlates in pancreatic cancer. *J Gastrointest Surg*. 2013;17(3):443–450.
248. Lindahl T, Barnes DE. Repair of endogenous DNA damage. *Cold Spring Harb Symp Quant Biol*. 2000;65:127–133.
249. Kastan MB, Bartek J. Cell-cycle checkpoints and cancer. *Nature*. 2004;432 (7015):316–323.
250. Thompson CB. Apoptosis in the pathogenesis and treatment of disease. *Science*. 1995;267(5203):1456–1462.
251. Bakkenist CJ, Kastan MB. DNA damage activates ATM through intermolecular autophosphorylation and dimer dissociation. *Nature*. 2003;421(6922):499–506.
252. Cimprich KA, Shin TB, Keith CT, Schreiber SL. cDNA cloning and gene mapping of a candidate human cell cycle checkpoint protein. *Proc Natl Acad Sci USA*. 1996;93 (7):2850–2855.
253. el-Deiry WS, Tokino T, Velculescu VE, et al. WAF1, a potential mediator of p53 tumor suppression. *Cell*. 1993;75(4):817–825.
254. Harper JW, Adami GR, Wei N, Keyomarsi K, Elledge SJ. The p21 Cdk-interacting protein Cip1 is a potent inhibitor of G1 cyclin-dependent kinases. *Cell*. 1993;75(4):805–816.
255. Rich T, Allen RL, Wyllie AH. Defying death after DNA damage. *Nature*. 2000;407 (6805):777–783.

256. Kang TH, Lindsey-Boltz LA, Reardon JT, Sancar A. Circadian control of XPA and excision repair of cisplatin-DNA damage by cryptochrome and HERC2 ubiquitin ligase. *Proc Natl Acad Sci USA*. 2010;107(11):4890–4895.
257. Gaddameedhi S, Selby CP, Kaufmann WK, Smart RC, Sancar A. Control of skin cancer by the circadian rhythm. *Proc Natl Acad Sci USA*. 2011;108(46):18790–18795.
258. Chaves I, Pokorny R, Byrdin M, et al. The cryptochromes: blue light photoreceptors in plants and animals. *Annu Rev Plant Biol*. 2011;62:335–364.
259. Barnes JW, Tischkau SA, Barnes JA, et al. Requirement of mammalian Timeless for circadian rhythmicity. *Science*. 2003;302(5644):439–442.
260. Oklejewicz M, Destici E, Tamanini F, Hut RA, Janssens R, van der Horst GT. Phase resetting of the mammalian circadian clock by DNA damage. *Curr Biol*. 2008;18 (4):286–291.
261. Gamsby JJ, Loros JJ, Dunlap JC. A phylogenetically conserved DNA damage response resets the circadian clock. *J Biol Rhythms*. 2009;24(3):193–202.
262. Warburg O. On the origin of cancer cells. *Science*. 1956;123(3191):309–314.
263. Ferreira LM, Hebrant A, Dumont JE. Metabolic reprogramming of the tumor. *Oncogene*. 2012;31(36):3999–4011.
264. Hsu PP, Sabatini DM. Cancer cell metabolism: Warburg and beyond. *Cell*. 2008;134 (5):703–707.
265. Kroemer G, Pouyssegur J. Tumor cell metabolism: cancer's Achilles' heel. *Cancer Cell*. 2008;13(6):472–482.
266. Vander Heiden MG, Cantley LC, Thompson CB. Understanding the Warburg effect: the metabolic requirements of cell proliferation. *Science*. 2009;324 (5930):1029–1033.
267. Tennant DA, Duran RV, Boulahbel H, Gottlieb E. Metabolic transformation in cancer. *Carcinogenesis*. 2009;30(8):1269–1280.
268. Bensaad K, Vousden KH. p53: new roles in metabolism. *Trends Cell Biol*. 2007;17 (6):286–291.
269. Levine AJ, Puzio-Kuter AM. The control of the metabolic switch in cancers by oncogenes and tumor suppressor genes. *Science*. 2010;330(6009):1340–1344.
270. Cairns RA, Harris IS, Mak TW. Regulation of cancer cell metabolism. *Nat Rev Cancer*. 2011;11(2):85–95.
271. Schulze A, Harris AL. How cancer metabolism is tuned for proliferation and vulnerable to disruption. *Nature*. 2012;491(7424):364–373.
272. Huang W, Ramsey KM, Marcheva B, Bass J. Circadian rhythms, sleep, and metabolism. *J Clin Invest*. 2011;121(6):2133–2141.
273. Maury E, Ramsey KM, Bass J. Circadian rhythms and metabolic syndrome: from experimental genetics to human disease. *Circ Res*. 2010;106(3):447–462.
274. Bass J, Takahashi JS. Circadian integration of metabolism and energetics. *Science*. 2010;330(6009):1349–1354.
275. Bass J. Circadian topology of metabolism. *Nature*. 2012;491(7424):348–356.
276. Koike N, Yoo SH, Huang HC. Transcriptional architecture and chromatin landscape of the core circadian clock in mammals. *Science*. 2012;338(6105):349–354.
277. Garbarino-Pico E, Green CB. Posttranscriptional regulation of mammalian circadian clock output. *Cold Spring Harb Symp Quant Biol*. 2007;72:145–156.
278. Damiola F, Le Minh N, Preitner N, Kornmann B, Fleury-Olela F, Schibler U. Restricted feeding uncouples circadian oscillators in peripheral tissues from the central pacemaker in the suprachiasmatic nucleus. *Genes Dev*. 2000;14(23):2950–2961.
279. Stokkan KA, Yamazaki S, Tei H, Sakaki Y, Menaker M. Entrainment of the circadian clock in the liver by feeding. *Science*. 2001;291(5503):490–493.
280. Rutter J, Reick M, Wu LC, McKnight SL. Regulation of clock and NPAS2 DNA binding by the redox state of NAD cofactors. *Science*. 2001;293(5529):510–514.

281. Ramsey KM, Yoshino J, Brace CS, et al. Circadian clock feedback cycle through NAMPT-mediated NAD+ biosynthesis. *Science*. 2009;324(5927):651–654.
282. Lamia KA, Sachdeva UM, DiTacchio L, et al. AMPK regulates the circadian clock by cryptochrome phosphorylation and degradation. *Science*. 2009;326(5951):437–440.
283. Yin L, Wu N, Curtin JC, et al. Rev-erbalpha, a heme sensor that coordinates metabolic and circadian pathways. *Science*. 2007;318(5857):1786–1789.
284. Dang CV. c-Myc target genes involved in cell growth, apoptosis, and metabolism. *Mol Cell Biol*. 1999;19(1):1–11.
285. Bensaad K, Tsuruta A, Selak MA, et al. TIGAR, a p53-inducible regulator of glycolysis and apoptosis. *Cell*. 2006;126(1):107–120.
286. Matoba S, Kang JG, Patino WD, et al. p53 regulates mitochondrial respiration. *Science*. 2006;312(5780):1650–1653.
287. Fernandez PC, Frank SR, Wang L, et al. Genomic targets of the human c-Myc protein. *Genes Dev*. 2003;17(9):1115–1129.
288. Orian A, Grewal SS, Knoepfler PS, Edgar BA, Parkhurst SM, Eisenman RN. Genomic binding and transcriptional regulation by the Drosophila Myc and Mnt transcription factors. *Cold Spring Harb Symp Quant Biol*. 2005;70:299–307.
289. Li Z, Van Calcar S, Qu C, Cavenee WK, Zhang MQ, Ren B. A global transcriptional regulatory role for c-Myc in Burkitt's lymphoma cells. *Proc Natl Acad Sci USA*. 2003;100(14):8164–8169.
290. Cawley S, Bekiranov S, Ng HH, et al. Unbiased mapping of transcription factor binding sites along human chromosomes 21 and 22 points to widespread regulation of noncoding RNAs. *Cell*. 2004;116(4):499–509.
291. Gao P, Tchernyshyov I, Chang TC, et al. c-Myc suppression of miR-23a/b enhances mitochondrial glutaminase expression and glutamine metabolism. *Nature*. 2009;458 (7239):762–765.
292. Cam H, Balciunaite E, Blais A, et al. A common set of gene regulatory networks links metabolism and growth inhibition. *Mol Cell*. 2004;16(3):399–411.
293. Morrish F, Giedt C, Hockenbery D. c-MYC apoptotic function is mediated by NRF-1 target genes. *Genes Dev*. 2003;17(2):240–255.
294. Hansson A, Hance N, Dufour E, et al. A switch in metabolism precedes increased mitochondrial biogenesis in respiratory chain-deficient mouse hearts. *Proc Natl Acad Sci USA*. 2004;101(9):3136–3141.
295. Vafa O, Wade M, Kern S, et al. c-Myc can induce DNA damage, increase reactive oxygen species, and mitigate p53 function: a mechanism for oncogene-induced genetic instability. *Mol Cell*. 2002;9(5):1031–1044.
296. Tanaka H, Matsumura I, Ezoe S, et al. E2F1 and c-Myc potentiate apoptosis through inhibition of NF-kappaB activity that facilitates MnSOD-mediated ROS elimination. *Mol Cell*. 2002;9(5):1017–1029.
297. Sagun KC, Carcamo JM, Golde DW. Antioxidants prevent oxidative DNA damage and cellular transformation elicited by the over-expression of c-MYC. *Mutat Res*. 2006;593(1–2):64–79.
298. Pelengaris S, Khan M, Evan G. c-MYC: more than just a matter of life and death. *Nat Rev Cancer*. 2002;2(10):764–776.
299. Breckenridge DG, Shore GC. Regulation of apoptosis by E1A and Myc oncoproteins. *Crit Rev Eukaryot Gene Expr*. 2000;10(3–4):273–280.
300. Blyth K, Terry A, O'Hara M, et al. Synergy between a human c-myc transgene and p53 null genotype in murine thymic lymphomas: contrasting effects of homozygous and heterozygous p53 loss. *Oncogene*. 1995;10(9):1717–1723.
301. Elson A, Deng C, Campos-Torres J, Donehower LA, Leder P. The MMTV/c-myc transgene and p53 null alleles collaborate to induce T-cell lymphomas, but not mammary carcinomas in transgenic mice. *Oncogene*. 1995;11(1):181–190.

302. Lowe SW, Cepero E, Evan G. Intrinsic tumour suppression. *Nature*. 2004;432 (7015):307–315.
303. Pan X, Zhang Y, Wang L, Hussain MM. Diurnal regulation of MTP and plasma triglyceride by CLOCK is mediated by SHP. *Cell Metab*. 2010;12(2):174–186.
304. Liu C, Li S, Liu T, Borjigin J, Lin JD. Transcriptional coactivator PGC-1alpha integrates the mammalian clock and energy metabolism. *Nature*. 2007;447(7143): 477–481.
305. Hofman MA, Swaab DF. Living by the clock: the circadian pacemaker in older people. *Ageing Res Rev*. 2006;5(1):33–51.
306. Kondratov RV. A role of the circadian system and circadian proteins in aging. *Ageing Res Rev*. 2007;6(1):12–27.
307. Asai M, Yoshinobu Y, Kaneko S, et al. Circadian profile of Per gene mRNA expression in the suprachiasmatic nucleus, paraventricular nucleus, and pineal body of aged rats. *J Neurosci Res*. 2001;66(6):1133–1139.
308. Lachman ME, Agrigoroaei S. Promoting functional health in midlife and old age: long-term protective effects of control beliefs, social support, and physical exercise. *PLoS One*. 2010;5(10):e13297.
309. Balducci L, Ershler WB. Cancer and ageing: a nexus at several levels. *Nat Rev Cancer*. 2005;5(8):655–662.
310. Jemal A, Siegel R, Xu J, Ward E. Cancer statistics, 2010. *CA Cancer J Clin*. 2010;60 (5):277–300.
311. Rose MR. The evolutionary biology of aging. New York: Oxford University Press. 1991. ix + 221 pp.
312. Pleiotropy GW. Natural selection, and senescence. *Evolution*. 1957;11:398–411.
313. Atwood CS, Bowen RL. The reproductive-cell cycle theory of aging: an update. *Exp Gerontol*. 2011;46(2–3):100–107.
314. Strehler BL. Genetic instability as the primary cause of human aging. *Exp Gerontol*. 1986;21(4–5):283–319.
315. Pawelec G, Derhovanessian E, Larbi A. Immunosenescence and cancer. *Crit Rev Oncol Hematol*. 2010;75(2):165–172.
316. Anisimov VN, Sikora E, Pawelec G. Relationships between cancer and aging: a multilevel approach. *Biogerontology*. 2009;10(4):323–338.
317. Lee DJ, Cha EK, Dubin JM, et al. Novel therapeutics for the management of castration-resistant prostate cancer (CRPC). *BJU Int*. 2012;109(7):968–985.
318. Minelli A, Bellezza I, Conte C, Culig Z. Oxidative stress-related aging: a role for prostate cancer? *Biochim Biophys Acta*. 2009;1795(2):83–91.
319. Best BP. Nuclear DNA, damage as a direct cause of aging. *Rejuvenation Res*. 2009;12 (3):199–208.
320. Espejel S, Martin M, Klatt P, Martin-Caballero J, Flores JM, Blasco MA. Shorter telomeres, accelerated ageing and increased lymphoma in DNA-PKcs-deficient mice. *EMBO Rep*. 2004;5(5):503–509.
321. de Boer J, Andressoo JO, de Wit J, et al. Premature aging in mice deficient in DNA repair and transcription. *Science*. 2002;296(5571):1276–1279.
322. Dolle ME, Busuttil RA, Garcia AM, et al. Increased genomic instability is not a prerequisite for shortened lifespan in DNA repair deficient mice. *Mutat Res*. 2006;596 (1–2):22–35.
323. Cao L, Li W, Kim S, Brodie SG, Deng CX. Senescence, aging, and malignant transformation mediated by p53 in mice lacking the Brca1 full-length isoform. *Genes Dev*. 2003;17(2):201–213.
324. Sager R. Senescence as a mode of tumor suppression. *Environ Health Perspect*. 1991;93:59–62.

325. Noureddine H, Gary-Bobo G, Alifano M, et al. Pulmonary artery smooth muscle cell senescence is a pathogenic mechanism for pulmonary hypertension in chronic lung disease. *Circ Res*. 2011;109(5):543–553.
326. Parrinello S, Coppe JP, Krtolica A, Campisi J. Stromal-epithelial interactions in aging and cancer: senescent fibroblasts alter epithelial cell differentiation. *J Cell Sci*. 2005;118 (Pt. 3):485–496.
327. Krtolica A, Parrinello S, Lockett S, Desprez PY, Campisi J. Senescent fibroblasts promote epithelial cell growth and tumorigenesis: a link between cancer and aging. *Proc Natl Acad Sci USA*. 2001;98(21):12072–12077.
328. Liu D, Hornsby PJ. Senescent human fibroblasts increase the early growth of xenograft tumors via matrix metalloproteinase secretion. *Cancer Res*. 2007;67(7):3117–3126.
329. Guney I, Wu S, Sedivy JM. Reduced c-Myc signaling triggers telomere-independent senescence by regulating Bmi-1 and p16(INK4a). *Proc Natl Acad Sci USA*. 2006;103 (10):3645–3650.
330. Campisi J, d'Adda di Fagagna F. Cellular senescence: when bad things happen to good cells. *Nat Rev Mol Cell Biol*. 2007;8(9):729–740.
331. Adams PD. Healing and hurting: molecular mechanisms, functions, and pathologies of cellular senescence. *Mol Cell*. 2009;36(1):2–14.
332. Collins CJ, Sedivy JM. Involvement of the INK4a/Arf gene locus in senescence. *Aging Cell*. 2003;2(3):145–150.
333. Rufini A, Tucci P, Celardo I, Melino G. Senescence and aging: the critical roles of p53. *Oncogene Advance Online Publication*. 2013; http://dx.doi.org/10.1038/onc.2012.640.
334. Hardeland R, Coto-Montes A, Poeggeler B. Circadian rhythms, oxidative stress, and antioxidative defense mechanisms. *Chronobiol Int*. 2003;20(6):921–962.
335. Longo VD, Kennedy BK. Sirtuins in aging and age-related disease. *Cell*. 2006;126 (2):257–268.
336. Vaziri H, Dessain SK, Ng Eaton E, et al. hSIR2(SIRT1) functions as an NAD-dependent p53 deacetylase. *Cell*. 2001;107(2):149–159.
337. Jeong J, Juhn K, Lee H, et al. SIRT1 promotes DNA repair activity and deacetylation of Ku70. *Exp Mol Med*. 2007;39(1):8–13.
338. Cho IR, Koh SS, Malilas W, et al. SIRT1 inhibits proliferation of pancreatic cancer cells expressing pancreatic adenocarcinoma up-regulated factor (PAUF), a novel oncogene, by suppression of beta-catenin. *Biochem Biophys Res Commun*. 2012;423 (2):270–275.
339. Kang TW, Yevsa T, Woller N, et al. Senescence surveillance of pre-malignant hepatocytes limits liver cancer development. *Nature*. 2011;479(7374):547–551.
340. Kalsbeek A, van der Spek R, Lei J, Endert E, Buijs RM, Fliers E. Circadian rhythms in the hypothalamo-pituitary-adrenal (HPA) axis. *Mol Cell Endocrinol*. 2012;349(1):20–29.
341. Rajagopal S, Rajagopal K, Lefkowitz RJ. Teaching old receptors new tricks: biasing seven-transmembrane receptors. *Nat Rev Drug Discov*. 2010;9(5):373–386.
342. Rosenbaum DM, Rasmussen SG, Kobilka BK. The structure and function of G-protein-coupled receptors. *Nature*. 2009;459(7245):356–363.
343. Janig W, Habler HJ. Specificity in the organization of the autonomic nervous system: a basis for precise neural regulation of homeostatic and protective body functions. *Prog Brain Res*. 2000;122:351–367.
344. Hanaoka BY, Peterson CA, Horbinski C, Crofford LJ. Implications of glucocorticoid therapy in idiopathic inflammatory myopathies. *Nat Rev Rheumatol*. 2012;8(8):448–457.
345. Samarasinghe RA, Witchell SF, DeFranco DB. Cooperativity and complementarity: synergies in non-classical and classical glucocorticoid signaling. *Cell Cycle*. 2012;11 (15):2819–2827.

346. Deblois G, Giguere V. Oestrogen-related receptors in breast cancer: control of cellular metabolism and beyond. *Nat Rev Cancer.* 2013;13(1):27–36.
347. Cui J, Shen Y, Li R. Estrogen synthesis and signaling pathways during aging: from periphery to brain. *Trends Mol Med.* 2013;19(3):197–209.
348. Hasan KM, Rahman MS, Arif KM, Sobhani ME. Psychological stress and aging: role of glucocorticoids (GCs). *Age (Dordr).* 2012;34(6):1421–1433.
349. Coto-Montes A, Boga JA, Rosales-Corral S, Fuentes-Broto L, Tan DX, Reiter RJ. Role of melatonin in the regulation of autophagy and mitophagy: a review. *Mol Cell Endocrinol.* 2012;361(1–2):12–23.
350. Greene MW. Circadian rhythms and tumor growth. *Cancer Lett.* 2012;318(2):115–123.
351. Reiter RJ, Calvo JR, Karbownik M, Qi W, Tan DX. Melatonin and its relation to the immune system and inflammation. *Ann N Y Acad Sci.* 2000;917:376–386.
352. Xu CX, Wang C, Krager SL, Bottum KM, Tischkau SA. Aryl hydrocarbon receptor activation attenuates Per1 gene induction and influences circadian clock resetting. *Toxicol Sci.* 2013;132(2):368–378.
353. Zhu H, Vadigepalli R, Rafferty R, Gonye GE, Weaver DR, Schwaber JS. Integrative gene regulatory network analysis reveals light-induced regional gene expression phase shift programs in the mouse suprachiasmatic nucleus. *PLoS One.* 2012;7(5):e37833.
354. Jakubcakova V, Oster H, Tamanini F, et al. Light entrainment of the mammalian circadian clock by a PRKCA-dependent posttranslational mechanism. *Neuron.* 2007;54 (5):831–843.
355. Hainich EC, Pizzio GA, Golombek DA. Constitutive activation of the ERK-MAPK pathway in the suprachiasmatic nuclei inhibits circadian resetting. *FEBS Lett.* 2006;580 (28–29):6665–6668.
356. Hirota T, Fukada Y. Resetting mechanism of central and peripheral circadian clocks in mammals. *Zoolog Sci.* 2004;21(4):359–368.
357. Shigeyoshi Y, Taguchi K, Yamamoto S, et al. Light-induced resetting of a mammalian circadian clock is associated with rapid induction of the mPer1 transcript. *Cell.* 1997;91 (7):1043–1053.
358. Takeuchi J, Shannon W, Aronin N, Schwartz WJ. Compositional changes of AP-1 DNA-binding proteins are regulated by light in a mammalian circadian clock. *Neuron.* 1993;11(5):825–836.
359. Obrietan K, Impey S, Storm DR. Light and circadian rhythmicity regulate MAP kinase activation in the suprachiasmatic nuclei. *Nat Neurosci.* 1998;1(8):693–700.
360. Ding JM, Faiman LE, Hurst WJ, Kuriashkina LR, Gillette MU. Resetting the biological clock: mediation of nocturnal CREB phosphorylation via light, glutamate, and nitric oxide. *J Neurosci.* 1997;17(2):667–675.
361. Hurst WJ, Mitchell JW, Gillette MU. Synchronization and phase-resetting by glutamate of an immortalized SCN cell line. *Biochem Biophys Res Commun.* 2002;298 (1):133–143.
362. Kon N, Hirota T, Kawamoto T, Kato Y, Tsubota T, Fukada Y. Activation of TGF-beta/activin signalling resets the circadian clock through rapid induction of Dec1 transcripts. *Nat Cell Biol.* 2008;10(12):1463–1469.
363. Gerber A, Esnault C, Aubert G, Treisman R, Pralong F, Schibler U. Blood-borne circadian signal stimulates daily oscillations in actin dynamics and SRF activity. *Cell.* 2013;152(3):492–503.
364. Lu NZ, Cidlowski JA. The origin and functions of multiple human glucocorticoid receptor isoforms. *Ann N Y Acad Sci.* 2004;1024:102–123.
365. Chung S, Son GH, Kim K. Adrenal peripheral oscillator in generating the circadian glucocorticoid rhythm. *Ann N Y Acad Sci.* 2011;1220:71–81.
366. Reddy AB, Maywood ES, Karp NA, et al. Glucocorticoid signaling synchronizes the liver circadian transcriptome. *Hepatology.* 2007;45(6):1478–1488.

367. Yamamoto T, Shimano H, Nakagawa Y, et al. SREBP-1 interacts with hepatocyte nuclear factor-4 alpha and interferes with PGC-1 recruitment to suppress hepatic gluconeogenic genes. *J Biol Chem*. 2004;279(13):12027–12035.
368. So AY, Bernal TU, Pillsbury ML, Yamamoto KR, Feldman BJ. Glucocorticoid regulation of the circadian clock modulates glucose homeostasis. *Proc Natl Acad Sci USA*. 2009;106(41):17582–17587.
369. Pezuk P, Mohawk JA, Wang LA, Menaker M. Glucocorticoids as entraining signals for peripheral circadian oscillators. *Endocrinology*. Oct 2012;153(10):4775–4783.
370. Furness JB. The organisation of the autonomic nervous system: peripheral connections. *Auton Neurosci*. 2006;130(1–2):1–5.
371. Shibata S. Neural regulation of the hepatic circadian rhythm. *Anat Rec A Discov Mol Cell Evol Biol*. 2004;280(1):901–909.
372. Vujovic N, Davidson AJ, Menaker M. Sympathetic input modulates, but does not determine, phase of peripheral circadian oscillators. *Am J Physiol Regul Integr Comp Physiol*. 2008;295(1):R355–R360.
373. Takekida S, Yan L, Maywood ES, Hastings MH, Okamura H. Differential adrenergic regulation of the circadian expression of the clock genes Period1 and Period2 in the rat pineal gland. *Eur J Neurosci*. 2000;12(12):4557–4561.
374. Logan RW, Arjona A, Sarkar DK. Role of sympathetic nervous system in the entrainment of circadian natural-killer cell function. *Brain Behav Immun*. 2011;25(1):101–109.
375. Mendez-Ferrer S, Lucas D, Battista M, Frenette PS. Haematopoietic stem cell release is regulated by circadian oscillations. *Nature*. 2008;452(7186):442–447.
376. Kang J, Shi Y, Xiang B, et al. A nuclear function of beta-arrestin1 in GPCR signaling: regulation of histone acetylation and gene transcription. *Cell*. 2005;123(5):833–847.
377. Tilley DG. G protein-dependent and G protein-independent signaling pathways and their impact on cardiac function. *Circ Res*. 2011;109(2):217–230.
378. Reiter E, Ahn S, Shukla AK, Lefkowitz RJ. Molecular mechanism of beta-arrestin-biased agonism at seven-transmembrane receptors. *Annu Rev Pharmacol Toxicol*. 2012;52:179–197.
379. DeFea KA. Beta-arrestins as regulators of signal termination and transduction: how do they determine what to scaffold? *Cell Signal*. 2011;23(4):621–629.
380. Verkaar F, van Rosmalen JW, Blomenrohr M, et al. G protein-independent cell-based assays for drug discovery on seven-transmembrane receptors. *Biotechnol Annu Rev*. 2008;14:253–274.
381. Ling J, Kumar R. Crosstalk between NFkB and glucocorticoid signaling: a potential target of breast cancer therapy. *Cancer Lett*. 2012;322(2):119–126.
382. Grossmann C, Gekle M. Interaction between mineralocorticoid receptor and epidermal growth factor receptor signaling. *Mol Cell Endocrinol*. 2012;350(2):235–241.
383. Schakman O, Gilson H, Kalista S, Thissen JP. Mechanisms of muscle atrophy induced by glucocorticoids. *Horm Res*. 2009;72(Suppl 1):36–41.
384. Mulholland DJ, Dedhar S, Coetzee GA, Nelson CC. Interaction of nuclear receptors with the Wnt/beta-catenin/Tcf signaling axis: Wnt you like to know? *Endocr Rev*. 2005;26(7):898–915.
385. Lang F, Perrotti N, Stournaras C. Colorectal carcinoma cells—regulation of survival and growth by SGK1. *Int J Biochem Cell Biol*. 2010;42(10):1571–1575.
386. Zhang JY, Selim MA. The role of the c-Jun N-terminal Kinase signaling pathway in skin cancer. *Am j cancer res*. 2012;2(6):691–698.
387. Romitti M, Ceolin L, Siqueira DR, Ferreira CV, Wajner SM, Maia AL. Signaling pathways in follicular cell-derived thyroid carcinomas (review). *Int J Oncol*. Jan 2013;42(1):19–28.
388. Servillo G, Della Fazia MA, Sassone-Corsi P. Coupling cAMP signaling to transcription in the liver: pivotal role of CREB and CREM. *Exp Cell Res*. 2002;275(2):143–154.

389. Vucenik I, Stains JP. Obesity and cancer risk: evidence, mechanisms, and recommendations. *Ann N Y Acad Sci*. Oct 2012;1271:37–43.
390. Rodon J, Dienstmann R, Serra V, Tabernero J. Development of PI3K inhibitors: lessons learned from early clinical trials. *Nat Rev Clin Oncol*. 2013;10(3):143–153.
391. Diaz-Meco MT, Moscat J. The atypical PKCs in inflammation: NF-kappaB and beyond. *Immunol Rev*. Mar 2012;246(1):154–167.
392. Steelman LS, Pohnert SC, Shelton JG, Franklin RA, Bertrand FE, McCubrey JA. JAK/STAT, Raf/MEK/ERK, PI3K/Akt and BCR-ABL in cell cycle progression and leukemogenesis. *Leukemia*. Feb 2004;18(2):189–218.
393. Dobrovolna J, Chinenov Y, Kennedy MA, Liu B, Rogatsky I. Glucocorticoid-dependent phosphorylation of the transcriptional coregulator GRIP1. *Mol Cell Biol*. Feb 2012;32(4):730–739.
394. Dahlman-Wright K, Qiao Y, Jonsson P, Gustafsson JA, Williams C, Zhao C. Interplay between AP-1 and estrogen receptor alpha in regulating gene expression and proliferation networks in breast cancer cells. *Carcinogenesis*. Sep 2012;33(9):1684–1691.
395. Fu L, Lee CC. The circadian clock: pacemaker and tumour suppressor. *Nat Rev Cancer*. 2003;3(5):350–361.
396. Rhind N, Russell P. Signaling pathways that regulate cell division. *Cold Spring Harb Perspect Biol*. 2012;4(10).
397. Gallego M, Virshup DM. Post-translational modifications regulate the ticking of the circadian clock. *Nat Rev Mol Cell Biol*. 2007;8(2):139–148.
398. Cao R, Obrietan K. mTOR signaling and entrainment of the mammalian circadian clock. *Mol Cell Pharmacol*. 2010;2(4):125–130.
399. Volden PA, Conzen SD. The influence of glucocorticoid signaling on tumor progression. *Brain Behav Immun*. 2013;30(Suppl):S26–S31.
400. Cole SW. Nervous system regulation of the cancer genome. *Brain Behav Immun*. 2012;30:S10–S18.
401. Anand P, Kunnumakkara AB, Sundaram C, et al. Cancer is a preventable disease that requires major lifestyle changes. *Pharm Res*. 2008;25(9):2097–2116.
402. Dunn GP, Old LJ, Schreiber RD. The three Es of cancer immunoediting. *Annu Rev Immunol*. 2004;22:329–360.
403. Mlecnik B, Bindea G, Pages F, Galon J. Tumor immunosurveillance in human cancers. *Cancer Metastasis Rev*. 2011;30(1):5–12.
404. Street SE, Cretney E, Smyth MJ. Perforin and interferon-gamma activities independently control tumor initiation, growth, and metastasis. *Blood*. 2001;97(1):192–197.
405. Street SE, Trapani JA, MacGregor D, Smyth MJ. Suppression of lymphoma and epithelial malignancies effected by interferon gamma. *J Exp Med*. 2002;196(1): 129–134.
406. van den Broek ME, Kagi D, Ossendorp F, et al. Decreased tumor surveillance in perforin-deficient mice. *J Exp Med*. 1996;184(5):1781–1790.
407. Shinkai Y, Rathbun G, Lam KP, et al. RAG-2-deficient mice lack mature lymphocytes owing to inability to initiate V(D)J rearrangement. *Cell*. 1992;68(5):855–867.
408. Shankaran V, Ikeda H, Bruce AT, et al. IFNgamma and lymphocytes prevent primary tumour development and shape tumour immunogenicity. *Nature*. 2001;410 (6832):1107–1111.
409. Khong HT, Restifo NP. Natural selection of tumor variants in the generation of "tumor escape" phenotypes. *Nat Immunol*. 2002;3(11):999–1005.
410. Mellman I, Coukos G, Dranoff G. Cancer immunotherapy comes of age. *Nature*. 2011;480(7378):480–489.
411. Scheiermann C, Kunisaki Y, Frenette PS. Circadian control of the immune system. *Nat Rev Immunol*. 2013;13(3):190–198.
412. Bellinger DL, Millar BA, Perez S, et al. Sympathetic modulation of immunity: relevance to disease. *Cell Immunol*. 2008;252(1–2):27–56.

413. Sloan EK, Capitanio JP, Cole SW. Stress-induced remodeling of lymphoid innervation. *Brain Behav Immun.* 2008;22(1):15–21.
414. Srinivasan V, Pandi-Perumal SR, Brzezinski A, Bhatnagar KP, Cardinali DP. Melatonin, immune function and cancer. *Recent Pat Endocr Metab Immune Drug Discov.* 2011;5(2):109–123.
415. Bellavance MA, Rivest S. The neuroendocrine control of the innate immune system in health and brain diseases. *Immunol Rev.* 2012;248(1):36–55.
416. Chighizola C, Meroni PL. The role of environmental estrogens and autoimmunity. *Autoimmun Rev.* 2012;11(6–7):A493–A501.
417. Arjona A, Boyadjieva N, Sarkar DK. Circadian rhythms of granzyme B, perforin, IFN-gamma, and NK cell cytolytic activity in the spleen: effects of chronic ethanol. *J Immunol.* 2004;172(5):2811–2817.
418. Arjona A, Sarkar DK. Evidence supporting a circadian control of natural killer cell function. *Brain Behav Immun.* Sep 2006;20(5):469–476.
419. Arjona A, Sarkar DK. Are circadian rhythms the code of hypothalamic-immune communication? Insights from natural killer cells. *Neurochem Res.* 2008;33(4):708–718.
420. Haus E, Smolensky MH. Biologic rhythms in the immune system. *Chronobiol Int.* 1999;16(5):581–622.
421. Silver AC, Arjona A, Hughes ME, Nitabach MN, Fikrig E. Circadian expression of clock genes in mouse macrophages, dendritic cells, and B cells. *Brain Behav Immun.* 2012;26(3):407–413.
422. Suzuki S, Toyabe S, Moroda T, et al. Circadian rhythm of leucocytes and lymphocytes subsets and its possible correlation with the function of the autonomic nervous system. *Clin Exp Immunol.* 1997;110(3):500–508.
423. Okamoto S, Ibaraki K, Hayashi S, Saito M. Ventromedial hypothalamus suppresses splenic lymphocyte activity through sympathetic innervation. *Brain Res.* 1996;739(1–2):308–313.
424. Maestroni GJ, Cosentino M, Marino F, et al. Neural and endogenous catecholamines in the bone marrow. Circadian association of norepinephrine with hematopoiesis? *Exp Hematol.* 1998;26(12):1172–1177.
425. Sephton S, Spiegel D. Circadian disruption in cancer: a neuroendocrine-immune pathway from stress to disease? *Brain Behav Immun.* Oct 2003;17(5):321–328.
426. Wang X, Reece SP, Van Scott MR, Brown JM. A circadian clock in murine bone marrow-derived mast cells modulates IgE-dependent activation in vitro. *Brain Behav Immun.* 2011;25(1):127–134.
427. Logan RW, Sarkar DK. Circadian nature of immune function. *Mol Cell Endocrinol.* 2012;349(1):82–90.
428. Tong X, Buelow K, Guha A, Rausch R, Yin L. USP2a protein deubiquitinates and stabilizes the circadian protein CRY1 in response to inflammatory signals. *J Biol Chem.* 2012;287(30):25280–25291.
429. Murphy BA, Vick MM, Sessions DR, Cook RF, Fitzgerald BP. Acute systemic inflammation transiently synchronizes clock gene expression in equine peripheral blood. *Brain Behav Immun.* 2007;21(4):467–476.
430. Kouri VP, Olkkonen J, Kaivosoja E, et al. Circadian timekeeping is disturbed in rheumatoid arthritis at molecular level. *PLoS One.* 2013;8(1):e54049.
431. Ohdo S, Koyanagi S, Suyama H, Higuchi S, Aramaki H. Changing the dosing schedule minimizes the disruptive effects of interferon on clock function. *Nat Med.* 2001;7(3):356–360.
432. Besedovsky H, del Rey A, Sorkin E, Dinarello CA. Immunoregulatory feedback between interleukin-1 and glucocorticoid hormones. *Science.* 1986;233(4764):652–654.
433. Pan A, Schernhammer ES, Sun Q, Hu FB. Rotating night shift work and risk of type 2 diabetes: two prospective cohort studies in women. *PLoS Med.* 2011;8(12):e1001141.

434. Husse J, Hintze SC, Eichele G, Lehnert H, Oster H. Circadian clock genes Per1 and Per2 regulate the response of metabolism-associated transcripts to sleep disruption. *PLoS One*. 2012;7(12):e52983.
435. Schiavo-Cardozo D, Lima MM, Pareja JC, Geloneze B. Appetite-regulating hormones from the upper gut: disrupted control of xenin and ghrelin in night workers. *Clin Endocrinol (Oxf)*. 2012; http://dx.doi.org/10.1111/cen.12114.
436. Morris CJ, Yang JN, Scheer FA. The impact of the circadian timing system on cardiovascular and metabolic function. *Prog Brain Res*. 2012;199:337–358.
437. Roenneberg T, Allebrandt KV, Merrow M, Vetter C. Social jetlag and obesity. *Curr Biol*. 2012;22(10):939–943.
438. Milagro FI, Gomez-Abellan P, Campion J, Martinez JA, Ordovas JM, Garaulet M. CLOCK, PER2 and BMAL1 DNA methylation: association with obesity and metabolic syndrome characteristics and monounsaturated fat intake. *Chronobiol Int*. 2012;29(9):1180–1194.
439. Garcia-Rios A, Perez-Martinez P, Delgado-Lista J, et al. A Period 2 genetic variant interacts with plasma SFA to modify plasma lipid concentrations in adults with metabolic syndrome. *J Nutr*. 2012;142(7):1213–1218.
440. Gomez-Abellan P, Madrid JA, Lujan JA, et al. Sexual dimorphism in clock genes expression in human adipose tissue. *Obes Surg*. 2012;22(1):105–112.
441. Gomez-Abellan P, Hernandez-Morante JJ, Lujan JA, Madrid JA, Garaulet M. Clock genes are implicated in the human metabolic syndrome. *Int J Obes (Lond)*. 2008;32 (1):121–128.
442. Williams SR, Zies D, Mullegama SV, Grotewiel MS, Elsea SH. Smith-Magenis syndrome results in disruption of CLOCK gene transcription and reveals an integral role for RAI1 in the maintenance of circadian rhythmicity. *Am J Hum Genet*. 2012;90 (6):941–949.
443. Garaulet M, Corbalan-Tutau MD, Madrid JA, et al. PERIOD2 variants are associated with abdominal obesity, psycho-behavioral factors, and attrition in the dietary treatment of obesity. *J Am Diet Assoc*. 2010;110(6):917–921.
444. Morton GJ, Cummings DE, Baskin DG, Barsh GS, Schwartz MW. Central nervous system control of food intake and body weight. *Nature*. 2006;443(7109):289–295.
445. Myers Jr MG, Olson DP. Central nervous system control of metabolism. *Nature*. 2012;491(7424):357–363.
446. Gonnissen HK, Hulshof T, Westerterp-Plantenga MS. Chronobiology, endocrinology, and energy- and food-reward homeostasis. *Obes Rev*. 2013;14(5):405–416.
447. Schibler U, Ripperger J, Brown SA. Peripheral circadian oscillators in mammals: time and food. *J Biol Rhythms*. 2003;18(3):250–260.
448. Verwey M, Amir S. Nucleus-specific effects of meal duration on daily profiles of Period1 and Period2 protein expression in rats housed under restricted feeding. *Neuroscience*. 2011;192:304–311.
449. Guilding C, Hughes AT, Brown TM, Namvar S, Piggins HD. A riot of rhythms: neuronal and glial circadian oscillators in the mediobasal hypothalamus. *Mol Brain*. 2009;2:28.
450. Mieda M, Williams SC, Richardson JA, Tanaka K, Yanagisawa M. The dorsomedial hypothalamic nucleus as a putative food-entrainable circadian pacemaker. *Proc Natl Acad Sci USA*. 2006;103(32):12150–12155.
451. Wyse CA, Coogan AN. Impact of aging on diurnal expression patterns of CLOCK and BMAL1 in the mouse brain. *Brain Res*. 2010;1337:21–31.
452. Abrahamson EE, Leak RK, Moore RY. The suprachiasmatic nucleus projects to posterior hypothalamic arousal systems. *Neuroreport*. 2001;12(2):435–440.
453. Saper CB, Scammell TE, Lu J. Hypothalamic regulation of sleep and circadian rhythms. *Nature*. 2005;437(7063):1257–1263.

454. Kalra SP, Kalra PS. Neuroendocrine control of energy homeostasis: update on new insights. *Prog Brain Res.* 2010;181:17–33.
455. Beaudry JL, Riddell MC. Effects of glucocorticoids and exercise on pancreatic beta-cell function and diabetes development. *Diabetes Metab Res Rev.* 2012;28(7):560–573.
456. Roberge C, Carpentier AC, Langlois MF, et al. Adrenocortical dysregulation as a major player in insulin resistance and onset of obesity. *Am J Physiol Endocrinol Metab.* 2007;293 (6):E1465–E1478.
457. Thorens B. Brain glucose sensing and neural regulation of insulin and glucagon secretion. *Diabetes Obes Metab.* 2011;13(Suppl 1):82–88.
458. Chandra R, Liddle RA. Recent advances in pancreatic endocrine and exocrine secretion. *Curr Opin Gastroenterol.* 2011;27(5):439–443.
459. Gale SM, Castracane VD, Mantzoros CS. Energy homeostasis, obesity and eating disorders: recent advances in endocrinology. *J Nutr.* 2004;134(2):295–298.
460. Gautron L, Elmquist JK. Sixteen years and counting: an update on leptin in energy balance. *J Clin Invest.* 2011;121(6):2087–2093.
461. Vansaun MN. Molecular pathways: adiponectin and Leptin signaling in cancer. *Clin Cancer Res.* 2013;19:1926–1932.
462. Hursting SD, Dunlap SM. Obesity, metabolic dysregulation, and cancer: a growing concern and an inflammatory (and microenvironmental) issue. *Ann N Y Acad Sci.* Oct 2012;1271:82–87.
463. Brugge J, Hung MC, Mills GB. A new mutational AKTivation in the PI3K pathway. *Cancer Cell.* 2007;12(2):104–107.
464. Boudny V, Kovarik J. JAK/STAT signaling pathways and cancer. Janus kinases/signal transducers and activators of transcription. *Neoplasma.* 2002;49(6):349–355.
465. Bodart JF. Extracellular-regulated kinase-mitogen-activated protein kinase cascade: unsolved issues. *J Cell Biochem.* 2010;109(5):850–857.
466. Okwan-Duodu D, Umpierrez GE, Brawley OW, Diaz R. Obesity-driven inflammation and cancer risk: role of myeloid derived suppressor cells and alternately activated macrophages. *Am J Cancer Res.* 2013;3(1):21–33.
467. Smorlesi A, Frontini A, Giordano A, Cinti S. The adipose organ: white-brown adipocyte plasticity and metabolic inflammation. *Obes Rev.* Dec 2012;13(Suppl 2):83–96.
468. Karin M. Nuclear factor-kappaB in cancer development and progression. *Nature.* 2006;441(7092):431–436.
469. Coussens LM, Werb Z. Inflammation and cancer. *Nature.* 2002;420(6917):860–867.
470. Allavena P, Sica A, Garlanda C, Mantovani A. The Yin-Yang of tumor-associated macrophages in neoplastic progression and immune surveillance. *Immunol Rev.* 2008;222:155–161.
471. Koki A, Khan NK, Woerner BM, et al. Cyclooxygenase-2 in human pathological disease. *Adv Exp Med Biol.* 2002;507:177–184.
472. Khan Z, Pillay V, Choonara YE, du Toit LC. Drug delivery technologies for chronotherapeutic applications. *Pharm Dev Technol.* 2009;14(6):602–612.
473. Firer MA, Gellerman G. Targeted drug delivery for cancer therapy: the other side of antibodies. *J Hematol Oncol.* 2012;5:70.
474. Lodish MB. Kinase inhibitors: adverse effects related to the endocrine system. *J Clin Endocrinol Metab.* 2013;98:1333–1342.
475. Philip AK, Philip B. Chronopharmaceuticals: hype or future of pharmaceutics. *Curr Pharm Des.* 2011;17(15):1512–1516.
476. Smaaland R. Circadian rhythm of cell division. *Prog Cell Cycle Res.* 1996;2:241–266.
477. Levi F, Okyar A, Dulong S, Innominato PF, Clairambault J. Circadian timing in cancer treatments. *Annu Rev Pharmacol Toxicol.* 2010;50:377–421.
478. Levi F, Focan C, Karaboue A, et al. Implications of circadian clocks for the rhythmic delivery of cancer therapeutics. *Adv Drug Deliv Rev.* 2007;59(9–10):1015–1035.

479. Innominato PF, Levi FA, Bjarnason GA. Chronotherapy and the molecular clock: clinical implications in oncology. *Adv Drug Deliv Rev*. 2010;62(9–10):979–1001.
480. Giacchetti S, Dugue PA, Innominato PF, et al. Sex moderates circadian chemotherapy effects on survival of patients with metastatic colorectal cancer: a meta-analysis. *Ann Oncol*. Dec 2012;23(12):3110–3116.
481. Rivard GE, Infante-Rivard C, Dresse MF, Leclerc JM, Champagne J. Circadian time-dependent response of childhood lymphoblastic leukemia to chemotherapy: a long-term follow-up study of survival. *Chronobiol Int*. Jun 1993;10(3):201–204.
482. Jemal A, Bray F, Center MM, Ferlay J, Ward E, Forman D. Global cancer statistics. *CA Cancer J Clin*. 2011;61(2):69–90.
483. Saunders NA, Simpson F, Thompson EW, et al. Role of intratumoural heterogeneity in cancer drug resistance: molecular and clinical perspectives. *EMBO Mol Med*. 2012;4 (8):675–684.
484. Stein CJ, Colditz GA. Modifiable risk factors for cancer. *Br J Cancer*. 2004;90 (2):299–303.
485. McPherson K, Steel CM, Dixon JM. ABC of breast diseases. Breast cancer-epidemiology, risk factors, and genetics. *BMJ*. 2000;321(7261):624–628.

CHAPTER TEN

Health Consequences of Circadian Disruption in Humans and Animal Models

Jennifer A. Evans, Alec J. Davidson
Department of Neurobiology, Morehouse School of Medicine, Atlanta, Georgia, USA

Contents

Abstract

Daily rhythms in behavior and physiology are programmed by a hierarchical collection of biological clocks located throughout the brain and body, known as the circadian system. Mounting evidence indicates that disruption of circadian regulation is associated with a wide variety of adverse health consequences, including increased risk for premature death, cancer, metabolic syndrome, cardiovascular dysfunction, immune dysregulation, reproductive problems, mood disorders, and learning deficits. Here we review the evidence for the pervasive effects of circadian disruption in humans and animal models, drawing from both environmental and genetic studies, and identify questions for future research.

Progress in Molecular Biology and Translational Science, Volume 119
ISSN 1877-1173
http://dx.doi.org/10.1016/B978-0-12-396971-2.00010-5

1. INTRODUCTION

Modern times demand much of the circadian system that coordinates myriad rhythms in our bodies. Nontraditional work schedules are becoming increasingly commonplace with the demand for goods and services at all hours of the day, rapid industrialization, and economic globalization. Schedules that include night shifts require employees to be awake during nighttime hours when their biological clocks program inactivity. Mounting evidence indicates that shift work is associated with a wide variety of serious physiological and psychological disorders, including cardiovascular disease, metabolic syndrome, obesity, immune dysfunction, increased risk for cancer, and reproductive complications. In fact, epidemiological studies may underestimate the adverse health consequences of shift work owing to self-selection and survival bias.

The three main factors thought to produce adverse effects on performance and health in shift workers are (1) circadian misalignment, (2) sleep deprivation, and (3) exposure to light at night. However, the specific role of circadian disruption can be difficult to assess in epidemiological reports because of variability in shift exposure conditions and the potential contribution of covariates (i.e., changes in diet, social stress, and the use of tobacco, alcohol, and prescription drugs). Epidemiological reports that control for some of these noncircadian variables do find effects of circadian disruption. Together with experimental evidence, these reports strongly indicate that circadian disruption and misalignment are key factors contributing to the adverse health consequences commonly observed in shift workers. Studies with animal models of circadian disruption are useful in that they afford stricter control to investigate the contribution of circadian misalignment in isolation of, or in combination with, these other factors.

This review summarizes the pervasive effects of circadian disruption on health in humans and animal models. First, we describe the circadian system and types of circadian disruption in humans along with the corresponding experimental manipulations used to simulate these conditions in humans and animal models. Second, we provide a comprehensive review of the health consequences that are evident under these conditions and discuss the factors that are thought to contribute to these disease states. Finally, we discuss future areas of research that can identify the specific causes and consequences, as well as the countermeasures for the health consequences of circadian disruption.

2. CIRCADIAN REGULATION

In mammals, the circadian system is a hierarchical collection of biological clocks controlled by a master pacemaker within the suprachiasmatic nucleus (SCN) of the anterior hypothalamus.[1] The SCN is synchronized by light and, in turn, coordinates downstream tissues so that they adopt specific temporal relationships to the external environment and to one another. In addition to monosynaptic and polysynaptic neural projections from the SCN to downstream targets, the SCN also controls overt rhythms such as body temperature, feeding, and hormone release, which can act to synchronize peripheral clocks. Individual cells within central and peripheral clocks contain molecular oscillators that generate circadian rhythms in protein expression and metabolic state.[2] Coordination at both the molecular and systems level is important for circadian function, as is synchronization to the 24 h environment and local time.[3]

The SCN receives a direct retinal projection from a population of intrinsically photoresponsive retinal ganglion cells expressing the photopigment melanopsin.[4] Circadian responses to light are time-dependent, with light at night causing the release of glutamate and PACAP to reset the phase of the SCN.[5] Light at night also produces acute effects, including changes in sleep, alertness, SCN electrical activity and gene expression, and melatonin levels. While the circadian visual system in humans has typically been viewed as having a relatively high photic threshold compared to the classic image-forming visual system, recent work indicates that the human circadian system can be affected by even low levels of illumination at night.[6,7] This suggests that even the use of low levels of artificial lighting at home and work may alter circadian function.

At the molecular level, circadian rhythms are controlled by autoregulatory feedback loops that control the transcription and translation of clock genes.[8] Within the core molecular loop, CLOCK and BMAL1 form a heterodimer that activates transcription of *period* (*per1*, *per2*) and *cryptochrome* (*cry1*, *cry2*) genes, which form proteins that complex with casein kinase 1ε and translocate back into the nucleus to inhibit CLOCK/BMAL1-mediated transcription. Degradation of PER/CRY repressor elements relieves the inhibition on CLOCK/BMAL1 transcription allowing the cycle to begin anew once every ~24 h, and this oscillatory process operates in nearly every cell of the body to control both ubiquitous and tissue-specific function. Additional loops involving other clock gene elements modulate clock gene/protein expression, influencing the pace and amplitude of circadian rhythms (e.g., REV-ERB and ROR). Moreover, these transcriptional–translational feedback loops interact with electrical and metabolic oscillations to influence cellular function.[9,10]

3. TYPES OF CIRCADIAN DISRUPTION IN HUMANS AND ANIMAL MODELS

3.1. Circadian disruption in the real world

3.1.1 Nontraditional work schedules

Nontraditional work schedules are defined as those that occur outside the hours of 9 am–5 pm, but this category encompasses a wide variety of work schedules, including fixed night shifts, early morning shifts, flex shifts, extended shifts, rotating shifts, and frequent international travel.[11] It is estimated that more than 15% of US employees (>20 million Americans) work on a nontraditional schedule,[12–15] leading to a higher incidence of sick leave, stress, decreased productivity, and accidents on and off the job with an estimated cost exceeding $200 billion a year.[16,17] While shift workers can force themselves to be awake at night and asleep during the day, this oftentimes entails a reduced alertness and sleep efficiency.[18] Despite the ability to meet expected behavioral demands through conscious efforts, the majority of shift workers do not display a corresponding shift in circadian function, and instead, remain in a state where the biological clock is programming sleep during the nighttime hours and wakefulness during the daytime hours.[19,20]

3.1.2 Transmeridian travel

After transmeridian travel, our biological clocks gradually realign to the new time zone, and the transient mismatch between internal timing and external conditions produces symptoms of jetlag that might last for days to weeks.[21] While jetlag is often viewed as a temporary inconvenience, the widespread and varied symptoms of jetlag are emblematic of the influence and importance of the circadian system for human health. Following transmeridian travel, it typically requires 1 day to adjust for each time zone crossed and the latency to recover is longer when traveling in the eastward direction,[21] although the extent and severity of jetlag can be affected by a host of personal and experiential factors. Moreover, large population surveys indicate that many people are affected by "social jetlag," a term that refers to a misalignment between social demands and the inherently preferred time as dictated by the circadian system of individuals with an early or late chronotype (i.e., morning larks and evening owls, respectively).[22]

3.1.3 Non-24 h sleep:wake disorder

Non-24 h sleep:wake disorder occurs when people fail to synchronize with the external light:dark (LD) cycle and instead "free-run" with an inherent period different from 24 h.[23] Since the daily rhythms of these people get progressively earlier or later each day, this results in intermittent misalignment between internal timing and external conditions. Non-24 h sleep:wake disorder occurs most often in individuals who lack photic input to the SCN,[24] but has been documented to occur in sighted individuals as well.[25] Patients with non-24 h sleep:wake disorder may be able to maintain a fixed diurnal schedule of wakefulness but experience cyclical patterns of daytime sleepiness and nighttime alertness, along with psychiatric symptoms and cognitive deficits.[23,26]

3.2. Models of circadian disruption in humans and animals

3.2.1 Simulated shift work

To simulate shift work, people can be brought into the laboratory and engaged in low-stress activities during nighttime hours to maintain wakefulness for several days or allowed to return home during daytime hours.[27,28] In nocturnal rodent models, rotating shift work has been simulated through the use of repeated shifts of the LD cycle (see later), whereas static night shift work has been simulated by imposing diurnal work schedules using running wheels, restricted food access, or operant training protocols.[29–31] In addition to producing an overt misalignment with the external environment, these protocols may also alter the temporal relationship between the SCN and peripheral tissues.[32,33]

3.2.2 Simulated jetlag

Simulated jetlag in humans and rodents is typically produced by shifting the LD cycle by 4–11 h in either an eastward direction (advance) or a westward direction (delay). Advances of the LD cycle typically require more time for adjustment in humans[34] and rodent models.[35,36] A 12 h shift of the LD cycle causes a complete inversion and commonly elicits resynchronization in the delay direction, but this may vary by species and individual. Research has uncovered that simulated jetlag produces various forms of circadian disruption, including a transient misalignment between the internal and external milieus,[21] misalignment among different tissues comprising the circadian system,[37] misalignment among neural oscillators within the SCN,[37–39] and misalignment among molecular components of the circadian clock.[40] In addition, a subset of Siberian hamsters provided a specific jetlag protocol will not resynchronize and, instead, will remain perpetually misaligned or become arrhythmic.[41] Unusual resetting behavior has also been reported to occur in humans[42] and diurnal rodent species.[43,44] These forms of odd resetting behavior may reflect that the circadian clock has been arrested by light occurring during mid-subjective night, a time in the circadian cycle that is predicted to be a phase singularity point.[45]

3.2.3 Non-24 h LD cycles

In the laboratory, misalignment between internal timing and external cycles can be produced through exposure to non-24 h LD cycles,[46,47] which may be similar to work schedules commonly employed in some occupations.[48] As the circadian system is able to entrain to non-24 h cycles only within a certain range (typically 24 ± 2 h), exposure to non-24 h LD cycles outside these limits causes rhythms to "free-run," with intermittent and temporary alignment between the internal and external milieu. In humans, exposure to non-24 h LD cycles can also cause internal desynchrony, where different rhythms free-run independently of one another (e.g., body temperature, melatonin, and sleep/wake rhythms).[49] Lastly, exposing rodents to non-24 h LD cycles can cause different SCN compartments to desynchronize such that one is synchronized to the external lighting environment and the other free-runs.[50]

3.2.4 Light at night

Exposure to light at night is a feature of environmental conditions found in both ecological and industrialized settings. Annual fluctuations in day length entail changes in duration of daytime light exposure, although use of artificial light in humans can serve to attenuate seasonal changes. That even low levels of light at night can reset circadian phase and suppress melatonin release in humans[6,7] suggests that the relatively common exposure to artificial light may in fact affect circadian function. One classic circadian manipulation is exposure to constant light conditions, which can produce internal desynchrony in humans.[51] In nocturnal rodents, exposure to constant light lengthens the circadian period and can cause arrhythmia or "splitting" of behavioral and physiological rhythms.[52,53] Recent work has also investigated the effects of light pollution at night by providing nocturnal rodents with low levels of light at night (typically ~5 lux) under an otherwise normal

LD cycle.[54,55] Lastly, non-24 h LD cycles that do not support entrainment entail aberrant exposure to light at night, with more light exposure occurring under ultra short LD cycles (e.g., LD cycles of 3.5 h of light and 3.5 h of darkness).[56,57]

3.2.5 Genetic models

Mutations that affect the molecular oscillator can produce deficits ranging from altered period to a complete lack of circadian coordination. The first genetic model to be discovered was the tau mutant hamster, which displays a 2 h reduction in period with each copy of the mutated gene,[58] due to a reduction in the phosphorylation capacity of casein kinase 1ε.[59] With the advance of molecular tools, a variety of mouse models have been generated that display altered circadian clock function.[60] In the most extreme, these mouse models are arrhythmic under constant dark conditions (e.g., the *bmal1* knockout mouse,[61] the *clock* mutant mouse,[62] the *per1* −/− *per2* −/− mouse,[63] and the *cry1* −/− *cry2* −/− mouse[64]). These mice models lack self-sustained circadian rhythms but may nevertheless display daily changes in behavior and physiology when maintained under 24 h LD cycles because of the masking effects of light and darkness. While deletions of only one clock gene paralog (e.g., *per1*) are typically insufficient to completely abolish circadian rhythms, this can produce a change in the free-running period and species-atypical alignment to the 24 h LD cycle. These latter models may be of particular interest for addressing the consequences of circadian disruption because overall clock function remains intact but is altered in a manner that disrupts the temporal relationship with the environment.

3.2.6 Surgical models

Brain lesions targeting the SCN have been used historically in rodent models to understand the consequences of complete loss of circadian rhythmicity.[65,66] However, recent work has focused on environmental and genetic manipulation of circadian regulation.

4. HEALTH CONSEQUENCES OF CIRCADIAN DISRUPTION

4.1. Longevity and aging

Shift workers display an increased mortality risk.[67,68] That proper alignment to environmental cycles has benefits for the fitness and longevity of invertebrate species has been suggested for a variety of organisms, including cyanobacteria,[69] plants,[70] and flies.[71,72] Within these studies, longevity or fitness is reduced if organisms are housed under LD cycles that differ from the inherent period of the organism. While the relatively long gestation and lifespan of mammals limits the number of comparable studies conducted, some evidence indicates that circadian misalignment imposes a biological cost that can decrease lifespan and accelerate the aging processes.[73]

4.1.1 Environmental models

Halberg and Cadotte first demonstrated that mortality was increased in mice provided weekly LD cycle inversions.[74] Similar results were obtained in a line of Syrian hamsters predisposed to develop cardiomyopathy: weekly inversions of the LD cycle led to a 11% decrease in median lifespan.[75] It was postulated that this result reflected the effect of circadian disruption rather than increased light exposure as constant light can produce therapeutic effects in cardiomyopathic hamsters.[76] Weekly 6 h shifts of the LD cycle can also have profound effects on lifespan, with aged mice displaying a higher mortality without increases in fecal corticosterone.[77] Simulated changes in season can also affect lifespan, with mouse lemurs held under accelerated seasonal cycles displaying a 30% reduction in average lifespan.[78] Similarly, rapid and repeated changes in day length decreased the lifespan of flies.[79]

4.1.2 Genetic models

Recent studies have provided evidence that molecular deficits in circadian clock function decrease survival and accelerate aging, despite earlier suggestions that health was not adversely affected.[61,62,80] A 38% decrease in median lifespan with pervasive organ dysfunction occurs when heterozygote tau mutant hamsters are housed under static 24 h LD cycles to which they entrain in a species-atypical manner.[81] Interestingly, homozygote tau mutant hamsters, which fail to entrain entirely, did not display a significantly increased mortality (9% decrease in median lifespan), prompting the suggestion that misalignment (like that which occurs in jetlag and shift work) may incur worse physiological costs than lack of entrainment (as in non-24 h sleep/wake disorder). In addition, it is now appreciated that *bmal1−/−* mice display an increased mortality and accelerated development of age-related pathologies starting at about 26 weeks of age.[82,83] Similarly, an accelerated rate of aging and a 13% decrease in average lifespan has been reported for *clock* mutant mice.[84,85] It is unclear whether pathology would emerge if *bmal1−/−* and *clock* mutant mice were permitted to live arrhythmic under constant conditions, but this question is of interest given previous work on the tau mutant hamster. Recently, it has been demonstrated that genetic disruption of the fly circadian clock through a *per* null mutation likewise increases mortality and leads to accelerated aging.[86]

4.1.3 SCN lesions

SCN lesions lead to arrhythmia at the overt level and if a lack of circadian rhythmicity is detrimental to health, then one would expect SCN lesions to decrease lifespan. While most studies have failed to find any pronounced effects (e.g., see Refs. 87,88), it is important to note that most laboratory conditions lack the rich ecological environment under which animals need to acquire food and avoid becoming food. Within the few reports where these variables are taken into account in seminaturalistic conditions, it has been reported that arrhythmic animals with SCN lesions display a decreased survival due to an increased predation risk[89] or some unspecified cause.[90] However, a recent study of mice with a *per2* null mutation held in seminaturalistic enclosures did not find long-term negative effects on fitness or survival.[91]

4.2. Cancer and oxidative stress

Recent studies indicate use of artificial light at night contributes to increasing rates of breast cancer in sighted women,[92–94] which is not evident in blind women.[95] Furthermore, shift work and/or frequent air travel is associated with higher risk for several forms of cancer,[96] including endometrial,[97] colorectal,[98] lymphatic,[99] prostate,[100–103] and breast cancer (reviewed in Ref. 104). Moreover, shift workers present epigenetic profiles indicative of an increased risk for tumorogenesis.[105,106] Melatonin suppression and/or the disruption of its proper timing under shift work conditions may be an important element in the increased risk of cancer development.[104,107] These epidemiological findings have prompted the decision to reclassify shift work from a possible to a probable human carcinogen (class 2A, International Agency for Research on Cancer) and to compensate Danish shift workers who developed breast cancer.[108]

4.2.1 Environmental models

Nocturnal rodents provided a fast-rotating simulated jetlag schedule display accelerated tumor growth relative to rodents that remain on static LD cycles.[109–112] Increased tumor growth is likewise observed under static LD cycles following SCN lesion.[113] Moreover, constant light increases the rate of tumor growth in rodent models[114–119] and adversely affects oxidative stress levels.[120,121] Light at night in rodent models is also associated with increased tumorigenesis[117,118] and changes in oxidative stress.[122] Given the role of melatonin suppression in mediating these effects, it is perhaps not surprising that manipulations of day length, which affect the duration of melatonin secretion, can influence tumor growth in rodent models.[119,123,124]

4.2.2 Genetic models

The circadian clock regulates myriad components of cell cycle and metabolic pathways and altered circadian gene expression has been detected in a variety of cancers (reviewed in Ref. 125). Of the circadian clock mutant models, *per* mutant and knockout mice display a pronounced phenotype, with molecular deregulation of cell cycle and metabolic pathways, increased risk for spontaneous tumor growth, and accelerated growth in induced tumor models.[110,126–129] Furthermore, *per* overexpression in tumor cell cultures retards growth while downregulation increases growth.[126,129–132] Knockout mouse models for *bmal1* and *cry* also display increased risk for spontaneous and induced tumor growth,[110] highlighting the overall importance of the circadian clock in the proper regulation of the cell cycle and tumor growth (but see Ref. 133). Furthermore, both *bmal1* and *cry* play a role in regulating oxidative stress, with *bmal1* deficiency increasing sensitivity and ROS levels[83,134,135] and *cry* deficiency reducing sensitivity in p53 knockout mice.[136] Associations between clock gene polymorphisms in humans and risk or severity of cancer have also been reported (*clock*,[137–139] *npas2*,[140,141] *per3*,[142,143] *cry1*,[139] *cry2*,[144,145] and *timeless*[146]).

4.3. Metabolic syndrome, obesity, and gastrointestinal problems

Metabolic function and feeding processes are regulated by the circadian clock.[147] Across a variety of cultures and occupations, shift workers display a higher incidence of a variety of risk factors for metabolic syndrome,[148–151] including increased weight gain and obesity,[152–162] altered free fatty acid, cholesterol, and triglyceride levels,[155,156,158,163–167] altered lipid and carbohydrate metabolism,[155,166,168–171] insulin resistance, glucose tolerance, and diabetes,[155,169,172–175] with changes in appetite-regulating hormones, growth hormones, and cortisol level.[152,176–179] In addition, shift workers display increased rates of several gastrointestinal disorders, including increased incidence of gastrointestinal discomfort, ulcers, and irritable bowel syndrome.[180–185] Shift workers also display altered feeding behavior,[164,186,187] which likely contributes to these pathologies. Lastly, recent work indicates that social jetlag is associated with obesity in humans.[22]

4.3.1 Environmental models

Simulated shift work in humans affects postprandial hormone levels and metabolic state in a manner that depends on the nutritional content of the meal.[188,189] Simulated conditions of circadian misalignment using forced desynchrony in humans provided isocaloric meals adversely affects metabolic state, with decreased leptin and increased glucose.[190] Adding sleep deprivation to this type of forced desynchrony protocol also produces adverse metabolic consequences in humans, with decreased resting metabolism and increased postprandial glucose levels.[191] Moreover, simulated eastward or westward travel in humans caused increased insulin and glucose levels, increased carbohydrate oxidation, and decreased protein oxidation.[192] Simulated shift work in rodent models is likewise linked to metabolic syndrome and obesity,[33,193,194] which may be related to food intake at inappropriate times.[195,196] Likewise, exposure of mice to non-24 h LD cycles results in accelerated weight gain, obesity, and higher glucose and insulin levels.[197,198] The adverse effects of circadian disruption are also evident during development, with repeated shifts in the LD cycle predisposing rats to sex-dependent metabolic consequences during adulthood, including increased adiposity, hyperinsulinemia, reduced glucose tolerance, and increased insulin secretion.[199] Moreover, rodents held under constant light display dysfunctional changes in metabolic state, adiposity, and feeding behaviors.[200–203] Light at night in mice also leads to a change in the time of food intake, increased body mass, and reduced glucose tolerance, which can be ameliorated with properly phased food intake.[54]

4.3.2 Genetic models

Mice with a dominant negative mutation in the *clock* gene display a change in the timing of food intake, increased body weight gain, increased adiposity, increased cholesterol, triglycerides, and insulin levels, reduced levels of orexin and ghrelin, and impaired pancreatic function.[204–206] A metabolic phenotype with hypotension and renal defects is also evident in *clock−/−* mice.[207] Moreover, *bmal1−/−* mice display a diabetic phenotype, insulin resistance, ectopic fat formation, increased circulating fatty acids, impaired pancreatic function, and a lowered genetic profile for adipose function,[83,203,205,206,208–210] with impaired metabolic function evident when *bmal1−/−* is restricted to the pancreas,[205,211] liver,[212,213] or adipose tissue.[214] One factor that may contribute to the manifestation of these phenotypes is that *clock* mutant and *bmal1*−/− mice display skeletal muscle and mitochondrial dysfunction.[215] Abnormal metabolic function, including altered lipid metabolism, impaired glucose tolerance, and/or insulin resistance can also be observed in mice with a null mutation in the *per* gene,[216–222] double *cry* knockout mice,[223,224] *cry1* transgenic mice,[225,226] and *rev-erb* knockout mice.[227,228] One question left unresolved in these models is whether the metabolic disruptions develop due to a disrupted clock *per se* or atypical synchronization under 24 h LD cycles. Recent genetic association studies have also provided evidence linking altered clock function with metabolic syndrome (*bmal1*,[229] *clock*,[230–234] *cry2*,[235,236] *per2*,[237] and *rev-erb*[238]).

4.4. Cardiovascular disorders

Hemodynamic, hemostatic, endothelial, and autonomic variables fluctuate over the course of the day, with consequences for the peak incidence of cardiac ischemic events such as angina, acute myocardial infarction, and sudden cardiac arrest (reviewed in Refs. 239,240). Shift workers display symptoms of cardiovascular disease,[68,241–244] including atherosclerosis,[245] coronary and ischemic heart disease,[158,169,171,246–248] myocardial infarction,[171,245] stroke,[68,249,250] atherosclerosis,[251] left ventricular hypertrophy,[167] and hypertension.[160,167,252–255]

4.4.1 Environmental models

In humans, light at night can acutely increase heart rate and cortisol levels.[256–263] Simulated conditions of circadian misalignment with and without sleep deprivation alters cardiovascular function in humans.[190,191] Following inversions of the LD cycle in rats, daily rhythms in systolic blood pressure, diastolic blood pressure, and heart rate rhythms resynchronize at a similar rate to the rhythm in locomotor activity.[264] In hypertensive, cardiomyopathic, and obese rodent models, chronic shifting of the LD cycle decreases survival, increases blood pressure, and increases expression of cardiovascular risk factors.[75,265,266]

4.4.2 Genetic models

When heterozygote tau mutant hamsters are housed under 24 h LD cycles that conflict with their inherent period length (22 h), they display multiple measures of cardiomyopathy, including hypotension, myocyte hypertrophy, interstitial fibrosis, and collagen deposition.[267] Interestingly, these pathologies are not evident under 22 h LD cycles, which suggests that the pathology is caused by the altered alignment to the external lighting condition and not the genetic mutation itself. Furthermore, pathology was not evident in SCN-lesioned hamsters, suggesting that no clock is better than a misaligned clock. Pathology was not evident in homozygote tau mutant hamsters, perhaps due to their free-running state under 24 h LD cycles. As further support for circadian disruption being associated with cardiovascular disease, recent work demonstrates that *bmal1−/−* mice display cardiomyopathy, endothelial dysfunction, and arteriosclerosis.[208,268–273] While *bmal1−/−* mice display a variety of pathologies that could contribute to cardiovascular dysfunction, the hypothesis that the circadian clock *per se* is required for cardiovascular health is supported by results obtained in other clock gene models.[207,224,268,271,273–275] Similar to metabolic syndrome, there is some evidence linking clock gene polymorphisms in humans with an increased risk for cardiovascular disease (*bmal1*[229] and *clock*[276]).

4.5. Reproduction issues

Reproductive cycles are regulated by circadian clocks, and reproductive function serves as a useful indicator of disease states that would limit the ability to support pregnancy and lactation.[277] Relative to day workers, rotating and night shift workers display a higher risk for a number of reproductive issues, including irregular menstrual cycles, endometriosis, preterm births, spontaneous abortions, and low birth weight.[278–300]

4.5.1 Environmental models

In rodents, the circadian system regulates the timing of reproduction so that it occurs at optimal times of the day and year (reviewed in Ref. 301), and the timing of behavioral and physiological reproductive events can be altered and disrupted by environmental stimuli that adjust the circadian phase.[302–304] Non-24 h schedules in rats produces a lower amplitude surge in luteinizing hormone likely due to lower phase coherence between SCN compartments.[305] Likewise, hamsters with bimodal activity rhythms under constant light display two daily surges in luteinizing hormone with lower amplitude peaks, presumably due to antiphase cycling of the bilateral SCN lobes.[306] Furthermore, constant light in rodents disrupts estrous cycles[307,308] and inhibits male reproductive function in South Indian gerbils.[309] Repeated 6 h phase advances of the LD cycle applied every 5 days after copulation severely compromised pregnancy outcomes in mice, with normal term births successful in 11/12 control mice compared to only 4/18 in advancing mice and 9/18 in delaying mice.[310] However, LD inversions every 3–4 days provided to pregnant rats did not compromise reproductive success despite the appearance of glucose intolerance and insulin resistance in adult offspring.[199] In contrast, exposure to constant light and non-24 h LD cycles (12 h) in rats reduced fertility and preterm fetus body weight.[311] Higher levels of embryonic reabsorption and pup mortality after birth are also evident in mice housed under non-24 h LD cycles (22 and 26 h).[312] Environmental circadian disruption also compromises reproduction in flies, with restricted food access to the typically inactive phase of the LD cycle causing female flies to produce fewer eggs.[313]

4.5.2 Genetic models

Clock gene rhythms are evident in a variety of central and peripheral reproductive tissues,[277,302] and proper temporal coordination within and between tissues is thought to be important for reproductive viability. Consistent with this view, homozygous *clock* mutant mice display poor reproductive success, longer and irregular estrous cycles, and an undetectable luteinizing hormone surge at the appropriate phase.[314–319] Irregular estrous cycles were further exacerbated when *clock* mutant mice were arrhythmic after release into constant darkness.[318] Furthermore, of the mice that successfully mated, *clock* mutant mice displayed higher rates of abnormalities during pregnancy, including reabsorption of embryos and fully formed fetuses, along with difficulties during labor under an LD cycle[317] or constant darkness.[318] However, these effects appear to be influenced by strain, with more severe effects observed with the *clock* mutation on the B6 than on the Balb/c background.[319] Some conflicting results have also been obtained with the *bmal1−/−* mouse model. While initial reports did not indicate the presence of reproductive defects,[61,320] subsequent studies report that *bmal1−/−* mice display irregular estrous cycles and impaired fertility.[321–323] Female mice lacking *per1* or *per2* also display irregular estrous cycles and reduced fertility during middle age (9–12 months of age) despite fecundity comparable to wild type mice at 2–6 months of age.[324] While there is little work investigating potential associations between reproductive complications and clock gene polymorphisms in humans, one study reported a relationship between *bmal1* and *npas2* variants and miscarriage.[325]

4.6. Mood disorders

Close links between circadian regulation and mood can be inferred from the fact that many psychological disorders are characterized by sleep and circadian disruptions, and that both physiological and circadian symptoms can be ameliorated with timed administration of bright light and/or melatonin (reviewed in Ref. 326). While jetlag and shift work do not necessarily induce psychotic or major affective disorders, circadian disruption in humans can lead to relapse of symptoms in people with a history of psychiatric illness.[327–330] Moreover, the chronotype and the magnitude of misalignment in humans correlates with the severity of symptoms in nonseasonal depression[331–337] and seasonal depression.[338,339]

4.6.1 Environmental models

Seasonal depression is associated with changes in circadian function caused by alterations in light exposure, and experimental work has focused on the nature of this relationship. Entrainment to short photoperiods and decreased light exposure is associated with depression in diurnal rodents,[340–343] and may be observed in some nocturnal rodent species.[344–346] However, whether day length influences affective responses of nocturnal rodents can be markedly influenced by species, strain, type of behavioral test employed, and time of day tested.[347] Increased exposure to light is also associated with changes in affective state in nocturnal rodents, with mice held under constant light displaying behavioral measures of depression and decreased anxiety.[348] Moreover, anxious- and depressive-like behavior is evident in mice exposed to light at night under both entrained[349–352] and free-running conditions.[56] Under free-running conditions with aberrant light at night, depression-like behavior can be ameliorated with fluoxetine administration or ablation of intrinsically photosensitive melanopsin-containing retinal ganglion cells.[56] Lastly, mice exposed to non-24 h LD cycles (20 h) display an increased impulsivity phenotype when tested under novel environmental conditions.[197] In contrast, anxiety or depressive behavior was not observed in mice exposed to a weekly 6 h LD shift.[353]

4.6.2 Genetic models

Clock gene expression rhythms are evident in a variety of structures important for mood, reward processing, and motivation,[354] and these local clocks may contribute to mood regulation.[355] To date, the strongest association between clock gene function and mood is provided by studies conducted with the *clock* mutant mouse, which has been proposed as a model for mania due to its hyperactivity, decreased depressive- and anxiety-like behavior, and increased reward-seeking and goal-directed behavior.[134,356–358] Adding to the face validity of the *clock* mutant model for mania, this behavioral profile can be ameliorated by lithium treatment or by *clock* rescue in the ventral tegmental area.[357] Interestingly, both manic- and depressive-like behaviors can be induced by RNAi knockdown of *clock* in the ventral tegmental area, which highlights the importance of clock gene expression within this specific brain region for proper mood regulation.[359] Decreased depression and anxiety are also reported in the *ror* knockout mouse, the *per2* mutant mice, and the *fbxl3* mutant mouse, which has a 27 h period due to altered *cry* degradation.[360–362] Moreover, there are a number of human genetic association studies that have implicated clock genes in the etiology of mood disorders (reviewed in Ref. 355), with most of the known clock genes implicated (*clock*, *npas2*, *rev-erb*, *ror*, *cry*, and *per*).

4.7. Learning and memory deficits

Cognitive processes in humans display circadian fluctuations and many of the key processes involved in learning and memory display daily fluctuations.[363] Environmental conditions with circadian disruption adversely affect many of these processes.[364–368] There is also evidence for actual neuropathological consequences of circadian disruption in humans, with temporal lobe atrophy and spatial cognitive deficits in long-term airline flight crews.[369,370]

4.7.1 Environmental models

In humans, non-24 h LD cycles that produce misalignment between sleep: wake cycles and melatonin secretion disrupt cognitive performance.[371] In rodents, exposure to non-24 h LD cycles can decrease cognitive function and produce remodeling of the prefrontal cortex.[197,372] Furthermore, simulated jetlag in rodents produces deficits in hippocampal-dependent forms of learning and memory and reductions in hippocampal neurogenesis.[373–379] In these studies, the deficits induced by jetlag can persist after adjustment to the shifted LD cycle and can be influenced by both procedural factors (e.g., shift magnitude and direction) and intrinsic factors (e.g., history of jetlag). Rats held under constant light display learning deficits associated with altered hippocampal function.[376,380,381] Learning deficits are also evident in rodent models provided light at night under both entrained[350,352] and free-running conditions.[56] In the latter context, these light-induced learning deficits are dependent on the presence of intrinsically photosensitive melanopsin-containing retinal ganglion cells.[56] Lastly, arrhythmic Siberian hamsters display learning deficits in a delayed novel-object recognition task.[382]

4.7.2 Genetic models

Clock gene expression rhythms are evident in a variety of structures important for learning, and these local clocks may contribute to memory processes.[354,383] The first indication that clock genes may contribute to learning processes derived from studies demonstrating that flies with a *per* null mutation displayed learning deficits in an experience-dependent courtship paradigm.[384] Building on this initial work, it has been demonstrated that mice carrying a *per2* null mutation display deficits in long-term potentiation and trace fear conditioning, with normal initial acquisition and short-term recall.[383] Similarly, *per1* knockout mice display blunted hippocampal clock gene rhythms and learning deficits in a radial arm maze task.[385] Furthermore, mice with null mutations of either *per1* or *per2* display altered cocaine-induced behavioral sensitization and place preference, with *per1* mutants and *per2* mutants displaying impoverished and enhanced responses, respectively.[386] However, the role of *per* for learning in general remains unclear, as mice deficient for either of these genes do not display deficits in cued fear conditioning, spatial learning in the Morris Water maze, or ethanol-induced learning.[383,387,388] In contrast, mice lacking the *clock* paralog *npas2* display deficits in cued fear conditioning, but not the Morris water maze.[389] While *bmal1*−/− mice, *clock* mutant mice, and double *cry*−/− mice display altered responses under a habituation protocol,[134] these models should be studied in the context of other learning paradigms.[390] In genetic association studies, cognitive function following sleep deprivation has been found to be related to a *per3* polymorphisms.[391–393]

4.8. Immune dysfunction

Immune function and response are regulated in a circadian manner with pronounced consequences (reviewed in Ref. 394). Relative to day workers, rotating and night shift workers display a higher risk for common infections,[395] multiple sclerosis,[396] and other autoimmune disorders.[397] Uncontrolled or chronic inflammation is a risk factor for several of the pathologies observed under conditions of circadian disruption, such as cardiovascular disease,[398] cancer,[399] and diabetes.[400] Thus, changes in the immune system may represent a common factor influencing the varied adverse health consequences associated with shift work.

4.8.1 Environmental models

In rats, chronic reversals of the LD cycle have been found to suppress immune function, with decreased cellular responses to concanavalin

A stimulation of peripheral blood and subcutaneous injection in a Popliteal Lymph Node Assay.[401] In contrast, chronic reversals of the LD cycle in mice were found to sensitize the inflammatory response to dextran sodium sulfate, a model of colitis.[402] Similarly, weekly 6 h shifts of the LD cycle in mice exacerbated the immune response to lipopolysaccharide, a model for sepsis.[353] Repeated LD shifts also suppressed the cytotoxic activity of Natural Killer cells and promoted the development of lung tumors.[111] Exposure to constant light, likewise, affects immune function in rodent and avian species.[403–405] Additionally, exposure to light at night suppresses immune function in Siberian hamsters, with chronic exposure producing reduced delayed-type hypersensitivity and suppressed blood plasma bactericidal activity following lipopolysaccharide challenge.[55] Seasonal changes in immune function and responses have been documented in the Siberian hamster.[406–409] Additionally, rapid and repeated changes in day length results in immune dysfunction in rats, mice, and flies.[79]

4.8.2 Genetic models

Clock gene expression rhythms are evident in various types of immune cells, with local clock function important for the temporal regulation of inflammatory responses (reviewed in Ref. 394). *Rev-erbα* has been identified as an important part of the mechanism by which the circadian clock regulates the temporal gating of immune function, with *rev-erbα−/−* mice displaying a hypersensitive response to the lipopolysaccharide challenge.[410] It has also been found that the clock gene *cry* negatively regulates cytokine production[411] and mice with deletions of both *cry* paralogs display an immune phenotype, with exacerbated cytokine production and joint swelling after arthritic induction.[412] In addition, *per2* mutant mice and *per1−/−* mice display altered cytokine and cytolytic expression rhythms in Natural Killer cells,[413–415] with reduced responses to the lipopolysaccharide challenge in *per2* mutant mice.[416] Deletion of *bmal1* in mice is associated with progressive corneal inflammation and decreased immune cell expression.[83,417,418] Lastly, mice with a *clock* null mutation display altered transcription of immune genes[419] and reduced activation of NF-κB activation in cultured fibroblasts and hepatocytes.[420] In genetic association studies examining the relationship between clock gene polymorphisms and immune disorders, a link has been identified between a specific *per3* polymorphism and the incidence or severity of inflammatory bowel disease.[421]

5. FUTURE RESEARCH QUESTIONS

Collectively, the epidemiological and experimental evidence accumulated thus far presents a compelling case that circadian disruption due to environmental and genetic means produces a wide variety of adverse health consequences. For people, the environment is the primary source of disruption of the circadian clock, although genetic polymorphisms altering clock function may also affect health. The use of artificial light and increased globalization facilitate and encourage the modern 24 h lifestyle, and nontraditional work schedules are becoming increasingly common. While there has been research into factors that influence worker productivity, there is comparatively little work exploring the potential short-term and long-term health consequences of scheduling strategies designed to provide staffing coverage across the day and night. Thus, there is a need for a better understanding of the effects of nontraditional schedules and the mechanisms through which these conditions affect health.

Shift work commonly involves misalignment of the circadian clock relative to external time, insufficient sleep, suppression of the hormone melatonin, psychosocial stress, and changes in other behavioral health variables such as feeding behavior, alcohol, and tobacco use. Animal models afford the opportunity to disentangle these variables in order to identify the relative consequences of each for specific health outcomes. For example, increased light exposure in mice can induce sleep[422] rather than produce sleep deprivation, thus enabling the results to distinguish between circadian- versus sleep-related mechanisms. Similarly, most mouse strains lack the enzymes required to synthesize melatonin,[423] but express functional receptors for this hormone. Furthermore, mice can be exposed to schedules that impose circadian disruption without inducing a stress response or changes in anxiety or depressive behavior,[77,353] although this may be schedule-specific.[56] Thus, the nocturnal mouse presents a useful model to further explore the contribution of circadian disruption *per se*.

The severity of health consequences across different work and light schedules remains a major outstanding question. It has been difficult to differentiate different types of shift schedules in epidemiological studies to date, thus the relative consequences of various shift work paradigms on sleep, cognitive, and health factors remain poorly understood.[424–426] Shift work can be fixed or rotating, with rotating shift work in the forward- (delaying) or backward-rotating (advancing) direction. Rotation speed and shift duration also vary by industry and work site. Studies in animal models have demonstrated the negative impact of shift sizes of 6 h,[77,111,353] 8 h,[109] and 12 h,[75,402] but the relative consequences of different schedules such as those used in industry represents an area for further investigation. The few studies that have addressed these types of questions suggest that health effects of circadian disruption are significantly modulated by schedule characteristics such as shift direction and exposure duration.[77,353,373–379] Additional research in both animal models and people comparing the effects of various schedules on specific health measures would aid in shift work scheduling. Of primary value would be additional research examining the role of exposure duration across schedule types, the persistence of health consequences, and individual susceptibility factors.

Also of interest remains the role of central versus local clock function in the effects of circadian disruption. In humans, clock gene polymorphisms would be expected to affect the function of all cells and most of the genetic models employed to date likewise have ubiquitous effects on cells in both central and peripheral clocks. Over the last few years, several reports have described health consequences of tissue-specific genetic manipulations,[205,211–214] supporting a role for local clock function in determining health outcomes. SCN-specific gene targeting strategies have yet to be developed, but environmental models that disrupt SCN function and system-level control provide useful models and may be highly representative of the specific conditions commonly producing circadian disruption in humans. With environmental lighting manipulations, it remains unclear the degree to which health consequences arise due to changes in central or peripheral function (i.e., loss of synchronization within or among peripheral clocks). Cancer research represents one area where these questions have begun to be addressed, using tissue-specific profiling during disease states[427] and independent manipulation of implanted cell and host genomes during tumor progression.[428] Recent advances in cell-specific *in vivo* expression technologies can facilitate research into the relative consequences of losing normal circadian timing in central versus peripheral sites.

One final area in need of additional insight is the role of a broken versus misaligned clock in disease genesis. Both environmental and nonarrhythmic genetic models provide excellent tools here as well; however, work thus far has failed at times to adequately describe the rhythmic state of the model at the gross (e.g., behavioral and hormonal) and fine (cellular) levels of analysis. For example, experiments using constant light should document the rhythmic state of animals, providing the opportunity to correlate health outcomes with the degree and timing of circadian disruption among animals. Further, it remains unclear how disease states in arrhythmic clock gene models (i.e., *bmal1*−/−, *clock* mutant, double *per*−/− or *cry*−/−) are affected by different entrainment and free-running conditions (e.g., LD versus DD conditions). Finally, the health consequences of circadian disruption can be further studied in rhythmic clock gene models (e.g., *per2* mutant) held under both 24 h LD cycles and LD cycles with a period closer to the inherent period of the mouse model. In this manner, research can address the relative consequences of misalignment versus arrhythmia, determine whether having the right clock for the environment is corrective, and whether there are situations where a broken clock may be advantageous.[69,267]

In conclusion, there is compelling evidence that circadian disruption is detrimental to health, and there is an increasing appreciation of the factor of circadian timing in the manifestation of disease states. In most of the research within this review, it would appear that the organism can compensate for circadian disruption when environmental conditions are devoid of challenge or stress; however, severe health consequences manifest when circadian regulation is compromised in the face of adversity. Collectively, this would suggest that circadian disruption *per se* does not cause disease, but can interact with disease states to increase outcome severity and/or accelerate disease progression. Uncovering the mechanistic relationships that underlie these synergistic and modulatory effects of circadian disruption on health are important areas for future research.

REFERENCES

1. Mohawk JA, Green CB, Takahashi JS. Central and peripheral circadian clocks in mammals. *Annu Rev Neurosci.* 2012;35:445–462.
2. Ko CH, Takahashi JS. Molecular components of the mammalian circadian clock. *Hum Mol Genet.* 2006;15:R271–R277.
3. Takahashi JS, Hong HK, Ko CH, McDearmon EL. The genetics of mammalian circadian order and disorder: implications for physiology and disease. *Nat Rev Genet.* 2008;9:764–775.
4. Berson DM. Strange vision: ganglion cells as circadian photoreceptors. *Trends Neurosci.* 2003;26:314–320.
5. Meijer JH, Michel S, Vansteensel MJ. Processing of daily and seasonal light information in the mammalian circadian clock. *Gen Comp Endocrinol.* 2007;152:159–164.
6. Boivin DB, Duffy JF, Kronauer RE, Czeisler CA. Dose-response relationships for resetting of human circadian clock by light. *Nature.* 1996;379:540–542.
7. Zeitzer JM, Dijk DJ, Kronauer R, Brown E, Czeisler C. Sensitivity of the human circadian pacemaker to nocturnal light: melatonin phase resetting and suppression. *J Physiol.* 2000;526(Pt. 3):695–702.

8. Dardente H, Cermakian N. Molecular circadian rhythms in central and peripheral clocks in mammals. *Chronobiol Int*. 2007;24:195–213.
9. Edgar RS, Green EW, Zhao Y, et al. Peroxiredoxins are conserved markers of circadian rhythms. *Nature*. 2012;485:459–464.
10. Wang TA, Yu YV, Govindaiah G, et al. Circadian rhythm of redox state regulates excitability in suprachiasmatic nucleus neurons. *Science*. 2012;337:839–842.
11. Costa G. The problem: shiftwork. *Chronobiol Int*. 1997;14:89–98.
12. Rosekind MR. Underestimating the societal costs of impaired alertness: safety, health and productivity risks. *Sleep Med*. 2005;6:S21–S25.
13. McMenamin TM. A time to work: recent trends in shift work and flexible schedules. *Mon Labor Rev*. 2007;130:9–11.
14. US Department of Labor. *Workers on flexible and shift schedules in 2004*. Available at: http://www.bls.gov/news.release/flex.nr0.htm.
15. US Department of Labor. *Beginning and ending hours: full-time wage and salary workers*. Available at: http://www.bls.gov/news.release/flex.t07.htm; 2004.
16. Kerin A, Aguirre A. Improving health, safety, and profits in extended hours operations (shiftwork). *Ind Health*. 2005;3:201–208 2004.
17. Folkard S, Tucker P. Shift work, safety and productivity. *Occup Med*. 2003;53: 95–101.
18. Akerstedt T, Wright Jr KP. Sleep loss and fatigue in shift work and shift work disorder. *Sleep Med Clin*. 2009;4:257–271.
19. Folkard S. Do permanent night workers show circadian adjustment? A review based on the endogenous melatonin rhythm. *Chronobiol Int*. 2008;25:215–224.
20. Dumont M, Lanctot V, Cadieux-Viau R, Paquet J. Melatonin production and light exposure of rotating night workers. *Chronobiol Int*. 2012;29:203–210.
21. Arendt J. Managing jet lag: some of the problems and possible new solutions. *Sleep Med Rev*. 2009;13:249–256.
22. Roenneberg T, Allebrandt KV, Merrow M, Vetter C. Social jetlag and obesity. *Curr Biol*. 2012;22:939–943.
23. Okawa M, Uchiyama M. Circadian rhythm sleep disorders: characteristics and entrainment pathology in delayed sleep phase and non-24-h sleep-wake syndrome. *Sleep Med Rev*. 2007;11:485–496.
24. Lockley SW, Arendt J, Skene DJ. Visual impairment and circadian rhythm disorders. *Dialogues Clin Neurosci*. 2007;9:301–314.
25. Hayakawa T, Uchiyama M, Kamei Y, et al. Clinical analyses of sighted patients with non-24-hour sleep-wake syndrome: a study of 57 consecutively diagnosed cases. *Sleep*. 2005;28:945–952.
26. Lockley SW, Dijk DJ, Kosti O, Skene DJ, Arendt J. Alertness, mood and performance rhythm disturbances associated with circadian sleep disorders in the blind. *J Sleep Res*. 2008;17:207–216.
27. Crowley SJ, Lee C, Tseng CY, Fogg LF, Eastman CI. Combinations of bright light, scheduled dark, sunglasses, and melatonin to facilitate circadian entrainment to night shift work. *J Biol Rhythms*. 2003;18:513–523.
28. Chapdelaine S, Paquet J, Dumont M. Effects of partial circadian adjustments on sleep and vigilance quality during simulated night work. *J Sleep Res*. 2012;21: 380–389.
29. Salgado-Delgado R, Nadia S, Angeles-Castellanos M, Buijs RM, Escobar C. In a rat model of night work, activity during the normal resting phase produces desynchrony in the hypothalamus. *J Biol Rhythms*. 2010;25:421–431.
30. Gritton HJ, Sutton BC, Martinez V, Sarter M, Lee TM. Interactions between cognition and circadian rhythms: attentional demands modify circadian entrainment. *Behav Neurosci*. 2009;123:937–948.

31. Murphy HM, Wideman CH, Nadzam GR. A laboratory animal model of human shift work. *Integr Physiol Behav Sci*. 2003;38:316–328.
32. Hoogerwerf WA, Hellmich HL, Cornelissen G, et al. Clock gene expression in the murine gastrointestinal tract: endogenous rhythmicity and effects of a feeding regimen. *Gastroenterology*. 2007;133:1250–1260.
33. Salgado-Delgado R, Angeles-Castellanos M, Buijs MR, Escobar C. Internal desynchronization in a model of night-work by forced activity in rats. *Neuroscience*. 2008;154:922–931.
34. Aschoff J, Hoffmann K, Pohl H, Wever R. Re-entrainment of circadian rhythms after phase-shifts of the Zeitgeber. *Chronobiologia*. 1975;2:23–78.
35. Zee PC, Rosenberg RS, Turek FW. Effects of aging on entrainment and rate of resynchronization of circadian locomotor activity. *Am J Physiol*. 1992;263: R1099–R1103.
36. Illnerova H, Vanecek J, Hoffmann K. Adjustment of the rat pineal N-acetyltransferase rhythm to eight-hour shifts of the light-dark cycle: advance of the cycle disturbs the rhythm more than delay. *Brain Res*. 1987;417:167–171.
37. Sellix MT, Evans JA, Leise TL, et al. Aging differentially affects the re-entrainment response of central and peripheral circadian oscillators. *J Neurosci*. 2012;32: 16193–16202.
38. Nagano M, Adachi A, Nakahama K, et al. An abrupt shift in the day/night cycle causes desynchrony in the mammalian circadian center. *J Neurosci*. 2003;23:6141–6151.
39. Rohling JH, Vanderleest HT, Michel S, Vansteensel MJ, Meijer JH. Phase resetting of the mammalian circadian clock relies on a rapid shift of a small population of pacemaker neurons. *PLoS One*. 2011;6:e25437.
40. Reddy AB, Field MD, Maywood ES, Hastings MH. Differential resynchronisation of circadian clock gene expression within the suprachiasmatic nuclei of mice subjected to experimental jet lag. *J Neurosci*. 2002;22:7326–7330.
41. Ruby NF, Joshi N, Heller HC. Phase shift magnitude and direction determine whether Siberian hamsters reentrain to the photocycle. *J Biol Rhythms*. 1998;13:506–517.
42. Jewett ME, Kronauer RE, Czeisler CA. Light-induced suppression of endogenous circadian amplitude in humans. *Nature*. 1991;350:59–62.
43. Honma S, Honma K. Light-induced uncoupling of multioscillatory circadian system in a diurnal rodent, Asian chipmunk. *Am J Physiol*. 1999;276:R1390–R1396.
44. Navaneethakannan K, Chandrashekaran MK. Light and dark pulse response curves in a day active palm squirrel *Funambulus palmarum*. *Exp Biol*. 1986;45:267–273.
45. Jewett ME, Forger DB, Kronauer RE. Revised limit cycle oscillator model of human circadian pacemaker. *J Biol Rhythms*. 1999;14:493–499.
46. Aschoff J, Pohl H. Phase relations between a circadian rhythm and its zeitgeber within the range of entrainment. *Naturwissenschaften*. 1978;65:80–84.
47. Gronfier C, Wright Jr KP, Kronauer RE, Czeisler CA. Entrainment of the human circadian pacemaker to longer-than-24-h days. *Proc Natl Acad Sci USA*. 2007;104: 9081–9086.
48. Kelly TL, Neri DF, Grill JT, et al. Nonentrained circadian rhythms of melatonin in submariners scheduled to an 18-hour day. *J Biol Rhythms*. 1999;14:190–196.
49. Dijk DJ, Duffy JF, Riel E, Shanahan TL, Czeisler CA. Ageing and the circadian and homeostatic regulation of human sleep during forced desynchrony of rest, melatonin and temperature rhythms. *J Physiol*. 1999;516:611–627.
50. de la Iglesia HO, Cambras T, Schwartz WJ, Diez-Noguera A. Forced desynchronization of dual circadian oscillators within the rat suprachiasmatic nucleus. *Curr Biol*. 2004;14:796–800.
51. Moore-Ede MC, Sulzman FM, Fuller C, eds. *The Clocks that Time Us: Physiology of the Circadian Timing System*. Cambridge: Harvard University Press; 1982.

52. Ohta H, Yamazaki S, McMahon DG. Constant light desynchronizes mammalian clock neurons. *Nat Neurosci*. 2005;8:267–269.
53. de la Iglesia HO, Meyer J, Carpino Jr A, Schwartz WJ. Antiphase oscillation of the left and right suprachiasmatic nuclei. *Science*. 2000;290:799–801.
54. Fonken LK, Workman JL, Walton JC, et al. Light at night increases body mass by shifting the time of food intake. *Proc Natl Acad Sci USA*. 2010;107:18664–18669.
55. Bedrosian TA, Fonken LK, Walton JC, Nelson RJ. Chronic exposure to dim light at night suppresses immune responses in Siberian hamsters. *Biol Lett*. 2011;7:468–471.
56. LeGates TA, Altimus CM, Wang H, et al. Aberrant light directly impairs mood and learning through melanopsin-expressing neurons. *Nature*. 2012;491:594–598.
57. Mrosovsky N. Masking: history, definitions, and measurement. *Chronobiol Int*. 1999;16:415–429.
58. Ralph MR, Menaker M. A mutation of the circadian system in golden hamsters. *Science*. 1988;241:1225–1227.
59. Lowrey PL, Shimomura K, Antoch MP, et al. Positional syntenic cloning and functional characterization of the mammalian circadian mutation tau. *Science*. 2000;288:483–492.
60. Siepka SM, Yoo SH, Park J, Lee C, Takahashi JS. Genetics and neurobiology of circadian clocks in mammals. *Cold Spring Harb Symp Quant Biol*. 2007;72:251–259.
61. Bunger MK, Wilsbacher LD, Moran SM, et al. *Mop3* is an essential component of the master circadian pacemaker in mammals. *Cell*. 2000;103:1009–1017.
62. King DP, Zhao Y, Sangoram AM, et al. Positional cloning of the mouse circadian clock gene. *Cell*. 1997;89:641–653.
63. Bae K, Jin X, Maywood ES, et al. Differential functions of *mPer1*, *mPer2*, and *mPer3* in the SCN circadian clock. *Neuron*. 2001;30:525–536.
64. van der Horst GT, Muijtjens M, Kobayashi K, et al. Mammalian *Cry1* and *Cry2* are essential for maintenance of circadian rhythms. *Nature*. 1999;398:627–630.
65. Stephan FK, Zucker I. Circadian rhythms in drinking behavior and locomotor activity of rats are eliminated by hypothalamic lesions. *Proc Natl Acad Sci USA*. 1972;69:1583–1586.
66. Moore RY, Eichler VB. Loss of a circadian adrenal corticosterone rhythm following suprachiasmatic lesions in the rat. *Brain Res*. 1972;42:201–206.
67. Knutsson A, Hammar N, Karlsson B. Shift workers' mortality scrutinized. *Chronobiol Int*. 2004;21:1049–1053.
68. Karlsson B, Alfredsson L, Knutsson A, Andersson E, Toren K. Total mortality and cause-specific mortality of Swedish shift- and dayworkers in the pulp and paper industry in 1952-2001. *Scand J Work Environ Health*. 2005;31:30–35.
69. Woelfle MA, Ouyang Y, Phanvijhitsiri K, Johnson CH. The adaptive value of circadian clocks: an experimental assessment in cyanobacteria. *Curr Biol*. 2004;14: 1481–1486.
70. Dodd AN, Salathia N, Hall A, et al. Plant circadian clocks increase photosynthesis, growth, survival, and competitive advantage. *Science*. 2005;309:630–633.
71. Aschoff J, von Saint Paul U, Wever R. Lifetime of flies under influence of time displacement. *Naturwissenschaften*. 1971;58:574.
72. Pittendrigh CS, Minis DH. Circadian systems: longevity as a function of circadian resonance in *Drosophila melanogaster*. *Proc Natl Acad Sci USA*. 1972;69:1537–1539.
73. Wyse CA, Coogan AN, Selman C, Hazlerigg DG, Speakman JR. Association between mammalian lifespan and circadian free-running period: the circadian resonance hypothesis revisited. *Biol Lett*. 2010;6:696–698.
74. Halberg F, Cadotte I. Increased mortality in mice exposed to weekly 180° shifts of lighting regimine LD 12:12 beginning at 1 year of age. *Chronobiologia*. 1975;2:26.

75. Penev PD, Kolker DE, Zee PC, Turek FW. Chronic circadian desynchronization decreases the survival of animals with cardiomyopathic heart disease. *Am J Physiol.* 1998;275:H2334–H2337.
76. Natelson BH, Ottenweller JE, Tapp WN, Bergen M, Soldan S. Phototherapeutic effects in hamsters with heart disease. *Physiol Behav.* 1996;60:463–468.
77. Davidson AJ, Sellix MT, Daniel J, et al. Chronic jet-lag increases mortality in aged mice. *Curr Biol.* 2006;16:R914–R916.
78. Perret M. Change in photoperiodic cycle affects life span in a prosimian primate (*Microcebus murinus*). *J Biol Rhythms.* 1997;12:136–145.
79. Li JC, Xu F. Influences of light-dark shifting on the immune system, tumor growth and life span of rats, mice and fruit flies as well as on the counteraction of melatonin. *Biol Signals.* 1997;6:77–89.
80. Menaker M, Vogelbaum MA. Mutant circadian period as a marker of suprachiasmatic nucleus function. *J Biol Rhythms.* 1993;8(Suppl.):S93–S98.
81. Hurd MW, Ralph MR. The significance of circadian organization for longevity in the golden hamster. *J Biol Rhythms.* 1998;13:430–436.
82. Bunger MK, Walisser JA, Sullivan R, et al. Progressive arthropathy in mice with a targeted disruption of the *Mop3/Bmal-1* locus. *Genesis.* 2005;41:122–132.
83. Kondratov RV, Kondratova AA, Gorbacheva VY, Vykhovanets OV, Antoch MP. Early aging and age-related pathologies in mice deficient in *BMAL1*, the core componentof the circadian clock. *Genes Dev.* 2006;20:1868–1873.
84. Antoch MP, Gorbacheva VY, Vykhovanets O, et al. Disruption of the circadian clock due to the *Clock* mutation has discrete effects on aging and carcinogenesis. *Cell Cycle.* 2008;7:1197–1204.
85. Dubrovsky YV, Samsa WE, Kondratov RV. Deficiency of circadian protein CLOCK reduces lifespan and increases age-related cataract development in mice. *Aging.* 2010;2:936–944.
86. Krishnan N, Rakshit K, Chow ES, et al. Loss of circadian clock accelerates aging in neurodegeneration-prone mutants. *Neurobiol Dis.* 2012;45:1129–1135.
87. Zucker I, Boshes M, Dark J. Suprachiasmatic nuclei influence circannual and circadian rhythms of ground squirrels. *Am J Physiol.* 1983;244:R472–R480.
88. Dark J, Pickard GE, Zucker I. Persistence of circannual rhythms in ground squirrels with lesions of the suprachiasmatic nuclei. *Brain Res.* 1985;332:201–207.
89. DeCoursey PJ, Krulas JR, Mele G, Holley DC. Circadian performance of suprachiasmatic nuclei (SCN)-lesioned antelope ground squirrels in a desert enclosure. *Physiol Behav.* 1997;62:1099–1108.
90. DeCoursey PJ, Krulas JR. Behavior of SCN-lesioned chipmunks in natural habitat: a pilot study. *J Biol Rhythms.* 1998;13:229–244.
91. Daan S, Spoelstra K, Albrecht U, et al. Lab mice in the field: unorthodox daily activity and effects of a dysfunctional circadian clock allele. *J Biol Rhythms.* 2011;26: 118–129.
92. Kloog I, Stevens RG, Haim A, Portnov BA. Nighttime light level co-distributes with breast cancer incidence worldwide. *Cancer Causes Control.* 2010;21:2059–2068.
93. Stevens RG. Light-at-night, circadian disruption and breast cancer: assessment of existing evidence. *Int J Epidemiol.* 2009;38:963–970.
94. Costa G, Haus E, Stevens R. Shift work and cancer—considerations on rationale, mechanisms, and epidemiology. *Scand J Work Environ Health.* 2010;36:163–179.
95. Hahn RA. Profound bilateral blindness and the incidence of breast cancer. *Epidemiology.* 1991;2:208–210.
96. World Health Organization International Agency for Research on Cancer. *IARC Monographs on the Evaluation of Carcinogenic Risks to Humans.* Vol. 98. Painting,

firefighting, and shiftwork. Available at: http://monographs.iarc.fr/ENG/Monographs/vol98/mono98.pdf; 2010.
97. Viswanathan AN, Hankinson SE, Schernhammer ES. Night shift work and the risk of endometrial cancer. *Cancer Res.* 2007;67:10618–10622.
98. Schernhammer ES, Laden F, Speizer FE, et al. Night-shift work and risk of colorectal cancer in the nurses' health study. *J Natl Cancer Inst.* 2003;95:825–828.
99. Lahti TA, Partonen T, Kyyronen P, Kauppinen T, Pukkala E. Night-time work predisposes to non-Hodgkin lymphoma. *Int J Cancer.* 2008;123:2148–2151.
100. Kubo T, Ozasa K, Mikami K, et al. Prospective cohort study of the risk of prostate cancer among rotating-shift workers: findings from the Japan collaborative cohort study. *Am J Epidemiol.* 2006;164:549–555.
101. Krstev S, Baris D, Stewart P, et al. Occupational risk factors and prostate cancer in U.S. blacks and whites. *Am J Ind Med.* 1998;34:421–430.
102. Demers PA, Checkoway H, Vaughan TL, et al. Cancer incidence among firefighters in Seattle and Tacoma, Washington (United States). *Cancer Causes Control.* 1994;5:129–135.
103. Conlon M, Lightfoot N, Kreiger N. Rotating shift work and risk of prostate cancer. *Epidemiology.* 2007;18:182–183.
104. Haus EL, Smolensky MH. Shift work and cancer risk: potential mechanistic roles of circadian disruption, light at night, and sleep deprivation. *Sleep Med Rev.* 2013;17:273–284.
105. Bollati V, Baccarelli A, Sartori S, et al. Epigenetic effects of shiftwork on blood DNA methylation. *Chronobiol Int.* 2010;27:1093–1104.
106. Zhu Y, Stevens RG, Hoffman AE, et al. Epigenetic impact of long-term shiftwork: pilot evidence from circadian genes and whole-genome methylation analysis. *Chronobiol Int.* 2011;28:852–861.
107. Blask DE, Brainard GC, Dauchy RT, et al. Melatonin-depleted blood from premenopausal women exposed to light at night stimulates growth of human breast cancer xenografts in nude rats. *Cancer Res.* 2005;65:11174–11184.
108. Wise J. Danish night shift workers with breast cancer awarded compensation. *BMJ.* 2009;338:b1152.
109. Filipski E, Delaunay F, King VM, et al. Effects of chronic jet lag on tumor progression in mice. *Cancer Res.* 2004;64:7879–7885.
110. Lee S, Donehower LA, Herron AJ, Moore DD, Fu L. Disrupting circadian homeostasis of sympathetic signaling promotes tumor development in mice. *PLoS One.* 2010;5:e10995.
111. Logan RW, Zhang C, Murugan S, et al. Chronic shift-lag alters the circadian clock of NK cells and promotes lung cancer growth in rats. *J Immunol.* 2012;188:2583–2591.
112. Filipski E, Innominato PF, Wu M, et al. Effects of light and food schedules on liver and tumor molecular clocks in mice. *J Natl Cancer Inst.* 2005;97:507–517.
113. Filipski E, King VM, Li X, et al. Host circadian clock as a control point in tumor progression. *J Natl Cancer Inst.* 2002;94:690–697.
114. van den Heiligenberg S, Depres-Brummer P, Barbason H, et al. The tumor promoting effect of constant light exposure on diethylnitrosamine-induced hepatocarcinogenesis in rats. *Life Sci.* 1999;64:2523–2534.
115. Blask DE, Dauchy RT, Sauer LA, Krause JA, Brainard GC. Growth and fatty acid metabolism of human breast cancer (MCF-7) xenografts in nude rats: impact of constant light-induced nocturnal melatonin suppression. *Breast Cancer Res Treat.* 2003;79:313–320.
116. Cos S, Mediavilla D, Martinez-Campa C, et al. Exposure to light-at-night increases the growth of DMBA-induced mammary adenocarcinomas in rats. *Cancer Lett.* 2006;235:266–271.

117. Vinogradova IA, Anisimov VN, Bukalev AV, Semenchenko AV, Zabezhinski MA. Circadian disruption induced by light-at-night accelerates aging and promotes tumorigenesis in rats. *Aging*. 2009;1:855–865.
118. Dauchy RT, Blask DE, Sauer LA, Brainard GC, Krause JA. Dim light during darkness stimulates tumor progression by enhancing tumor fatty acid uptake and metabolism. *Cancer Lett.* 1999;144:131–136.
119. Shah PN, Mhatre MC, Kothari LS. Effect of melatonin on mammary carcinogenesis in intact and pinealectomized rats in varying photoperiods. *Cancer Res.* 1984;44:3403–3407.
120. Baydas G, Ercel E, Canatan H, Donder E, Akyol A. Effect of melatonin on oxidative status of rat brain, liver and kidney tissues under constant light exposure. *Cell Biochem Funct.* 2001;19:37–41.
121. Cruz A, Padillo FJ, Granados J, et al. Effect of melatonin on cholestatic oxidative stress under constant light exposure. *Cell Biochem Funct.* 2003;21:377–380.
122. Reiter RJ, Tan DX, Osuna C, Gitto E. Actions of melatonin in the reduction of oxidative stress. A review. *J Biomed Sci.* 2000;7:444–458.
123. Nelson RJ, Blom JM. Photoperiodic effects on tumor development and immune function. *J Biol Rhythms.* 1994;9:233–249.
124. Bartsch H, Bartsch C. Effect of melatonin on experimental tumors under different photoperiods and times of administration. *J Neural Transm.* 1981;52:269–279.
125. Gery S, Koeffler HP. The role of circadian regulation in cancer. *Cold Spring Harb Symp Quant Biol.* 2007;72:459–464.
126. Fu L, Pelicano H, Liu J, Huang P, Lee C. The circadian gene *Period2* plays an important role in tumor suppression and DNA damage response in vivo. *Cell.* 2002;111:41–50.
127. Lee CC. Tumor suppression by the mammalian *Period* genes. *Cancer Causes Control.* 2006;17:525–530.
128. Yang X, Wood PA, Ansell CM, et al. The circadian clock gene *Per1* suppresses cancer cell proliferation and tumor growth at specific times of day. *Chronobiol Int.* 2009;26:1323–1339.
129. Wood PA, Yang X, Taber A, et al. *Period 2* mutation accelerates $Apc^{Min/+}$ tumorigenesis. *Mol Cancer Res.* 2008;6:1786–1793.
130. Gery S, Virk RK, Chumakov K, Yu A, Koeffler HP. The clock gene *Per2* links the circadian system to the estrogen receptor. *Oncogene.* 2007;26:7916–7920.
131. Xiang S, Coffelt SB, Mao L, et al. Period-2: a tumor suppressor gene in breast cancer. *J Circadian Rhythms.* 2008;6:4.
132. Yang X, Wood PA, Oh EY, et al. Down regulation of circadian clock gene *Period 2* accelerates breast cancer growth by altering its daily growth rhythm. *Breast Cancer Res Treat.* 2009;117:423–431.
133. Gauger MA, Sancar A. Cryptochrome, circadian cycle, cell cycle checkpoints, and cancer. *Cancer Res.* 2005;65:6828–6834.
134. Kondratova AA, Dubrovsky YV, Antoch MP, Kondratov RV. Circadian clock proteins control adaptation to novel environment and memory formation. *Aging.* 2010;2:285–297.
135. Khapre RV, Kondratova AA, Susova O, Kondratov RV. Circadian clock protein BMAL1 regulates cellular senescence in vivo. *Cell Cycle.* 2011;10:4162–4169.
136. Ozturk N, Lee JH, Gaddameedhi S, Sancar A. Loss of cryptochrome reduces cancer risk in *p53* mutant mice. *Proc Natl Acad Sci USA.* 2009;106:2841–2846.
137. Hoffman AE, Yi CH, Zheng T, et al. CLOCK in breast tumorigenesis: genetic, epigenetic, and transcriptional profiling analyses. *Cancer Res.* 2010;70:1459–1468.
138. Zhou F, He X, Liu H, et al. Functional polymorphisms of circadian positive feedback regulation genes and clinical outcome of Chinese patients with resected colorectal cancer. *Cancer.* 2012;118:937–946.

139. Dai H, Zhang L, Cao M, et al. The role of polymorphisms in circadian pathway genes in breast tumorigenesis. *Breast Cancer Res Treat.* 2011;127:531–540.
140. Zhu Y, Stevens RG, Leaderer D, et al. Non-synonymous polymorphisms in the circadian gene NPAS2 and breast cancer risk. *Breast Cancer Res Treat.* 2008;107:421–425.
141. Monsees GM, Kraft P, Hankinson SE, Hunter DJ, Schernhammer ES. Circadian genes and breast cancer susceptibility in rotating shift workers. *Int J Cancer.* 2012;131:2547–2552.
142. Zhu Y, Brown HN, Zhang Y, Stevens RG, Zheng T. *Period3* structural variation: a circadian biomarker associated with breast cancer in young women. *Cancer Epidemiol Biomarkers Prev.* 2005;14:268–270.
143. Zhao B, Lu J, Yin J, et al. A functional polymorphism in *PER3* gene is associated with prognosis in hepatocellular carcinoma. *Liver Int.* 2012;32:1451–1459.
144. Hoffman AE, Zheng T, Stevens RG, et al. Clock-cancer connection in non-Hodgkin's lymphoma: a genetic association study and pathway analysis of the circadian gene cryptochrome 2. *Cancer Res.* 2009;69:3605–3613.
145. Hoffman AE, Zheng T, Yi CH, et al. The core circadian gene *Cryptochrome 2* influences breast cancer risk, possibly by mediating hormone signaling. *Cancer Prev Res.* 2010;3:539–548.
146. Fu A, Leaderer D, Zheng T, et al. Genetic and epigenetic associations of circadian gene *TIMELESS* and breast cancer risk. *Mol Carcinog.* 2012;51:923–929.
147. Li MD, Li CM, Wang Z. The role of circadian clocks in metabolic disease. *Yale J Biol Med.* 2012;85:387–401.
148. Li Y, Sato Y, Yamaguchi N. Shift work and the risk of metabolic syndrome: a nested case-control study. *Int J Occup Environ Health.* 2011;17:154–160.
149. Lin YC, Hsiao TJ, Chen PC. Persistent rotating shift-work exposure accelerates development of metabolic syndrome among middle-aged female employees: a five-year follow-up. *Chronobiol Int.* 2009;26:740–755.
150. De Bacquer D, Van Risseghem M, Clays E, et al. Rotating shift work and the metabolic syndrome: a prospective study. *Int J Epidemiol.* 2009;38:848–854.
151. Wang XS, Armstrong ME, Cairns BJ, Key TJ, Travis RC. Shift work and chronic disease: the epidemiological evidence. *Occup Med.* 2011;61:78–89.
152. Schiavo-Cardozo D, Lima MM, Pareja JC, Geloneze B. Appetite-regulating hormones from the upper gut: disrupted control of xenin and ghrelin in night workers. *Clin Endocrinol.* 2013; http://dx.doi.org/10.1111.
153. Antunes Lda C, Jornada MN, Ramalho L, Hidalgo MP. Correlation of shift work and waist circumference, body mass index, chronotype and depressive symptoms. *Arq Bras Endocrinol Metabol.* 2010;54:652–656.
154. Niedhammer I, Lert F, Marne MJ. Prevalence of overweight and weight gain in relation to night work in a nurses' cohort. *Int J Obes Relat Metab Disord.* 1996;20:625–633.
155. Karlsson B, Knutsson A, Lindahl B. Is there an association between shift work and having a metabolic syndrome? Results from a population based study of 27,485 people. *Occup Environ Med.* 2001;58:747–752.
156. Karlsson BH, Knutsson AK, Lindahl BO, Alfredsson LS. Metabolic disturbances in male workers with rotating three-shift work. Results of the WOLF study. *Int Arch Occup Environ Health.* 2003;76:424–430.
157. van Amelsvoort LG, Schouten EG, Kok FJ. Duration of shiftwork related to body mass index and waist to hip ratio. *Int J Obes Relat Metab Disord.* 1999;23:973–978.
158. Biggi N, Consonni D, Galluzzo V, Sogliani M, Costa G. Metabolic syndrome in permanent night workers. *Chronobiol Int.* 2008;25:443–454.

159. Macagnan J, Pattussi MP, Canuto R, et al. Impact of nightshift work on overweight and abdominal obesity among workers of a poultry processing plant in southern Brazil. *Chronobiol Int.* 2012;29:336–343.
160. Suwazono Y, Dochi M, Sakata K, et al. A longitudinal study on the effect of shift work on weight gain in male Japanese workers. *Obesity.* 2008;16:1887–1893.
161. Di Lorenzo L, De Pergola G, Zocchetti C, et al. Effect of shift work on body mass index: results of a study performed in 319 glucose-tolerant men working in a Southern Italian industry. *Int J Obes Relat Metab Disord.* 2003;27:1353–1358.
162. Morikawa Y, Nakagawa H, Miura K, et al. Effect of shift work on body mass index and metabolic parameters. *Scand J Work Environ Health.* 2007;33:45–50.
163. Lund J, Arendt J, Hampton SM, English J, Morgan LM. Postprandial hormone and metabolic responses amongst shift workers in Antarctica. *J Endocrinol.* 2001;171:557–564.
164. Esquirol Y, Bongard V, Mabile L, et al. Shift work and metabolic syndrome: respective impacts of job strain, physical activity, and dietary rhythms. *Chronobiol Int.* 2009;26:544–559.
165. Ghiasvand M, Heshmat R, Golpira R, et al. Shift working and risk of lipid disorders: a cross-sectional study. *Lipids Health Dis.* 2006;5:9.
166. Romon M, Nuttens MC, Fievet C, et al. Increased triglyceride levels in shift workers. *Am J Med.* 1992;93:259–262.
167. Ekstrand K, Bostrom PA, Arborelius M, Nilsson JA, Lindell SE. Cardiovascular risk factors in commercial flight aircrew officers compared with those in the general population. *Angiology.* 1996;47:1089–1094.
168. Al-Naimi S, Hampton SM, Richard P, Tzung C, Morgan LM. Postprandial metabolic profiles following meals and snacks eaten during simulated night and day shift work. *Chronobiol Int.* 2004;21:937–947.
169. Morgan L, Hampton S, Gibbs M, Arendt J. Circadian aspects of postprandial metabolism. *Chronobiol Int.* 2003;20:795–808.
170. Knutsson A. Relationships between serum triglycerides and gamma-glutamyltransferase among shift and day workers. *J Intern Med.* 1989;226:337–339.
171. Kawachi I, Colditz GA, Stampfer MJ, et al. Prospective study of shift work and risk of coronary heart disease in women. *Circulation.* 1995;92:3178–3182.
172. Mikuni E, Ohoshi T, Hayashi K, Miyamura K. Glucose intolerance in an employed population. *Tohoku J Exp Med.* 1983;141(Suppl.):251–256.
173. Van Cauter E, Polonsky KS, Scheen AJ. Roles of circadian rhythmicity and sleep in human glucose regulation. *Endocr Rev.* 1997;18:716–738.
174. Pan A, Schernhammer ES, Sun Q, Hu FB. Rotating night shift work and risk of type 2 diabetes: two prospective cohort studies in women. *PLoS Med.* 2011;8:e1001141.
175. Nagaya T, Yoshida H, Takahashi H, Kawai M. Markers of insulin resistance in day and shift workers aged 30-59 years. *Int Arch Occup Environ Health.* 2002;75:562–568.
176. Van Cauter E, Plat L, Leproult R, Copinschi G. Alterations of circadian rhythmicity and sleep in aging: endocrine consequences. *Horm Res.* 1998;49:147–152.
177. Manenschijn L, van Kruysbergen RG, de Jong FH, Koper JW, van Rossum EF. Shift work at young age is associated with elevated long-term cortisol levels and body mass index. *J Clin Endocrinol Metab.* 2011;96:E1862–E1865.
178. Lac G, Chamoux A. Biological and psychological responses to two rapid shiftwork schedules. *Ergonomics.* 2004;47:1339–1349.
179. Hennig J, Kieferdorf P, Moritz C, Huwe S, Netter P. Changes in cortisol secretion during shiftwork: implications for tolerance to shiftwork? *Ergonomics.* 1998;41:610–621.
180. Knutsson A, Boggild H. Gastrointestinal disorders among shift workers. *Scand J Work Environ Health.* 2010;36:85–95.

181. Nojkov B, Rubenstein JH, Chey WD, Hoogerwerf WA. The impact of rotating shift work on the prevalence of irritable bowel syndrome in nurses. *Am J Gastroenterol.* 2010;105:842–847.
182. Lu WZ, Gwee KA, Moochhalla S, Ho KY. Melatonin improves bowel symptoms in female patients with irritable bowel syndrome: a double-blind placebo-controlled study. *Aliment Pharmacol Ther.* 2005;22:927–934.
183. Segawa K, Nakazawa S, Tsukamoto Y, et al. Peptic ulcer is prevalent among shift workers. *Dig Dis Sci.* 1987;32:449–453.
184. Drake CL, Roehrs T, Richardson G, Walsh JK, Roth T. Shift work sleep disorder: prevalence and consequences beyond that of symptomatic day workers. *Sleep.* 2004;27:1453–1462.
185. Saberi HR, Moravveji AR. Gastrointestinal complaints in shift-working and day-working nurses in Iran. *J Circadian Rhythms.* 2010;8:9.
186. Pasqua IC, Moreno CR. The nutritional status and eating habits of shift workers: a chronobiological approach. *Chronobiol Int.* 2004;21:949–960.
187. Waterhouse J, Buckley P, Edwards B, Reilly T. Measurement of, and some reasons for, differences in eating habits between night and day workers. *Chronobiol Int.* 2003;20:1075–1092.
188. Hampton SM, Morgan LM, Lawrence N, et al. Postprandial hormone and metabolic responses in simulated shift work. *J Endocrinol.* 1996;151:259–267.
189. Ribeiro DC, Hampton SM, Morgan L, Deacon S, Arendt J. Altered postprandial hormone and metabolic responses in a simulated shift work environment. *J Endocrinol.* 1998;158:305–310.
190. Scheer FA, Hilton MF, Mantzoros CS, Shea SA. Adverse metabolic and cardiovascular consequences of circadian misalignment. *Proc Natl Acad Sci USA.* 2009;106:4453–4458.
191. Buxton OM, Cain SW, O'Connor SP, et al. Adverse metabolic consequences in humans of prolonged sleep restriction combined with circadian disruption. *Sci Transl Med.* 2012;4:129ra143.
192. Gonnissen HK, Rutters F, Mazuy C, et al. Effect of a phase advance and phase delay of the 24-h cycle on energy metabolism, appetite, and related hormones. *Am J Clin Nutr.* 2012;96:689–697.
193. Barclay JL, Husse J, Bode B, et al. Circadian desynchrony promotes metabolic disruption in a mouse model of shiftwork. *PLoS One.* 2012;7:e37150.
194. Salgado-Delgado R, Angeles-Castellanos M, Saderi N, Buijs RM, Escobar C. Food intake during the normal activity phase prevents obesity and circadian desynchrony in a rat model of night work. *Endocrinology.* 2010;151:1019–1029.
195. Arble DM, Bass J, Laposky AD, Vitaterna MH, Turek FW. Circadian timing of food intake contributes to weight gain. *Obesity.* 2009;17:2100–2102.
196. Bray MS, Tsai JY, Villegas-Montoya C, et al. Time-of-day-dependent dietary fat consumption influences multiple cardiometabolic syndrome parameters in mice. *Int J Obes.* 2010;34:1589–1598.
197. Karatsoreos IN, Bhagat S, Bloss EB, Morrison JH, McEwen BS. Disruption of circadian clocks has ramifications for metabolism, brain, and behavior. *Proc Natl Acad Sci USA.* 2011;108:1657–1662.
198. Oishi K. Disrupted light-dark cycle induces obesity with hyperglycemia in genetically intact animals. *Neuro Endocrinol Lett.* 2009;30:458–461.
199. Varcoe TJ, Wight N, Voultsios A, Salkeld MD, Kennaway DJ. Chronic phase shifts of the photoperiod throughout pregnancy programs glucose intolerance and insulin resistance in the rat. *PLoS One.* 2011;6:e18504.
200. Dauchy RT, Dauchy EM, Tirrell RP, et al. Dark-phase light contamination disrupts circadian rhythms in plasma measures of endocrine physiology and metabolism in rats. *Comp Med.* 2010;60:348–356.

201. Wideman CH, Murphy HM. Constant light induces alterations in melatonin levels, food intake, feed efficiency, visceral adiposity, and circadian rhythms in rats. *Nutr Neurosci*. 2009;12:233–240.
202. Mustonen AM, Nieminen P, Hyvarinen H. Effects of continuous light and melatonin treatment on energy metabolism of the rat. *J Endocrinol Invest*. 2002;25:716–723.
203. Shi SQ, Ansari TS, McGuinness OP, Wasserman DH, Johnson CH. Circadian disruption leads to insulin resistance and obesity. *Curr Biol*. 2013;23:372–381.
204. Turek FW, Joshu C, Kohsaka A, et al. Obesity and metabolic syndrome in circadian *Clock* mutant mice. *Science*. 2005;308:1043–1045.
205. Marcheva B, Ramsey KM, Buhr ED, et al. Disruption of the clock components CLOCK and BMAL1 leads to hypoinsulinaemia and diabetes. *Nature*. 2010;466: 627–631.
206. Rudic RD, McNamara P, Curtis AM, et al. BMAL1 and CLOCK, two essential components of the circadian clock, are involved in glucose homeostasis. *PLoS Biol*. 2004;2: e377.
207. Zuber AM, Centeno G, Pradervand S, et al. Molecular clock is involved in predictive circadian adjustment of renal function. *Proc Natl Acad Sci USA*. 2009;106:16523–16528.
208. Hemmeryckx B, Van Hove CE, Fransen P, et al. Progression of the prothrombotic state in aging Bmal1-deficient mice. *Arterioscler Thromb Vasc Biol*. 2011;31:2552–2559.
209. Shimba S, Ishii N, Ohta Y, et al. Brain and muscle Arnt-like protein-1 (BMAL1), a component of the molecular clock, regulates adipogenesis. *Proc Natl Acad Sci USA*. 2005;102:12071–12076.
210. Shimba S, Ogawa T, Hitosugi S, et al. Deficient of a clock gene, brain and muscle Arnt-like protein-1 (BMAL1), induces dyslipidemia and ectopic fat formation. *PLoS One*. 2011;6:e25231.
211. Sadacca LA, Lamia KA, deLemos AS, Blum B, Weitz CJ. An intrinsic circadian clock of the pancreas is required for normal insulin release and glucose homeostasis in mice. *Diabetologia*. 2011;54:120–124.
212. Lamia KA, Storch KF, Weitz CJ. Physiological significance of a peripheral tissue circadian clock. *Proc Natl Acad Sci USA*. 2008;105:15172–15177.
213. Kornmann B, Schaad O, Bujard H, Takahashi JS, Schibler U. System-driven and oscillator-dependent circadian transcription in mice with a conditionally active liver clock. *PLoS Biol*. 2007;5:e34.
214. Paschos GK, Ibrahim S, Song WL, et al. Obesity in mice with adipocyte-specific deletion of clock component *Arntl*. *Nat Med*. 2012;18:1768–1777.
215. Andrews JL, Zhang X, McCarthy JJ, et al. CLOCK and BMAL1 regulate *MyoD* and are necessary for maintenance of skeletal muscle phenotype and function. *Proc Natl Acad Sci USA*. 2010;107:19090–19095.
216. Carvas JM, Vukolic A, Yepuri G, et al. *Period2* gene mutant mice show compromised insulin-mediated endothelial nitric oxide release and altered glucose homeostasis. *Front Physiol*. 2012;3:337.
217. Bhatwadekar AD, Yan Y, Qi X, et al. *Per2* mutation recapitulates the vascular phenotype of diabetes in the retina and bone marrow. *Diabetes*. 2013;62:273–282.
218. Stow LR, Richards J, Cheng KY, et al. The circadian protein period 1 contributes to blood pressure control and coordinately regulates renal sodium transport genes. *Hypertension*. 2012;59:1151–1156.
219. Bae K, Lee K, Seo Y, et al. Differential effects of two *Period* genes on the physiology and proteomic profiles of mouse anterior tibialis muscles. *Mol Cells*. 2006;22: 275–284.
220. Grimaldi B, Bellet MM, Katada S, et al. PER2 controls lipid metabolism by direct regulation of PPARgamma. *Cell Metab*. 2010;12:509–520.

221. Zhao Y, Zhang Y, Zhou M, et al. Loss of *mPer2* increases plasma insulin levels by enhanced glucose-stimulated insulin secretion and impaired insulin clearance in mice. *FEBS Lett.* 2012;586:1306–1311.
222. Yang S, Liu A, Weidenhammer A, et al. The role of *mPer2* clock gene in glucocorticoid and feeding rhythms. *Endocrinology.* 2009;150:2153–2160.
223. Lamia KA, Papp SJ, Yu RT, et al. Cryptochromes mediate rhythmic repression of the glucocorticoid receptor. *Nature.* 2011;480:552–556.
224. Doi M, Takahashi Y, Komatsu R, et al. Salt-sensitive hypertension in circadian clock-deficient Cry-null mice involves dysregulated adrenal Hsd3b6. *Nat Med.* 2010;16:67–74.
225. Okano S, Hayasaka K, Igarashi M, et al. Non-obese early onset diabetes mellitus in mutant cryptochrome1 transgenic mice. *Eur J Clin Invest.* 2010;40:1011–1017.
226. Okano S, Akashi M, Hayasaka K, Nakajima O. Unusual circadian locomotor activity and pathophysiology in mutant CRY1 transgenic mice. *Neurosci Lett.* 2009;451:246–251.
227. Bugge A, Feng D, Everett LJ, et al. Rev-erbalpha and Rev-erbbeta coordinately protect the circadian clock and normal metabolic function. *Genes Dev.* 2012;26:657–667.
228. Cho H, Zhao X, Hatori M, et al. Regulation of circadian behaviour and metabolism by REV-ERB-alpha and REV-ERB-beta. *Nature.* 2012;485:123–127.
229. Woon PY, Kaisaki PJ, Braganca J, et al. Aryl hydrocarbon receptor nuclear translocator-like (BMAL1) is associated with susceptibility to hypertension and type 2 diabetes. *Proc Natl Acad Sci USA.* 2007;104:14412–14417.
230. Garaulet M, Esteban Tardido A, Lee YC, et al. SIRT1 and CLOCK 3111T>C combined genotype is associated with evening preference and weight loss resistance in a behavioral therapy treatment for obesity. *Int J Obes.* 2012;36:1436–1441.
231. Garaulet M, Sanchez-Moreno C, Smith CE, et al. Ghrelin, sleep reduction and evening preference: relationships to CLOCK 3111 T/C SNP and weight loss. *PLoS One.* 2011;6:e17435.
232. Tsuzaki K, Kotani K, Sano Y, et al. The association of the Clock 3111 T/C SNP with lipids and lipoproteins including small dense low-density lipoprotein: results from the Mima study. *BMC Med Genet.* 2010;11:150.
233. Monteleone P, Tortorella A, Docimo L, et al. Investigation of 3111T/C polymorphism of the CLOCK gene in obese individuals with or without binge eating disorder: association with higher body mass index. *Neurosci Lett.* 2008;435:30–33.
234. Sookoian S, Castano G, Gemma C, Gianotti TF, Pirola CJ. Common genetic variations in CLOCK transcription factor are associated with nonalcoholic fatty liver disease. *World J Gastroenterol.* 2007;13:4242–4248.
235. Dupuis J, Langenberg C, Prokopenko I, et al. New genetic loci implicated in fasting glucose homeostasis and their impact on type 2 diabetes risk. *Nat Genet.* 2010;42:105–116.
236. Kelly MA, Rees SD, Hydrie MZ, et al. Circadian gene variants and susceptibility to type 2 diabetes: a pilot study. *PLoS One.* 2012;7:e32670.
237. Garcia-Rios A, Perez-Martinez P, Delgado-Lista J, et al. A Period 2 genetic variant interacts with plasma SFA to modify plasma lipid concentrations in adults with metabolic syndrome. *J Nutr.* 2012;142:1213–1218.
238. Goumidi L, Grechez A, Dumont J, et al. Impact of REV-ERB alpha gene polymorphisms on obesity phenotypes in adult and adolescent samples. *Int J Obes.* 2013;37:666–672.
239. Chatham JC, Young ME. Regulation of myocardial metabolism by the cardiomyocyte circadian clock. *J Mol Cell Cardiol.* 2013;55:139–146.
240. Morris CJ, Yang JN, Scheer FA. The impact of the circadian timing system on cardiovascular and metabolic function. *Prog Brain Res.* 2012;199:337–358.

241. Ruger M, Scheer FA. Effects of circadian disruption on the cardiometabolic system. *Rev Endocr Metab Disord.* 2009;10:245–260.
242. Wolk R, Gami AS, Garcia-Touchard A, Somers VK. Sleep and cardiovascular disease. *Curr Probl Cardiol.* 2005;30:625–662.
243. Knutsson A, Boggild H. Shiftwork and cardiovascular disease: review of disease mechanisms. *Rev Environ Health.* 2000;15:359–372.
244. Ha M, Park J. Shiftwork and metabolic risk factors of cardiovascular disease. *J Occup Health.* 2005;47:89–95.
245. Knutsson A, Hallquist J, Reuterwall C, Theorell T, Akerstedt T. Shiftwork and myocardial infarction: a case-control study. *Occup Environ Med.* 1999;56:46–50.
246. Knutsson A, Akerstedt T, Jonsson BG, Orth-Gomer K. Increased risk of ischaemic heart disease in shift workers. *Lancet.* 1986;2:89–92.
247. Knutsson A, Akerstedt T, Jonsson BG. Prevalence of risk factors for coronary artery disease among day and shift workers. *Scand J Work Environ Health.* 1988;14:317–321.
248. Tenkanen L, Sjoblom T, Harma M. Joint effect of shift work and adverse life-style factors on the risk of coronary heart disease. *Scand J Work Environ Health.* 1998;24:351–357.
249. Fujino Y, Iso H, Tamakoshi A, et al. A prospective cohort study of shift work and risk of ischemic heart disease in Japanese male workers. *Am J Epidemiol.* 2006;164:128–135.
250. Brown DL, Feskanich D, Sanchez BN, et al. Rotating night shift work and the risk of ischemic stroke. *Am J Epidemiol.* 2009;169:1370–1377.
251. Puttonen S, Kivimaki M, Elovainio M, et al. Shift work in young adults and carotid artery intima-media thickness: the cardiovascular risk in young Finns study. *Atherosclerosis.* 2009;205:608–613.
252. Lieu SJ, Curhan GC, Schernhammer ES, Forman JP. Rotating night shift work and disparate hypertension risk in African-Americans. *J Hypertens.* 2012;30:61–66.
253. Oishi M, Suwazono Y, Sakata K, et al. A longitudinal study on the relationship between shift work and the progression of hypertension in male Japanese workers. *J Hypertens.* 2005;23:2173–2178.
254. Ohira T, Tanigawa T, Iso H, et al. Effects of shift work on 24-hour ambulatory blood pressure and its variability among Japanese workers. *Scand J Work Environ Health.* 2000;26:421–426.
255. Yamasaki F, Schwartz JE, Gerber LM, Warren K, Pickering TG. Impact of shift work and race/ethnicity on the diurnal rhythm of blood pressure and catecholamines. *Hypertension.* 1998;32:417–423.
256. Scheer FA, van Doornen LJ, Buijs RM. Light and diurnal cycle affect human heart rate: possible role for the circadian pacemaker. *J Biol Rhythms.* 1999;14:202–212.
257. Ruger M, Gordijn MC, Beersma DG, de Vries B, Daan S. Time-of-day-dependent effects of bright light exposure on human psychophysiology: comparison of daytime and nighttime exposure. *Am J Physiol Regul Integr Comp Physiol.* 2006;290: R1413–R1420.
258. Tsunoda M, Endo T, Hashimoto S, Honma S, Honma KI. Effects of light and sleep stages on heart rate variability in humans. *Psychiatry Clin Neurosci.* 2001;55:285–286.
259. Burgess HJ, Sletten T, Savic N, Gilbert SS, Dawson D. Effects of bright light and melatonin on sleep propensity, temperature, and cardiac activity at night. *J Appl Physiol.* 2001;91:1214–1222.
260. Scheer FA, Van Doornen LJ, Buijs RM. Light and diurnal cycle affect autonomic cardiac balance in human: possible role for the biological clock. *Auton Neurosci.* 2004;110:44–48.
261. Cajochen C, Munch M, Kobialka S, et al. High sensitivity of human melatonin, alertness, thermoregulation, and heart rate to short wavelength light. *J Clin Endocrinol Metab.* 2005;90:1311–1316.

262. Scheer FA, Buijs RM. Light affects morning salivary cortisol in humans. *J Clin Endocrinol Metab*. 1999;84:3395–3398.
263. Badia P, Myers B, Boecker M, Culpepper J, Harsh JR. Bright light effects on body temperature, alertness, EEG and behavior. *Physiol Behav*. 1991;50:583–588.
264. Campos LA, Plehm R, Cipolla-Neto J, Bader M, Baltatu OC. Altered circadian rhythm reentrainment to light phase shifts in rats with low levels of brain angiotensinogen. *Am J Physiol Regul Integr Comp Physiol*. 2006;290:R1122–R1127.
265. Oishi K, Ohkura N. Chronic circadian clock disruption induces expression of the cardiovascular risk factor plasminogen activator inhibitor-1 in mice. *Blood Coagul Fibrinolysis*. 2013;24:106–108.
266. Kishi T, Sunagawa K. Experimental 'jet lag' causes sympathoexcitation via oxidative stress through AT1 receptor in the brainstem. *Conf Proc IEEE Eng Med Biol Soc*. 2011;2011:1969–1972.
267. Martino TA, Oudit GY, Herzenberg AM, et al. Circadian rhythm disorganization produces profound cardiovascular and renal disease in hamsters. *Am J Physiol Regul Integr Comp Physiol*. 2008;294:R1675–R1683.
268. Cheng B, Anea CB, Yao L, et al. Tissue-intrinsic dysfunction of circadian clock confers transplant arteriosclerosis. *Proc Natl Acad Sci USA*. 2011;108:17147–17152.
269. Lefta M, Campbell KS, Feng HZ, Jin JP, Esser KA. Development of dilated cardiomyopathy in *Bmal1*-deficient mice. *Am J Physiol Heart Circ Physiol*. 2012;303:H475–H485.
270. Somanath PR, Podrez EA, Chen J, et al. Deficiency in core circadian protein *Bmal1* is associated with a prothrombotic and vascular phenotype. *J Cell Physiol*. 2011;226:132–140.
271. Anea CB, Zhang M, Stepp DW, et al. Vascular disease in mice with a dysfunctional circadian clock. *Circulation*. 2009;119:1510–1517.
272. Anea CB, Cheng B, Sharma S, et al. Increased superoxide and endothelial NO synthase uncoupling in blood vessels of *Bmal1*-knockout mice. *Circ Res*. 2012;111:1157–1165.
273. Anea CB, Ali MI, Osmond JM, et al. Matrix metalloproteinase 2 and 9 dysfunction underlie vascular stiffness in circadian clock mutant mice. *Arterioscler Thromb Vasc Biol*. 2010;30:2535–2543.
274. Wang CY, Wen MS, Wang HW, et al. Increased vascular senescence and impaired endothelial progenitor cell function mediated by mutation of circadian gene *Per2*. *Circulation*. 2008;118:2166–2173.
275. Viswambharan H, Carvas JM, Antic V, et al. Mutation of the circadian clock gene *Per2* alters vascular endothelial function. *Circulation*. 2007;115:2188–2195.
276. Scott EM, Carter AM, Grant PJ. Association between polymorphisms in the *Clock* gene, obesity and the metabolic syndrome in man. *Int J Obes*. 2008;32:658–662.
277. Boden MJ, Varcoe TJ, Kennaway DJ. Circadian regulation of reproduction: from gamete to offspring. *Prog Biophys Mol Biol*. 2013; http://dx.doi.org/10.1016.
278. Abeysena C, Jayawardana P, DAS R. Maternal sleep deprivation is a risk factor for small for gestational age: a cohort study. *Aust N Z J Obstet Gynaecol*. 2009;49:382–387.
279. Mahoney MM. Shift work, jet lag, and female reproduction. *Int J Endocrinol*. 2010;2010:813764.
280. Uehata T, Sasakawa N. The fatigue and maternity disturbances of night workwomen. *J Hum Ergol*. 1982;11:465–474.
281. Xu X, Ding M, Li B, Christiani DC. Association of rotating shiftwork with preterm births and low birth weight among never smoking women textile workers in China. *Occup Environ Med*. 1994;51:470–474.
282. Cone JE, Vaughan LM, Huete A, Samuels SJ. Reproductive health outcomes among female flight attendants: an exploratory study. *J Occup Environ Med*. 1998;40:210–216.
283. Ahlborg Jr G, Axelsson G, Bodin L. Shift work, nitrous oxide exposure and subfertility among Swedish midwives. *Int J Epidemiol*. 1996;25:783–790.

284. Axelsson G, Ahlborg Jr G, Bodin L. Shift work, nitrous oxide exposure, and spontaneous abortion among Swedish midwives. *Occup Environ Med.* 1996;53:374–378.
285. Axelsson G, Lutz C, Rylander R. Exposure to solvents and outcome of pregnancy in university laboratory employees. *Br J Ind Med.* 1984;41:305–312.
286. Aspholm R, Lindbohm ML, Paakkulainen H, et al. Spontaneous abortions among Finnish flight attendants. *J Occup Environ Med.* 1999;41:486–491.
287. Axelsson G, Rylander R, Molin I. Outcome of pregnancy in relation to irregular and inconvenient work schedules. *Br J Ind Med.* 1989;46:393–398.
288. McDonald AD, McDonald JC, Armstrong B, et al. Prematurity and work in pregnancy. *Br J Ind Med.* 1988;45:56–62.
289. Marino JL, Holt VL, Chen C, Davis S. Shift work, hCLOCK T3111C polymorphism, and endometriosis risk. *Epidemiology.* 2008;19:477–484.
290. Lohstroh PN, Chen J, Ba J, et al. Bone resorption is affected by follicular phase length in female rotating shift workers. *Environ Health Perspect.* 2003;111:618–622.
291. Bisanti L, Olsen J, Basso O, Thonneau P, Karmaus W. Shift work and subfecundity: a European multicenter study. European Study Group on Infertility and Subfecundity. *J Occup Environ Med.* 1996;38:352–358.
292. Knutsson A. Health disorders of shift workers. *Occup Med.* 2003;53:103–108.
293. Labyak S, Lava S, Turek F, Zee P. Effects of shiftwork on sleep and menstrual function in nurses. *Health Care Women Int.* 2002;23:703–714.
294. Lawson CC, Whelan EA, Lividoti Hibert EN, et al. Rotating shift work and menstrual cycle characteristics. *Epidemiology.* 2011;22:305–312.
295. Su SB, Lu CW, Kao YY, Guo HR. Effects of 12-hour rotating shifts on menstrual cycles of photoelectronic workers in Taiwan. *Chronobiol Int.* 2008;25:237–248.
296. Lawson CC, Whelan EA, Hibert EN, et al. Occupational factors and risk of preterm birth in nurses. *Am J Obstet Gynecol.* 2009;200:51.e51–51.e58.
297. Whelan EA, Lawson CC, Grajewski B, et al. Work schedule during pregnancy and spontaneous abortion. *Epidemiology.* 2007;18:350–355.
298. Zhu JL, Hjollund NH, Andersen AM, Olsen J. Shift work, job stress, and late fetal loss: the national birth cohort in Denmark. *J Occup Environ Med.* 2004;46:1144–1149.
299. Zhu JL, Hjollund NH, Olsen J. Shift work, duration of pregnancy, and birth weight: the national birth cohort in Denmark. *Am J Obstet Gynecol.* 2004;191:285–291.
300. Bonzini M, Palmer KT, Coggon D, et al. Shift work and pregnancy outcomes: a systematic review with meta-analysis of currently available epidemiological studies. *BJOG.* 2011;118:1429–1437.
301. Ikegami K, Yoshimura T. Circadian clocks and the measurement of daylength in seasonal reproduction. *Mol Cell Endocrinol.* 2012;349:76–81.
302. Kennaway DJ, Boden MJ, Varcoe TJ. Circadian rhythms and fertility. *Mol Cell Endocrinol.* 2012;349:56–61.
303. Young Janik L, Janik D. Nonphotic phase shifting in female Syrian hamsters: interactions with the estrous cycle. *J Biol Rhythms.* 2003;18:307–317.
304. Legan SJ, Franklin KM, Peng XL, Duncan MJ. Novel wheel running blocks the preovulatory luteinizing hormone surge and advances the hamster circadian pacemaker. *J Biol Rhythms.* 2010;25:450–459.
305. Smarr BL, Morris E, de la Iglesia HO. The dorsomedial suprachiasmatic nucleus times circadian expression of *Kiss1* and the luteinizing hormone surge. *Endocrinology.* 2012;153:2839–2850.
306. Swann JM, Turek FW. Multiple circadian oscillators regulate the timing of behavioral and endocrine rhythms in female golden hamsters. *Science.* 1985;228:898–900.
307. Campbell CS, Ryan KD, Schwartz NB. Estrous cycles in the mouse: relative influence of continuous light and the presence of a male. *Biol Reprod.* 1976;14:292–299.

308. Hagino N, Sako T, Nakamoto O, Kunz Y, Saito H. Prevention of continuous light-induced anovulation in rats by early exposure to continuous light. *Biol Reprod.* 1983;29:355–361.
309. Thomas BB, Oommen MM, Ashadevi JS. Constant light and blinding effects on reproduction of male South Indian gerbils. *J Exp Zool.* 2001;289:59–65.
310. Summa KC, Vitaterna MH, Turek FW. Environmental perturbation of the circadian clock disrupts pregnancy in the mouse. *PLoS One.* 2012;7:e37668.
311. Gozeri E, Celik H, Ozercan I, et al. The effect of circadian rhythm changes on fetal and placental development (experimental study). *Neuro Endocrinol Lett.* 2008;29:87–90.
312. Endo A, Watanabe T. Effects of non-24-hour days on reproductive efficacy and embryonic development in mice. *Gamete Res.* 1989;22:435–441.
313. Xu K, DiAngelo JR, Hughes ME, Hogenesch JB, Sehgal A. The circadian clock interacts with metabolic physiology to influence reproductive fitness. *Cell Metab.* 2011;13:639–654.
314. Low-Zeddies SS, Takahashi JS. Chimera analysis of the *Clock* mutation in mice shows that complex cellular integration determines circadian behavior. *Cell.* 2001;105: 25–42.
315. Herzog ED, Grace MS, Harrer C, et al. The role of *Clock* in the developmental expression of neuropeptides in the suprachiasmatic nucleus. *J Comp Neurol.* 2000; 424:86–98.
316. Chappell PE, White RS, Mellon PL. Circadian gene expression regulates pulsatile gonadotropin-releasing hormone (GnRH) secretory patterns in the hypothalamic GnRH-secreting GT1-7 cell line. *J Neurosci.* 2003;23:11202–11213.
317. Miller BH, Olson SL, Turek FW, et al. Circadian clock mutation disrupts estrous cyclicity and maintenance of pregnancy. *Curr Biol.* 2004;14:1367–1373.
318. Dolatshad H, Campbell EA, O'Hara L, et al. Developmental and reproductive performance in circadian mutant mice. *Hum Reprod.* 2006;21:68–79.
319. Kennaway DJ, Boden MJ, Voultsios A. Reproductive performance in female Clock Delta19 mutant mice. *Reprod Fertil Dev.* 2004;16:801–810.
320. Cowden KD, Simon MC. The bHLH/PAS factor MOP3 does not participate in hypoxia responses. *Biochem Biophys Res Commun.* 2002;290:1228–1236.
321. Boden MJ, Varcoe TJ, Voultsios A, Kennaway DJ. Reproductive biology of female *Bmal1* null mice. *Reproduction.* 2010;139:1077–1090.
322. Ratajczak CK, Boehle KL, Muglia LJ. Impaired steroidogenesis and implantation failure in $Bmal1^{-/-}$ mice. *Endocrinology.* 2009;150:1879–1885.
323. Alvarez JD, Hansen A, Ord T, et al. The circadian clock protein BMAL1 is necessary for fertility and proper testosterone production in mice. *J Biol Rhythms.* 2008;23:26–36.
324. Pilorz V, Steinlechner S. Low reproductive success in *Per1* and *Per2* mutant mouse females due to accelerated ageing? *Reproduction.* 2008;135:559–568.
325. Kovanen L, Saarikoski ST, Aromaa A, Lonnqvist J, Partonen T. *ARNTL (BMAL1)* and *NPAS2* gene variants contribute to fertility and seasonality. *PLoS One.* 2010;5:e10007.
326. Lanfumey L, Mongeau R, Hamon M. Biological rhythms and melatonin in mood disorders and their treatments. *Pharmacol Ther.* 2013;138:176–184.
327. Katz G, Knobler HY, Laibel Z, Strauss Z, Durst R. Time zone change and major psychiatric morbidity: the results of a 6-year study in Jerusalem. *Compr Psychiatry.* 2002;43:37–40.
328. Skipper Jr JK, Jung FD, Coffey LC. Nurses and shiftwork: effects on physical health and mental depression. *J Adv Nurs.* 1990;15:835–842.
329. Jauhar P, Weller MP. Psychiatric morbidity and time zone changes: a study of patients from Heathrow airport. *Br J Psychiatry.* 1982;140:231–235.
330. Young DM. Psychiatric morbidity in travelers to Honolulu, Hawaii. *Compr Psychiatry.* 1995;36:224–228.

331. Drennan MD, Klauber MR, Kripke DF, Goyette LM. The effects of depression and age on the Horne–Ostberg morningness-eveningness score. *J Affect Disord.* 1991;23: 93–98.
332. Chelminski I, Ferraro FR, Petros TV, Plaud JJ. An analysis of the "eveningness-morningness" dimension in "depressive" college students. *J Affect Disord.* 1999;52:19–29.
333. Emens J, Lewy A, Kinzie JM, Arntz D, Rough J. Circadian misalignment in major depressive disorder. *Psychiatry Res.* 2009;168:259–261.
334. Hasler BP, Buysse DJ, Kupfer DJ, Germain A. Phase relationships between core body temperature, melatonin, and sleep are associated with depression severity: further evidence for circadian misalignment in non-seasonal depression. *Psychiatry Res.* 2010;178:205–207.
335. Levandovski R, Dantas G, Fernandes LC, et al. Depression scores associate with chronotype and social jetlag in a rural population. *Chronobiol Int.* 2011;28:771–778.
336. Wood J, Birmaher B, Axelson D, et al. Replicable differences in preferred circadian phase between bipolar disorder patients and control individuals. *Psychiatry Res.* 2009;166:201–209.
337. Mansour HA, Wood J, Chowdari KV, et al. Circadian phase variation in bipolar I disorder. *Chronobiol Int.* 2005;22:571–584.
338. Lewy AJ, Emens JS, Songer JB, et al. Winter Depression: integrating mood, circadian rhythms, and the sleep/wake and light/dark cycles into a bio-psycho-social-environmental model. *Sleep Med Clin.* 2009;4:285–299.
339. Lewy AJ, Lefler BJ, Emens JS, Bauer VK. The circadian basis of winter depression. *Proc Natl Acad Sci USA.* 2006;103:7414–7419.
340. Krivisky K, Ashkenazy T, Kronfeld-Schor N, Einat H. Antidepressants reverse short-photoperiod-induced, forced swim test depression-like behavior in the diurnal fat sand rat: further support for the utilization of diurnal rodents for modeling affective disorders. *Neuropsychobiology.* 2011;63:191–196.
341. Ashkenazy T, Einat H, Kronfeld-Schor N. Effects of bright light treatment on depression- and anxiety-like behaviors of diurnal rodents maintained on a short daylight schedule. *Behav Brain Res.* 2009;201:343–346.
342. Ashkenazy T, Einat H, Kronfeld-Schor N. We are in the dark here: induction of depression- and anxiety-like behaviours in the diurnal fat sand rat, by short daylight or melatonin injections. *Int J Neuropsychopharmacol.* 2009;12:83–93.
343. Ashkenazy-Frolinger T, Kronfeld-Schor N, Juetten J, Einat H. It is darkness and not light: depression-like behaviors of diurnal unstriped Nile grass rats maintained under a short photoperiod schedule. *J Neurosci Methods.* 2010;186:165–170.
344. Prendergast BJ, Nelson RJ. Affective responses to changes in day length in Siberian hamsters (*Phodopus sungorus*). *Psychoneuroendocrinology.* 2005;30:438–452.
345. Prendergast BJ, Kay LM. Affective and adrenocorticotrophic responses to photoperiod in Wistar rats. *J Neuroendocrinol.* 2008;20:261–267.
346. Monje FJ, Cabatic M, Divisch I, et al. Constant darkness induces IL-6-dependent depression-like behavior through the NF-kappaB signaling pathway. *J Neurosci.* 2011;31:9075–9083.
347. Flaisher-Grinberg S, Gampetro DR, Kronfeld-Schor N, Einat H. Inconsistent effects of photoperiod manipulations in tests for affective-like changes in mice: implications for the selection of appropriate model animals. *Behav Pharmacol.* 2011;22:23–30.
348. Fonken LK, Finy MS, Walton JC, et al. Influence of light at night on murine anxiety- and depressive-like responses. *Behav Brain Res.* 2009;205:349–354.
349. Bedrosian TA, Weil ZM, Nelson RJ. Chronic dim light at night provokes reversible depression-like phenotype: possible role for TNF. *Mol Psychiatry.* 2012; http://dx.doi.org/10.1038.

350. Bedrosian TA, Fonken LK, Walton JC, Haim A, Nelson RJ. Dim light at night provokes depression-like behaviors and reduces CA1 dendritic spine density in female hamsters. *Psychoneuroendocrinology*. 2011;36:1062–1069.
351. Fonken LK, Nelson RJ. Dim light at night increases depressive-like responses in male C3H/HeNHsd mice. *Behav Brain Res*. 2013;243C:74–78.
352. Fonken LK, Kitsmiller E, Smale L, Nelson RJ. Dim nighttime light impairs cognition and provokes depressive-like responses in a diurnal rodent. *J Biol Rhythms*. 2012;27:319–327.
353. Castanon-Cervantes O, Wu M, Ehlen JC, et al. Dysregulation of inflammatory responses by chronic circadian disruption. *J Immunol*. 2010;185:5796–5805.
354. Guilding C, Piggins HD. Challenging the omnipotence of the suprachiasmatic timekeeper: are circadian oscillators present throughout the mammalian brain? *Eur J Neurosci*. 2007;25:3195–3216.
355. McCarthy MJ, Welsh DK. Cellular circadian clocks in mood disorders. *J Biol Rhythms*. 2012;27:339–352.
356. Easton A, Arbuzova J, Turek FW. The circadian *Clock* mutation increases exploratory activity and escape-seeking behavior. *Genes Brain Behav*. 2003;2:11–19.
357. Roybal K, Theobold D, Graham A, et al. Mania-like behavior induced by disruption of *CLOCK*. *Proc Natl Acad Sci USA*. 2007;104:6406–6411.
358. Naylor E, Bergmann BM, Krauski K, et al. The circadian *clock* mutation alters sleep homeostasis in the mouse. *J Neurosci*. 2000;20:8138–8143.
359. Mukherjee S, Coque L, Cao JL, et al. Knockdown of *Clock* in the ventral tegmental area through RNA interference results in a mixed state of mania and depression-like behavior. *Biol Psychiatry*. 2010;68:503–511.
360. Masana MI, Sumaya IC, Becker-Andre M, Dubocovich ML. Behavioral characterization and modulation of circadian rhythms by light and melatonin in C_3H/HeN mice homozygous for the RORbeta knockout. *Am J Physiol Regul Integr Comp Physiol*. 2007;292:R2357–R2367.
361. Hampp G, Ripperger JA, Houben T, et al. Regulation of monoamine oxidase A by circadian-clock components implies clock influence on mood. *Curr Biol*. 2008;18:678–683.
362. Keers R, Pedroso I, Breen G, et al. Reduced anxiety and depression-like behaviours in the circadian period mutant mouse afterhours. *PLoS One*. 2012;7:e38263.
363. Wright KP, Lowry CA, Lebourgeois MK. Circadian and wakefulness-sleep modulation of cognition in humans. *Front Mol Neurosci*. 2012;5:50.
364. Santhi N, Horowitz TS, Duffy JF, Czeisler CA. Acute sleep deprivation and circadian misalignment associated with transition onto the first night of work impairs visual selective attention. *PLoS One*. 2007;2:e1233.
365. Dijk DJ, Duffy JF, Czeisler CA. Circadian and sleep/wake dependent aspects of subjective alertness and cognitive performance. *J Sleep Res*. 1992;1:112–117.
366. Goel N, Rao H, Durmer JS, Dinges DF. Neurocognitive consequences of sleep deprivation. *Semin Neurol*. 2009;29:320–339.
367. Akerstedt T. Shift work and disturbed sleep/wakefulness. *Sleep Med Rev*. 1998;2:117–128.
368. Akerstedt T, Fredlund P, Gillberg M, Jansson B. A prospective study of fatal occupational accidents—relationship to sleeping difficulties and occupational factors. *J Sleep Res*. 2002;11:69–71.
369. Cho K, Ennaceur A, Cole JC, Suh CK. Chronic jet lag produces cognitive deficits. *J Neurosci*. 2000;20:RC66.
370. Cho K. Chronic 'jet lag' produces temporal lobe atrophy and spatial cognitive deficits. *Nat Neurosci*. 2001;4:567–568.

371. Wright Jr KP, Hull JT, Hughes RJ, Ronda JM, Czeisler CA. Sleep and wakefulness out of phase with internal biological time impairs learning in humans. *J Cogn Neurosci.* 2006;18:508–521.
372. Neto SP, Carneiro BT, Valentinuzzi VS, Araujo JF. Dissociation of the circadian rhythm of locomotor activity in a 22 h light-dark cycle impairs passive avoidance but not object recognition memory in rats. *Physiol Behav.* 2008;94:523–527.
373. Gibson EM, Wang C, Tjho S, Khattar N, Kriegsfeld LJ. Experimental 'jet lag' inhibits adult neurogenesis and produces long-term cognitive deficits in female hamsters. *PLoS One.* 2010;5:e15267.
374. Kott J, Leach G, Yan L. Direction-dependent effects of chronic "jet-lag" on hippocampal neurogenesis. *Neurosci Lett.* 2012;515:177–180.
375. Devan BD, Goad EH, Petri HL, et al. Circadian phase-shifted rats show normal acquisition but impaired long-term retention of place information in the water task. *Neurobiol Learn Mem.* 2001;75:51–62.
376. Fekete M, van Ree JM, Niesink RJ, de Wied D. Disrupting circadian rhythms in rats induces retrograde amnesia. *Physiol Behav.* 1985;34:883–887.
377. Tapp WN, Holloway FA. Phase shifting circadian rhythms produces retrograde amnesia. *Science.* 1981;211:1056–1058.
378. Loh DH, Navarro J, Hagopian A, et al. Rapid changes in the light/dark cycle disrupt memory of conditioned fear in mice. *PLoS One.* 2010;5:e12546.
379. Craig LA, McDonald RJ. Chronic disruption of circadian rhythms impairs hippocampal memory in the rat. *Brain Res Bull.* 2008;76:141–151.
380. Fujioka A, Fujioka T, Tsuruta R, et al. Effects of a constant light environment on hippocampal neurogenesis and memory in mice. *Neurosci Lett.* 2011;488:41–44.
381. Ma WP, Cao J, Tian M, et al. Exposure to chronic constant light impairs spatial memory and influences long-term depression in rats. *Neurosci Res.* 2007;59: 224–230.
382. Ruby NF, Hwang CE, Wessells C, et al. Hippocampal-dependent learning requires a functional circadian system. *Proc Natl Acad Sci USA.* 2008;105:15593–15598.
383. Wang LM, Dragich JM, Kudo T, et al. Expression of the circadian clock gene *Period2* in the hippocampus: possible implications for synaptic plasticity and learned behaviour. *ASN Neuro.* 2009;1:e00012.
384. Sakai T, Tamura T, Kitamoto T, Kidokoro Y. A clock gene, *period,* plays a key role in long-term memory formation in *Drosophila. Proc Natl Acad Sci USA.* 2004;101: 16058–16063.
385. Jilg A, Lesny S, Peruzki N, et al. Temporal dynamics of mouse hippocampal clock gene expression support memory processing. *Hippocampus.* 2010;20:377–388.
386. Abarca C, Albrecht U, Spanagel R. Cocaine sensitization and reward are under the influence of circadian genes and rhythm. *Proc Natl Acad Sci USA.* 2002;99: 9026–9030.
387. Zueger M, Urani A, Chourbaji S, et al. *mPer1* and *mPer2* mutant mice show regular spatial and contextual learning in standardized tests for hippocampus-dependent learning. *J Neural Transm.* 2006;113:347–356.
388. Zghoul T, Abarca C, Sanchis-Segura C, et al. Ethanol self-administration and reinstatement of ethanol-seeking behavior in $Per1^{Brdm1}$ mutant mice. *Psychopharmacology (Berl).* 2007;190:13–19.
389. Garcia JA, Zhang D, Estill SJ, et al. Impaired cued and contextual memory in NPAS2-deficient mice. *Science.* 2000;288:2226–2230.
390. Crawley JN, Paylor R. A proposed test battery and constellations of specific behavioral paradigms to investigate the behavioral phenotypes of transgenic and knockout mice. *Horm Behav.* 1997;31:197–211.

391. Lo JC, Groeger JA, Santhi N, et al. Effects of partial and acute total sleep deprivation on performance across cognitive domains, individuals and circadian phase. *PLoS One*. 2012;7:e45987.
392. Groeger JA, Viola AU, Lo JC, et al. Early morning executive functioning during sleep deprivation is compromised by a *PERIOD3* polymorphism. *Sleep*. 2008;31: 1159–1167.
393. Viola AU, Archer SN, James LM, et al. *PER3* polymorphism predicts sleep structure and waking performance. *Curr Biol*. 2007;17:613–618.
394. Scheiermann C, Kunisaki Y, Frenette PS. Circadian control of the immune system. *Nat Rev Immunol*. 2013;13:190–198.
395. Mohren DC, Jansen NW, Kant IJ, et al. Prevalence of common infections among employees in different work schedules. *J Occup Environ Med*. 2002;44:1003–1011.
396. Hedstrom AK, Akerstedt T, Hillert J, Olsson T, Alfredsson L. Shift work at young age is associated with increased risk for multiple sclerosis. *Ann Neurol*. 2011;70:733–741.
397. Magrini A, Pietroiusti A, Coppeta L, et al. Shift work and autoimmune thyroid disorders. *Int J Immunopathol Pharmacol*. 2006;19:31–36.
398. Ryan MJ. The pathophysiology of hypertension in systemic lupus erythematosus. *Am J Physiol Regul Integr Comp Physiol*. 2009;296:R1258–R1267.
399. Bojarski C. Malignant transformation in inflammatory bowel disease: prevention, surveillance and treatment—new techniques in endoscopy. *Dig Dis*. 2009;27:571–575.
400. Goldberg RB. Cytokine and cytokine-like inflammation markers, endothelial dysfunction, and imbalanced coagulation in development of diabetes and its complications. *J Clin Endocrinol Metab*. 2009;94:3171–3182.
401. Kort WJ, Weijma JM. Effect of chronic light-dark shift stress on the immune response of the rat. *Physiol Behav*. 1982;29:1083–1087.
402. Preuss F, Tang Y, Laposky AD, et al. Adverse effects of chronic circadian desynchronization in animals in a "challenging" environment. *Am J Physiol Regul Integr Comp Physiol*. 2008;295:R2034–R2040.
403. Moore CB, Siopes TD. Effects of lighting conditions and melatonin supplementation on the cellular and humoral immune responses in Japanese quail *Coturnix coturnix japonica*. *Gen Comp Endocrinol*. 2000;119:95–104.
404. Kirby JD, Froman DP. Research note: evaluation of humoral and delayed hypersensitivity responses in cockerels reared under constant light or a twelve hour light:twelve hour dark photoperiod. *Poult Sci*. 1991;70:2375–2378.
405. Oishi K, Shibusawa K, Kakazu H, et al. Extended light exposure suppresses nocturnal increases in cytotoxic activity of splenic natural killer cells in rats. *Biol Rhythm Res*. 2006;37:21–35.
406. Prendergast BJ. Behavioral tolerance to endotoxin is enhanced by adaptation to winter photoperiods. *Psychoneuroendocrinology*. 2008;33:540–545.
407. Prendergast BJ, Bilbo SD, Nelson RJ. Short day lengths enhance skin immune responses in gonadectomised Siberian hamsters. *J Neuroendocrinol*. 2005;17:18–21.
408. Wen JC, Dhabhar FS, Prendergast BJ. Pineal-dependent and -independent effects of photoperiod on immune function in Siberian hamsters (*Phodopus sungorus*). *Horm Behav*. 2007;51:31–39.
409. Wen JC, Prendergast BJ. Photoperiodic regulation of behavioral responsiveness to proinflammatory cytokines. *Physiol Behav*. 2007;90:717–725.
410. Gibbs JE, Blaikley J, Beesley S, et al. The nuclear receptor REV-ERBalpha mediates circadian regulation of innate immunity through selective regulation of inflammatory cytokines. *Proc Natl Acad Sci USA*. 2012;109:582–587.
411. Narasimamurthy R, Hatori M, Nayak SK, et al. Circadian clock protein cryptochrome regulates the expression of proinflammatory cytokines. *Proc Natl Acad Sci USA*. 2012;109:12662–12667.

412. Hashiramoto A, Yamane T, Tsumiyama K, et al. Mammalian clock gene *Cryptochrome* regulates arthritis via proinflammatory cytokine TNF-alpha. *J Immunol.* 2010;184:1560–1565.
413. Logan RW, Wynne O, Levitt D, Price D, Sarkar DK. Altered circadian expression of cytokines and cytolytic factors in splenic natural killer cells of *Per1*$^{-/-}$ mutant mice. *J Interferon Cytokine Res.* 2013;33:108–114.
414. Arjona A, Sarkar DK. The circadian gene *mPer2* regulates the daily rhythm of IFN-gamma. *J Interferon Cytokine Res.* 2006;26:645–649.
415. Arjona A, Sarkar DK. Evidence supporting a circadian control of natural killer cell function. *Brain Behav Immun.* 2006;20:469–476.
416. Liu J, Malkani G, Shi X, et al. The circadian clock *Period 2* gene regulates gamma interferon production of NK cells in host response to lipopolysaccharide-induced endotoxic shock. *Infect Immun.* 2006;74:4750–4756.
417. Sun Y, Yang Z, Niu Z, et al. The mortality of *MOP3* deficient mice with a systemic functional failure. *J Biomed Sci.* 2006;13:845–851.
418. Sun Y, Yang Z, Niu Z, et al. *MOP3*, a component of the molecular clock, regulates the development of B cells. *Immunology.* 2006;119:451–460.
419. Oishi K, Miyazaki K, Kadota K, et al. Genome-wide expression analysis of mouse liver reveals CLOCK-regulated circadian output genes. *J Biol Chem.* 2003;278:41519–41527.
420. Spengler ML, Kuropatwinski KK, Comas M, et al. Core circadian protein CLOCK is a positive regulator of NF-kappaB-mediated transcription. *Proc Natl Acad Sci USA.* 2012;109:E2457–E2465.
421. Mazzoccoli G, Palmieri O, Corritore G, et al. Association study of a polymorphism in clock gene *PERIOD3* and risk of inflammatory bowel disease. *Chronobiol Int.* 2012;29:994–1003.
422. Altimus CM, Guler AD, Villa KL, et al. Rods-cones and melanopsin detect light and dark to modulate sleep independent of image formation. *Proc Natl Acad Sci USA.* 2008;105:19998–20003.
423. Ebihara S, Marks T, Hudson DJ, Menaker M. Genetic control of melatonin synthesis in the pineal gland of the mouse. *Science.* 1986;231:491–493.
424. Pilcher JJ, Lambert BJ, Huffcutt AI. Differential effects of permanent and rotating shifts on self-report sleep length: a meta-analytic review. *Sleep.* 2000;23:155–163.
425. Bambra CL, Whitehead MM, Sowden AJ, Akers J, Petticrew MP. Shifting schedules: the health effects of reorganizing shift work. *Am J Prev Med.* 2008;34:427–434.
426. Sallinen M, Kecklund G. Shift work, sleep, and sleepiness—differences between shift schedules and systems. *Scand J Work Environ Health.* 2010;36:121–133.
427. Chen ST, Choo KB, Hou MF, et al. Deregulated expression of the *PER1*, *PER2* and *PER3* genes in breast cancers. *Carcinogenesis.* 2005;26:1241–1246.
428. Zeng ZL, Wu MW, Sun J, et al. Effects of the biological clock gene *Bmal1* on tumour growth and anti-cancer drug activity. *J Biochem.* 2010;148:319–326.

CHAPTER ELEVEN

Sleep and Circadian Rhythm Disruption in Social Jetlag and Mental Illness

Russell G. Foster*, Stuart N. Peirson*, Katharina Wulff*, Eva Winnebeck†, Céline Vetter†, Till Roenneberg†

*Nuffield Department of Clinical Neurosciences, Nuffield Laboratory of Ophthalmology, Oxford, United Kingdom

†Institute for Medical Psychology, University of Munich, Munich, Germany

Contents

Abstract

Sleep and wake represent two profoundly different states of physiology that arise within the brain from a complex interaction between multiple neural circuits and neurotransmitter systems. These neural networks are, in turn, adjusted by three key drivers that collectively determine the duration, quality, and efficiency of sleep. Two of these drivers are endogenous, namely, the circadian system and a homeostatic hourglass oscillator, while the third is exogenous—our societal structure (social time).

In this chapter, we outline the neuroscience of sleep and highlight the links between sleep, mood, cognition, and mental health. We emphasize that the complexity of sleep/wake generation and regulation makes this behavioral cycle very vulnerable to disruption and then explore this concept by examining sleep and circadian rhythm disruption (SCRD) when the exogenous and endogenous drivers of sleep are in conflict.

Progress in Molecular Biology and Translational Science, Volume 119
ISSN 1877-1173
http://dx.doi.org/10.1016/B978-0-12-396971-2.00011-7

SCRD can be particularly severe when social timing forces an abnormal pattern of sleep and wake upon our endogenous sleep biology.

SCRD is also very common in mental illness, and although well known, this association is poorly understood or treated. Recent studies suggest that the generation of sleep and mental health shares overlapping neural mechanisms such that defects in these endogenous pathways result in pathologies to both behaviors. The evidence for this association is examined in some detail.

We conclude this review by suggesting that the emerging understanding of the neurobiology of sleep/wake behavior, and of the health consequences of sleep disruption, will provide new ways to decrease the conflict between biological and societal timing in both the healthy and individuals with mental illness.

1. THE BIOLOGY OF SLEEP

Sleep is a highly complex behavior arising from an interaction between multiple neural circuits, neurotransmitters, and hormones, none of which are exclusive to the generation of sleep.[1] The major brain structures and neurotransmitter systems involved in the sleep/wake cycle are summarized in Fig. 11.1.

The sleep systems illustrated in Fig. 11.1 are controlled by three key drivers that interact and collectively determine the duration, quality, and efficiency of sleep. Two of these drivers are endogenous, referred to as the circadian system and a homeostatic hourglass oscillator, while the third is exogenous and the product of our societal temporal structure (e.g., school and work times). Normal versus abnormal sleep has, in turn, a major impact upon mood, cognition, and mental and physical health. These relationships are summarized in Fig. 11.2.

2. SLEEP AND CIRCADIAN RHYTHM DISRUPTION ARISING FROM SOCIAL TIMING

As described above, efficient sleep depends upon complex biological drivers in the form of the circadian system and the homeostatic hourglass oscillator (Fig. 11.2). In real life, however, an additional critical factor—social timing—influences sleep opportunities and thereby sleep duration, quality, and efficiency, potentially resulting in sleep and circadian rhythm disruption (SCRD). A large database ($\approx$150,000 entries) has been established on sleep-activity behavior in the general population[2] by using an internet-based questionnaire (the Munich ChronoType Questionnaire,

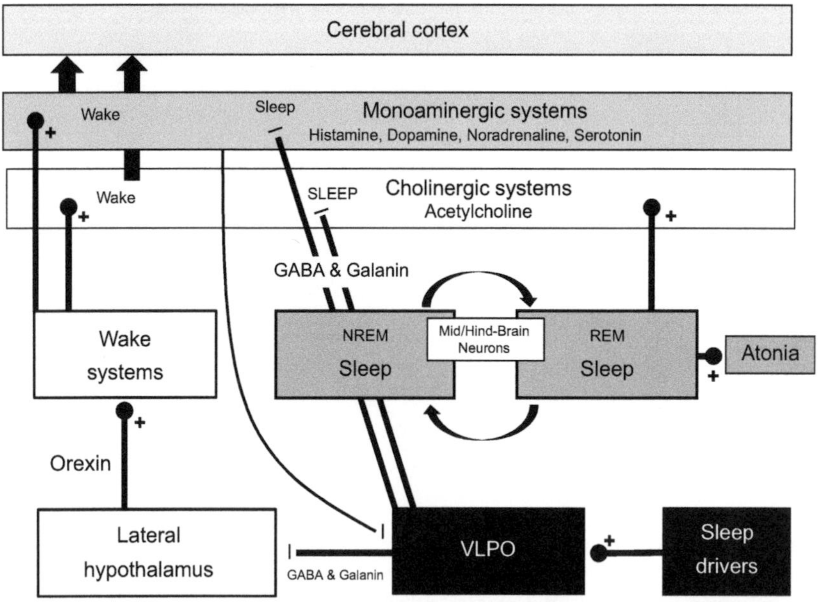

Figure 11.1 During wake orexin neurons in the lateral hypothalamus project to and excite (+) monoaminergic and cholinergic neurons. These neurotransmitters drive wakefulness and consciousness within the cortex. The monoaminergic neurons also inhibit (−) the ventrolateral preoptic nuclei (VLPO). During sleep, circadian and homeostatic sleep drivers (Fig. 11.2) activate the VLPO which releases GABA and galanin to inhibit the orexin neurons in the lateral hypothalamus and the aminergic and cholinergic neurons (−). The NREM–REM flip/flop every 70–90 min is driven by a network of neurons in the mid- and hind-brain. During REM sleep aminergic neurons remain inhibited, but cholinergic neurons are activated (+). REM-on neurons project to the spinal cord and drive muscle paralysis (atonia).

MCTQ[3]; see "chronotype study" at www.theWep.org). The MCTQ assesses sleep–wake behavior separately for work and free days. This epidemiological approach has shown that both sleep timing and duration are greatly challenged by work and school schedules or other social events. To align their sleep and wake times with social obligations, 80% of the population uses alarm clocks on workdays,[2] and a growing number of people use sleep medication at night and stimulants to drive wakefulness during the day.[4]

How individuals accommodate their endogenous sleep need between social constraints and biological timing depends on how their circadian clock embeds itself into the light–dark cycle of the environment (entrainment),[5] producing distinct "chronotypes".[6,7] Chronotype can be determined by an

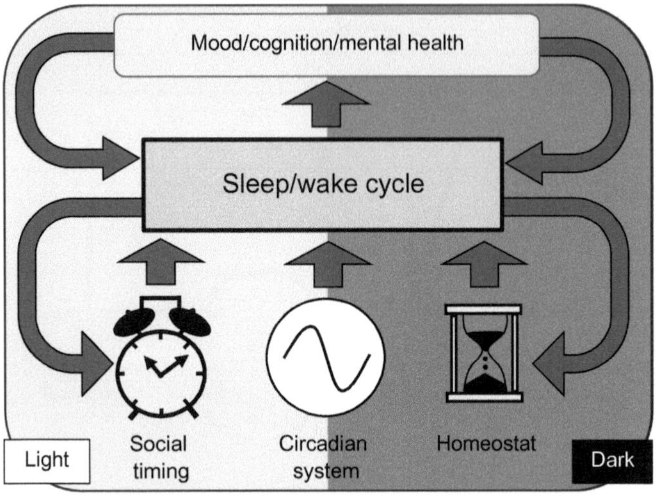

Figure 11.2 The key components in the regulation and maintenance of the sleep/wake cycle and its relationship to mood/cognition and mental health. The sleep/wake cycle is regulated by three key drivers which are illustrated here: (i) The endogenous 24-h body clock/circadian system, via the master circadian pacemaker located within the suprachiasmatic nuclei (SCN) drives wakefulness throughout the day and sleep during the night. The SCN regulates directly multiple neurotransmitter systems that either drive or modulate sleep, including the hypothalamo–pituitary–adrenal (HPA) axis and melatonin from the pineal gland. (ii) Wake-dependent homeostatic drivers (such as adenosine) build-up within the brain to generate a sleep pressure during the day. This "hourglass oscillator" drives increased sleep pressure during wake, which is counteracted by the circadian drive for wakefulness. Sleep pressure is dissipated during sleep, and sleep is maintained by the circadian drive for sleep. In general terms, wake occurs when sleep pressure is low and the circadian drive for sleep ceases. (iii) Social timing frequently forces an exogenous sleep/wake pattern on an individual's behavior. The light/dark cycle plays a key role in sleep regulation acting to (a) entrain the SCN to the 24-h cycle of dawn and dusk; (b) alter the production or different hormones associated with activity and rest, most notably melatonin; (c) modulate the HPA "stress" axis; and (d) elevate or suppress levels of mood and cognitive function. Social behaviors will also drive an individual's exposure to the light/dark cycle, and hence circadian entrainment and alertness. The duration, quality, and efficiency of sleep have a direct impact upon mood, cognitive processing, and mental health. (See Color Insert.)

individual's sleep–wake behavior, for example, by actimetry, sleep-logs, or questionnaires such as the MCTQ. The MCTQ[3] assesses chronotype based on sleep behavior on work-free days (the time of mid-sleep, MSF—the midpoint between sleep onset and sleep end), which is then corrected for "oversleep" on free days (MSF_{sc}); sleeping in on free days reflects compensation for accumulated sleep debt during the workweek (see below). The detailed evaluation of the MCTQ and the computation of all derived variables are given in the supplementary online material of Ref. 2.

MSF is normally distributed in the general population, but with a slight overrepresentation of late types. The differences between the extremes of this distribution (extreme "larks" vs. extreme "owls") are as much as 12 h. Chronotype depends on genetic disposition[8] age and sex[9] and on light exposure.[10] Chronotype becomes progressively later throughout puberty and adolescence, thereafter advancing again until the elderly become as early as children.[9] Social time has surprisingly little influence on free-day sleep timing. Within Germany, chronotype (MSF_{sc}; MSF adjusted for age and sex) becomes later by 4 min for every longitude from east to west, that is, by the same time that the sun needs to cross one longitude. Thus, despite experiencing similar social times (work times, evening news programs, etc.), the internal timing system of the German population is still adjusted to sun time.

The strong influence of social timing on human sleep becomes apparent when comparing sleep–wake behavior on work and free days. The vast majority of the population shows clear differences in both sleep duration (Fig. 11.3) and timing (Fig. 11.4). When expressed as weekly average, sleep

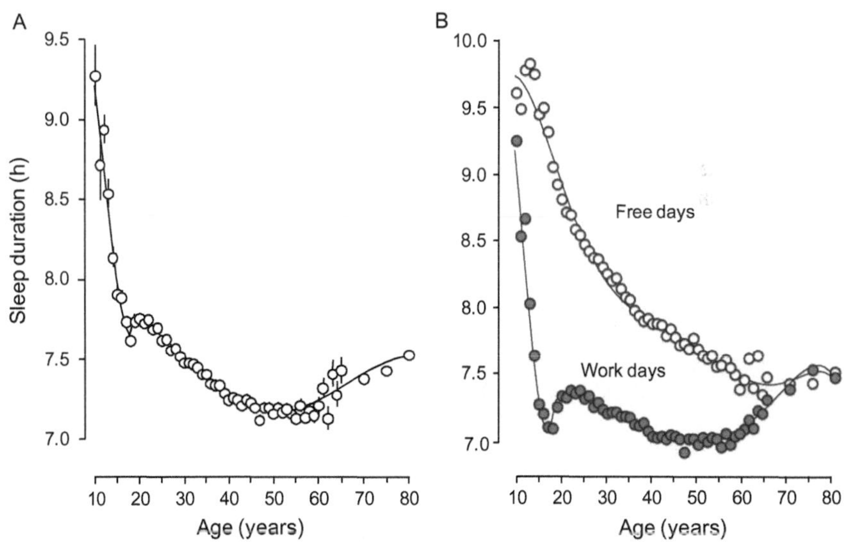

Figure 11.3 Average sleep duration as a function of age. (A) Weekly averages of sleep duration. (B) Sleep duration separately for free days (open circles) and for workdays (gray dots). Curves are polynomial fits; vertical lines represent standard errors of the mean (±SEM; in most cases they are smaller than the respective symbols). *Redrawn after Ref. 2.*

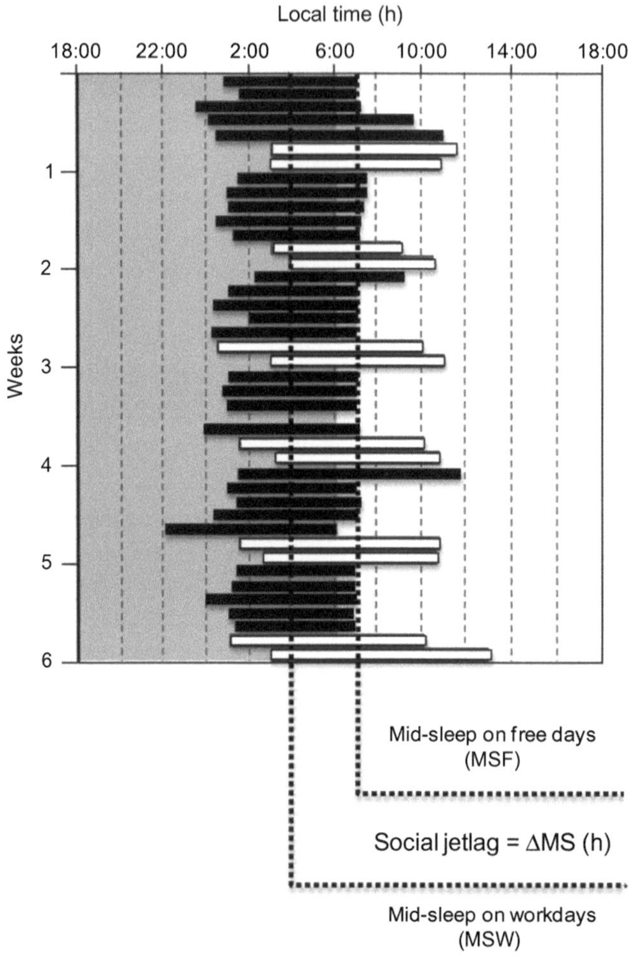

Figure 11.4 Social Jetlag. Six-week long sleep-log of an extremely late chronotype (MSF ≈ 7 a.m.), exemplifying the typical scalloping between sleep time on workdays (black bars) and on weekends (white bars). Local time is indicated at the top of the graph and the consecutive days of the sleep-log on the left. The difference between the mid-sleep point on free days (MSF) and that on workdays (MSW) is used to quantify the amount of "social jetlag." Note that sleep on workdays in this late type is interrupted by the alarm clock (constant sleep end at around 7 a.m., corresponding to the MSF point of this subject). Although this is an extreme example of social jetlag (due to the late chronotype in combination with an early work start), the majority of the population shows similar patterns. *Redrawn after Ref. 2.* (See Color Insert.)

duration shortens drastically (by almost 2 h) from the age of 10 to 17 (Fig. 11.3A). The separate analysis for work and free days reveals, however, that this sharp decline is mainly due to work- or school-day sleep (>2 h), while sleep duration on free days becomes only gradually shorter from the age of 10 to retirement. The profound mismatch between the "enforced" sleep duration on school/work days and free-day sleep duration demonstrates the immense impact of social times on our daily sleep behavior. Note that the mismatch disappears completely at around the average age of retirement (65 years).

The differences in sleep between work and free days not only relate to sleep duration but also to sleep timing (Fig. 11.4). Although one can sleep outside the temporal window provided by the circadian clock (e.g., naps), sleep is more efficient, and SCRD minimized, when it coincides with this window.[11] The sleep-log example in Fig. 11.4 illustrates the premature interruption of sleep by the alarm clock on workdays (black bars in Fig. 11.4). Thus, workday sleep—especially in the later chronotypes—can lead to a substantial sleep loss (see also Fig. 11.3). The main reason for this sleep deficit is that the circadian clock strongly influences when one can fall asleep (independent of sleep pressure, i.e., the homeostatic drive) while the alarm clock artificially puts an abrupt end to sleep. To compensate for this workday sleep debt, people commonly oversleep on free days (white bars in Fig. 11.4). As the difference in sleep timing between work and free days resembles the situation of traveling across several time zones to the West on Friday evenings and "flying" back on Monday mornings, the term "social jetlag" was coined to describe these weekly changes in sleep timing.[12] The symptoms of jetlag (e.g., problems in sleep, digestion, and performance) are manifestations of a misaligned circadian system. In travel-induced jetlag, they are transient until the clock reentrains. By contrast, social jetlag is chronic throughout a working career.

The developmental changes in circadian timing discussed above, in combination with the fact that school start times are suited to the predominantly late chronotype of teenagers,[13] lead to a peak of social jetlag at around the end of adolescence (Fig. 11.5). This is why teenagers show the largest discrepancy in sleep duration between free days and workdays compared to all other ages (Fig. 11.3).

Only 13% of the population represented in the MCTQ database is free of social jetlag, 69% experience at least 1 h, and a third suffers from 2 h or more. The discrepancy between biological (circadian) time and social time is a major reason for sleep deprivation (SD). About 5% of the database

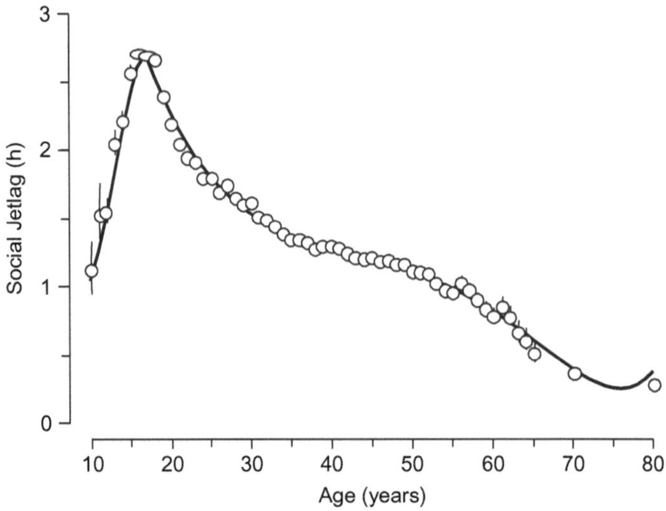

Figure 11.5 Age dependency of social jetlag. Circles represent average values of age groups (ages 10–65: 1-year bins; >65: 5-year bins; note that the age group "75" has no entries for social jetlag). Curves are polynomial fits; vertical lines represent standard errors of the mean (±SEM; in most cases they are smaller than the respective symbols). *Redrawn after Ref. 2.*

population sleep at least 20% shorter than their weekly average (an estimate for individual sleep need), that is, lose an entire night every week. An additional 35% sleep up to 10% shorter on workdays, missing half a night's sleep every week. Only a quarter of the population gets at least as much sleep on workdays as their weekly average. Many early chronotypes suffer from sleep loss on free days (rather than on workdays) for reasons that are not linked to alarm clocks but to other social constraints. They routinely stay up in the evenings beyond their circadian bed time due to the social pressure of the majority of late types in the population. Yet, their circadian clock wakes them up early in the morning (according to their chronotype). The pressure to stay up late is even stronger on evenings before work-free days.

The poor health consequences arising from SCRD are summarized in Table 11.1 (see references for details), and a few examples are highlighted here. There is a striking association between SCRD and smoking. For example, independent of social background and region, the number of smokers in the population increases with the greater the social jetlag. Further, the consumption of alcohol and caffeine increases with social jetlag (T. Roenneberg, unpublished). Based upon the scores from the Beck Depression Inventory,[16] the tendency toward depression increases when

Table 11.1 The impact of sleep and circadian rhythms disruption (SCRD) arising from social jetlag and shift work are summarized here and illustrate the severe health consequences of working against biological time (for full references, see Pritchett *et al.*[14])

Emotional responses	Cognitive responses	Somatic responses
• Exhaustion	• Reduced concentration	• Drowsiness
• Increased irritability	• Reduced performance	• Microsleeps
• Mood fluctuations	• Reduced attention	• Unintended sleep
• Anxiety	• Decreased memory	• Bodily sensations of pain
• Depressed mood	• Reduced recall of events	• Bodily sensations of cold
• Frustration	• Reduced multitasking	• Cardiovascular disease
• Anger	• Reduced decision making	• Risk of cancer
• Increased impulsivity	• Reduced creativity	• Metabolic abnormalities
• Decreased motor skills	• Reduced productivity	• Weight gain
• Increased stimulant use	• Reduced socialization	• Risk of diabetes II
• Increased sedative use	• Reduced communication	• Reduced immunity
• Alcohol use/misuse		• Disorders of the HPA

Associations between SCRD and poor health (this table) have long been a concern for shift workers, who suffer from the most extreme form of social jetlag. Shift-work schedules have been simulated in carefully controlled laboratory studies and result in the impairments to both cognitive and metabolic systems listed here. Subjects develop, for example, imbalanced glucose regulation resembling metabolic syndrome or type II diabetes.[15]

work times are not compatible with circadian sleep times.[17] Finally, being forced to live against ones circadian clock has metabolic consequences. Many studies have reported that short sleep duration is associated with an increased body mass index (BMI; for review, see Ref. 18), and more recent studies have shown that social jetlag also contributes, over and above sleep duration, to BMI.[2] With every hour of social jetlag, the probability of being overweight or obese increases by 30%.

3. SCRD AND PSYCHOSES

SCRD is a common comorbidity in numerous psychiatric disorders.[1] The greatest focus has been on mood disorders, especially unipolar and seasonal affective subtypes, yet SCRD is also prominent in the more severe, psychotic disorders such as schizophrenia.[19–21] The relationship between schizophrenia and abnormal sleep was first described in the late nineteenth century by the German psychiatrist Emil Kraepelin.[22] Today, SCRD is reported in 30–80% of patients with schizophrenia and is increasingly recognized as one of the most common features of the disorder.[23] Sleep disturbances in schizophrenia include increased sleep latency, and reductions in total sleep time, sleep efficiency, REM sleep latency, REM sleep density, and slow-wave sleep duration.[22,23] As illustrated in Fig. 11.6, schizophrenia is also associated with significant circadian disruption, including abnormal phasing, instability, and fragmentation in rest–activity cycles.[24–27]

Critically, schizophrenia patients with SCRD score badly on many quality-of-life clinical subscales, highlighting the human cost of SCRD in schizophrenia.[23,28,29] Significantly, schizophrenia patients often comment that an improvement in sleep is one of their highest priorities during treatment.[30] It is also becoming clear that SCRD impacts upon the onset, outcome, and relapse of mental illness.[31–33] These findings suggest that there are causal relationships between SCRD and psychoses, perhaps mediated via common (or overlapping) mechanisms.[20]

4. A CONCEPTUAL FRAMEWORK FOR SCRD IN PSYCHIATRIC ILLNESS

The association of mental illness and SCRD has until recently been considered to arise from exogenous factors including social isolation, antipsychotic medication, and/or activation of the stress axis.[1] Such a linear relationship between psychosis and SCRD is illustrated in Fig. 11.7A. Some recent studies[20] have addressed this hypothesis by examining SCRD in patients with schizophrenia and unemployed control subjects (Fig. 11.6) and showed that severe SCRD exists in schizophrenia and persists independently of antipsychotic medication. Further, sleep disruption cannot be explained on the basis of lack of employment as unemployed individuals had stable sleep/wake patterns.[20] These results are consistent with an alternative hypothesis, which suggests that psychoses and SCRD may share common and overlapping mechanistic pathways.[20]

As discussed above, the sleep and circadian timing system is the product of a complex interaction between multiple brain regions, neurotransmitters, and modulatory hormones (Figs. 11.1 and 11.2). As a consequence, abnormalities in any of the underlying neurotransmitter systems potentially impinge upon sleep/circadian timing at multiple levels. Similarly, psychoses involve several distributed brain circuits, affecting a range of neurotransmitter systems, many of which overlap with those underlying sleep and circadian rhythms.[1,34] Viewed in this context, it is no surprise that SCRD is

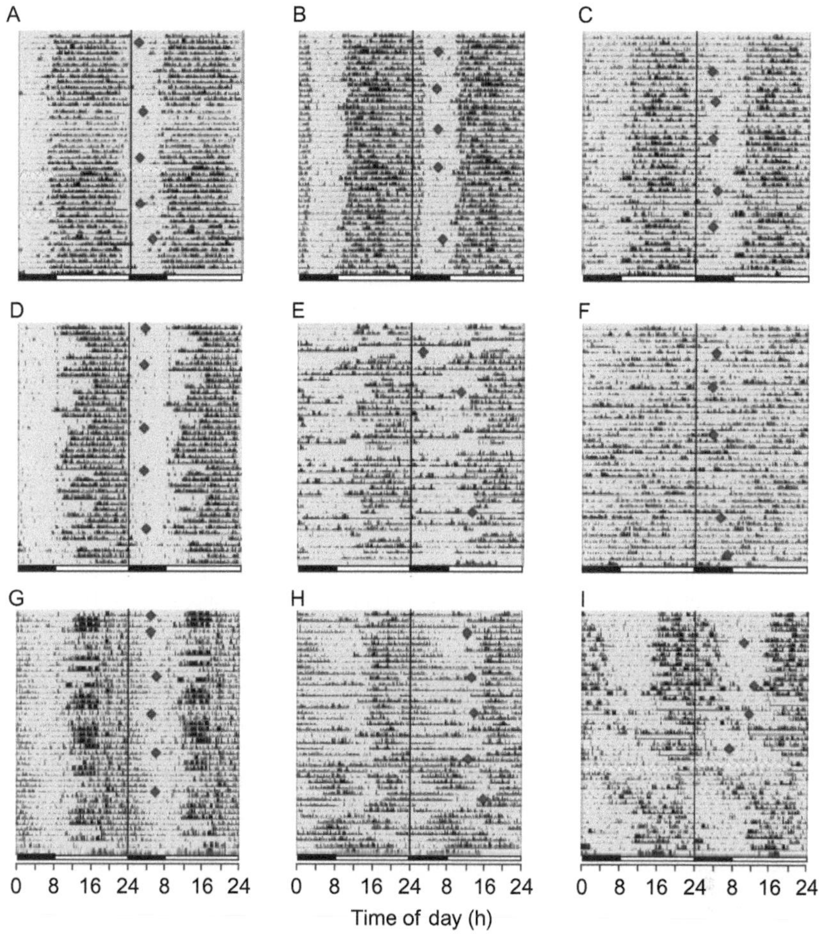

Figure 11.6 Six weeks of rest-activity patterns (actograms) and melatonin peak times (dots) from subjects' in their home environment. Activity data are 48-h double plotted. Black bars on the *x*-axis indicate nighttime, open bars indicate daytime, and the midline indicates midnight. (A–C) Unemployed individuals showing clear entrainment to the day/night cycle and low nighttime activity. (D–I) Actograms of patients with schizophrenia showing disrupted sleep patterns ranging from: (D) entrained but with variable sleep onsets and periods; (E) disrupted, and (F) highly irregular rest–activity cycles; (G) highly delayed rest–activity periods with bedtimes around 04:00; (H) reversed rest–activity cycles alternating with free-running periods; and (I) delayed rest–activity periods alternating with free-running periods. *From Ref. 20.* (See Color Insert.)

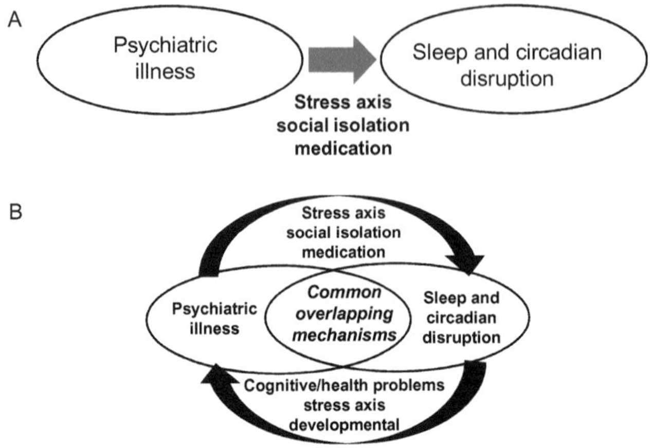

Figure 11.7 Diagram illustrating the possible relationships between psychiatric illness and sleep and circadian rhythm disruption (SCRD). (A) A straightforward and linear relationship whereby psychiatric illness results in SCRD as a result of activation of components of the stress axis, an absence of social constraints and social isolation and/or is the product of antipsychotic medication. (B) An alternative hypothesis which suggests that psychiatric illness and SCRD share common and overlapping mechanisms. Thus, aberrant functioning or synchrony of specific neural circuits, affecting several neurotransmitter systems that predispose an individual to psychiatric illness, will have a parallel effect upon the sleep/circadian system. Disruption of sleep will, likewise, impact upon multiple aspects of brain function, including activation of the stress axis, exacerbating or driving a range of health problems (Table 11.1), and in the young, may have developmental consequences. Medication, substance abuse, social isolation, and/or activation of the stress axis associated with psychiatric illness will certainly impinge upon the sleep and circadian systems, but are depicted here not as the primary cause of SCRD in neuropsychiatric illness.

common in psychoses, or that disruption of sleep/circadian biology will, in turn, have widespread effects, ranging across many aspects of neural and neuroendocrine function as outlined in Table 11.1. Significantly, many of the pathologies caused by SCRD are reported routinely as comorbid with neuropsychiatric illness but are rarely linked to the disruption of sleep. Furthermore, the consequences of SCRD result in abnormal light exposure and atypical patterns of social behavior (Fig. 11.2), closing a vicious cycle to further destabilize sleep/circadian physiology.[35,36] The common and overlapping mechanisms of psychosis and SCRD are illustrated in Fig. 11.7B. Critically, these relationships explain how relatively small changes in either the exogenous social time or endogenous brain neurotransmitters will be amplified by feedbacks to increase an individual's vulnerability to neuropsychiatric illness and comorbid health problems.[1]

The conceptual framework outlined in Fig. 11.7B allows four explicit predictions: (A) genes linked to mental illness will play a role in sleep and circadian rhythm generation and regulation; (B) genes that generate and regulate sleep and circadian rhythms will play a role in mental health and illness; (C) that SCRD will precede mental illness under some circumstances; and (D) that SCRD amelioration will have a positive impact upon mental illness. The evidence supporting these predictions is summarized below.

4.1. Genes linked to mental illnesses that also play a role in sleep and circadian rhythm generation and regulation

A number of genes have been linked with schizophrenia as a result of genetic associations and predictions based upon plausible biological mechanisms. Those most established in the "candidate gene" literature are *Nrg1*,[37–39] *Akt1*,[40] *Disc1*,[41,42] *Grm3*,[43] *Dao*,[44,45] *Comt*,[46] *Dtnbp1*,[47] and *ErbB4*.[38] More recent additions include *Snap-25*,[48] *Vipr2*,[49] *Gsk3b*,[50] *Pde4d*,[51–53] *Tcf4*,[54] *MIR137*,[55] and *ZNF804A*.[56]

Sleep and circadian functions have been analyzed in knockout or mutant models of just three of these genes, namely, *Snap-25*,[57] *Vipr2*,[58] and *Nrg1*.[59] Recent evidence has shown that a microduplication of the vasoactive intestinal polypeptide (VIP) receptor 2 (*Vipr2*) confers a significant risk for schizophrenia.[49] VIPR2 is known to play a critical role in the SCN in synchronizing neuronal oscillations,[60] and *Vipr2* knockout mice have reduced clock gene expression, blunted SCN electrical activity, and show a range of circadian phenotypes, from relatively normal to arrhythmic.[58] Further evidence comes from work on the *Blind-drunk* mutant mouse, which carries a mutation in *Snap25*, a gene encoding an exocytotic synaptic protein. This mouse model of schizophrenia displays behavioral phenotypes, which are modulated by environmental stress and corrected by antipsychotics.[61,62] In addition, these mice display phase advanced and fragmented circadian rhythms in locomotor activity.[57] Furthermore, plausible mechanistic evidence suggests that many more genes associated with schizophrenia may be linked to SCRD, including *Nrg1*, *Tcf4*, *Pde4d*, and *Cckar* (see Ref. 14). Neuregulin 1 (NRG1) is a growth factor involved in neurodevelopment and plasticity, which has been associated with both schizophrenia[37–39] and schizotypal personality disorder.[63] There is some evidence that its expression is increased in the brains of schizophrenia patients.[37,64] Mice heterozygous for a disruption in the *Nrg1* gene show disrupted rest/activity rhythms,[59] while wheel-running activity is inhibited by the long-term infusion of NRG1 into the third ventricle of the hamster brain.[65] NRG1 is expressed in the SCN and retinal ganglion cells,[66,67] consistent with its proposed involvement in circadian function.

4.2. Genes that generate and regulate sleep and circadian rhythms that also play a role in normal mental health

At the level of genetic association, several studies have identified associations between schizophrenia and clock genes (reviewed in Ref. 68). There is some evidence of an association with *Per3* and *Tim* with schizophrenia and schizoaffective disorder, and *Cry1* has been suggested as a candidate gene for schizophrenia based on its location near a linkage hot spot for schizophrenia on chromosome 12q24. A transmission bias for the *Clock* 3111C/T polymorphism has been suggested in a population of 145 Japanese schizophrenia patients. This polymorphism has been associated with aberrant dopaminergic neurotransmission to the SCN. Elevated *Per1* expression has also been shown in the temporal lobe of postmortem schizophrenic brains.[68]

Mechanistic evidence also exists for a link between the circadian system and bipolar disorder. First, the serine/threonine protein kinase GSK3B is known to phosphorylate elements of the transcriptional–translational feedback loop (TTFL) and is a target of the mood-stabilizer lithium.[69] Additional evidence has come from behavioral phenotyping of *Clock* mutant mice. Mice carrying a mutation of the TTFL component *Clock* show hyperactivity, reduced sleep, lower anxiety, increased risk-taking behavior, and increased reward value for cocaine, sucrose, or medial forebrain stimulation. Chronic administration of lithium returned these responses to wild-type levels. These behaviors are analogous to those seen in patients with mania. The *Clock* mutant animals also showed elevated dopamine function in the ventral tegmental area (VTA), and the behavioral phenotype could be rescued by expressing functional CLOCK protein in this region alone.[70] Moreover, *in vivo* gene silencing in the VTA alone has been shown to produce such mania-related behaviors.[71] It is possible that the role of CLOCK in the VTA that gives rise to the mania phenotype described in these studies may occur as a result of a deficit in circadian function, but could also occur via an unrelated role of this protein in the VTA. Certainly, silencing this gene in the VTA results in changes in the expression of other genes, including ion channels and genes involved in dopamine synthesis.[71] Further evidence comes from the *Myshkin* mouse mutant, which is caused by a mutation in the neuron-specific sodium, potassium ATPase α3 (*Atp1a3*). Similar to *Clock* mutant mice, these animals, both a mania and SCRD phenotype, with a reduction in REM and NREM sleep, reduced sleep bouts and increased waking.[72] Work on the *Afterhours* (*Afh*) mutant again shows a phenotype consistent with mania, reduced anxiety, and depression. Moreover, this study also showed an association between variation in *Fbxl3* and bipolar disorder in three separate human data sets.[73] Remarkably, *Clock*, *Myshkin*, and *Afh* mutants all show a long circadian period (27, 25, and 27 h, respectively), suggesting that a long circadian period may be associated with the mania-related behavior observed in these animal models.

There is now quite compelling genetic evidence for links between bipolar disorder and SCRD. Based on analysis of 46 SNPs of 8 clock genes (*Bmal1*, *Clock*, *Per1-3*, *Cry1-2*, *Tim*), *Bmal1* and *Tim* were found to be associated with bipolar disorder or schizophrenia.[74] The association of *Bmal1* has been confirmed in a separate study.[75] Further evidence comes from patients carrying the long allele variant of the clock gene *Per3* ($PER3^{5/5}$), which has been linked to early onset of bipolar disorder.[76]

4.3. Where SCRD precedes mental illness

In the few studies undertaken, SCRD appears to be a good predictor of mental illness in those individuals identified as "at risk" of developing some form of mental illness. It is worth stressing from the outset, however, that there are no studies that have undertaken longitudinal tracking of individuals using high-resolution sleep phenotyping techniques. The problem of falling asleep, early morning awakenings, and the decreased need for sleep (often termed insomnia) are the most often reported symptoms preceding the onset of bipolar disorder.[77,78] Prodromal sleep disturbances have also been reported in children with childhood-onset schizophrenia.[79] Another study focused on adolescents at risk of developing psychosis, but lacking mania or depressive symptoms, and showed that 37% of subjects scored high on sleep disturbances with females scoring higher than males.[80] A very recent study examined drug-naïve undergraduate students with a hypomania phenotype (defined by extremely high scores on the Mood Disorder Questionnaire), and as having a high risk of developing bipolar disorder. In these individuals, high-resolution sleep/wake analysis was undertaken. The results showed that during sleep, high-risk individuals compared to low-risk individuals showed significantly elevated levels of restlessness during the least active 5 h (2–7 a.m.) of sleep. Once again this behavior was more pronounced in women compared to men.[81] Collectively these few studies suggest that the structure of sleep/wake timing is beginning to break down before any clinical diagnosis of mental illness. Such findings raise the possibility that SCRD may be an important factor in the development, outcome, and treatment of individuals with mental illness.

4.4. SCRD stabilization and its impact on mental illness

Individuals with mood disorders often show a sleep/wake pattern that is not appropriately aligned to environmental or social time. These phenotypes prompted the use of chronotherapies, including bright light therapy (BLT), alone or in combination with exogenous melatonin, SD, and sleep–wake schedule intervention to stabilize sleep.[82] Such interventions are gaining acceptance and are frequently utilized treatment options for mood disorders such as seasonal affective disorders (SADs, "winter depression"), unipolar and bipolar depression, antepartum depression, and premenstrual depression. Rosenthal and colleagues[83] pioneered the use of appropriately timed artificial bright light to reduce the symptoms of SAD in individuals during the winter months. Timed light exposure is now being used for the treatment of a broad range of depressive illnesses. For example, in patients with mild to severe depression, light treatment achieves rapid remission rates of 40–67%.[84,85]

SD is widely recognized as an effective rapid-onset antidepressant therapy. Unfortunately, however, depressive symptoms return quickly after recovery sleep. As a result, SD is used in combination with mood stabilizers such as lithium or other antidepressants. SD has also been used effectively in combination with other chronotherapeutics, primarily BLT and scheduled sleep–wake, to accelerate and sustain antidepressant efficacy in conditions such as bipolar disorder.[86] Significantly, cognitive behavioral therapy for insomnia (CBT-I) has been shown to not only improve the initiation and maintenance of sleep but also reduce the severity of depressive symptoms and suicidal ideation.[87] CBT-I comprises behavioral components (relaxation techniques, stress management, sleep restriction, stimulus control), cognitive components (correcting unhelpful beliefs and attitudes about sleep, attentional bias), and psychoeducation (sleep biology and sleep hygiene) and was originally developed for primary insomnia but is now being used increasingly in individuals with psychiatric disorders comorbid with sleep disturbances.[88] In a recent study, the effect of four sessions of CBT-I was assessed in patients with schizophrenia and exhibiting persistent persecutory delusions. Participants self-reported significant reductions in sleep disturbances following the CBT-I intervention, and approximately half the patients showed a clinically significant decrease in paranoid thinking. In addition, CBT-I also reduced some of the emotional mediators, namely, anxiety, depression, and anomalies of experience.[89] Collectively the emerging data suggest that sleep stabilization can be an effective means to reduce the symptoms in a number of mental illnesses.

5. CONCLUSIONS

In this chapter, we have considered the exogenous and endogenous origins of SCRD with reference to social time and mental illness. The association between SCRD and mental illness is well recognized, but the causes, correlates, and effects of SCRD in mental illness have been poorly understood and its treatment often neglected. Recent advances in our appreciation of the endogenous mechanisms that generate sleep and circadian rhythms, and in the emerging understanding of the neurobiology of psychiatric disorders, suggest that there are common and overlapping mechanisms for the generation of normal mental health and sleep/wake behavior. Further, the possibility that SCRD may be important in the development, outcome, and treatment of mental illness provides an additional and urgent impetus for the study of these links. Significantly, the imposition of a structured social time upon individuals with mental illness and abnormal sleep/wake behavior improves both in parallel. These emerging results provide an unprecedented scientific opportunity to reevaluate the connections between these important domains of human health and to develop evidence-based interventions to ameliorate or correct abnormalities in these behaviors.

While the imposition of a structured social time may improve SCRD in psychiatric patients, societal structure is imposing an abnormal pattern of sleep and wake upon much of the population. As a result, SCRD dominates the lives of millions of individuals in the developed nations. The circadian sleep window of the vast majority of the population ends well past the time when people have to get up on workdays, which has made the use of alarm clocks reach epidemic levels. The result is a substantial discrepancy between endogenous and exogenous time. Social jetlag is a form of SCRD, with all the known health consequences (Table 11.1). With this growing body of evidence, society must address these conflicts with more flexible school and work times and/or more natural dynamic lighting in buildings that promotes the alignment of biological and solar time.

REFERENCES

1. Wulff K, Gatti S, Wettstein JG, Foster RG. Sleep and circadian rhythm disruption in psychiatric and neurodegenerative disease. *Nat Rev Neurosci*. 2010;11(8):589–599.
2. Roenneberg T, Allebrandt KV, Merrow M, Vetter C. Social jetlag and obesity. *Curr Biol*. 2012;22(10):939–943.
3. Roenneberg T, Wirz-Justice A, Merrow M. Life between clocks: daily temporal patterns of human chronotypes. *J Biol Rhythms*. 2003;18(1):80–90.
4. Foster RG, Wulff K. The rhythm of rest and excess. *Nat Rev Neurosci*. 2005;6(5):407–414.
5. Roenneberg T, Daan S, Merrow M. The art of entrainment. *J Biol Rhythms*. 2003;18(3):183–194.
6. Roenneberg T. What is chronotype? *Sleep Biol. Rhythms*. 2012;10(2):75–76.
7. Roenneberg T, Kantermann T, Juda M, Vetter C, Allebrandt KV. Light and the Human Circadian Clock. *Circadian Clocks*: Springer; in press.
8. Toh KL, Jones CR, He Y, et al. An hPer2 phosphorylation site mutation in familial advanced sleep phase syndrome. *Science*. 2001;291(5506):1040–1043.
9. Roenneberg T, Kuehnle T, Pramstaller PP, et al. A marker for the end of adolescence. *Curr Biol*. 2004;14(24):R1038–R1039.
10. Roenneberg T, Kumar CJ, Merrow M. The human circadian clock entrains to sun time. *Curr Biol*. 2007;17(2):R44–R45.
11. Wyatt JK, Ritz-De Cecco A, Czeisler CA, Dijk DJ. Circadian temperature and melatonin rhythms, sleep, and neurobehavioral function in humans living on a 20-h day. *Am J Physiol*. 1999;277(4 Pt 2):R1152–R1163.
12. Wittmann M, Dinich J, Merrow M, Roenneberg T. Social jetlag: misalignment of biological and social time. *Chronobiol Int*. 2006;23(1–2):497–509.
13. Carskadon MA, Labyak SE, Acebo C, Seifer R. Intrinsic circadian period of adolescent humans measured in conditions of forced desynchrony. *Neurosci Lett*. 1999;260(2): 129–132.
14. Pritchett D, Wulff K, Oliver PL, et al. Evaluating the links between schizophrenia and sleep and circadian rhythm disruption. *J Neural Transm*. 2012;119(10):1061–1075.
15. Scheer FA, Hilton MF, Mantzoros CS, Shea SA. Adverse metabolic and cardiovascular consequences of circadian misalignment. *Proc Natl Acad Sci USA*. 2009;106(11):4453–4458.
16. Beck AT, Steer RA, Ball R, Ranieri W. Comparison of Beck Depression Inventories - IA and -II in psychiatric outpatients. *J Pers Assess*. 1996;67(3):588–597.
17. Levandovski R, Dantas G, Fernandes LC, et al. Depression scores associate with chronotype and social jetlag in a rural population. *Chronobiol Int*. 2011;28(9):771–778.
18. Mitchell NS, Catenacci VA, Wyatt HR, Hill JO. Obesity: overview of an epidemic. *Psychiatr Clin North Am*. 2011;34(4):717–732.
19. Benca RM, Obermeyer WH, Thisted RA, Gillin JC. Sleep and psychiatric disorders. A meta-analysis. *Arch Gen Psychiatry*. 1992;49(8):651–668, discussion 669–670.
20. Wulff K, Middleton B, Dijk D-J, Foster RG, Joyce E. Sleep and circadian rhythm disruption in schizophrenia. *Br J Psychiatry*. 2012;200:1–9.
21. NIH . National Institutes of Health State of the Science Conference statement on Manifestations and Management of Chronic Insomnia in Adults, June 13–15, 2005. *Sleep*. 2005;28(9):1049–1057.
22. Manoach DS, Stickgold R. Does abnormal sleep impair memory consolidation in schizophrenia? *Front Hum Neurosci*. 2009;3:21.
23. Cohrs S. Sleep disturbances in patients with schizophrenia: impact and effect of antipsychotics. *CNS Drugs*. 2008;22(11):939–962.
24. Martin J, Jeste DV, Caliguiri MP, Patterson T, Heaton R, Ancoli-Israel S. Actigraphic estimates of circadian rhythms and sleep/wake in older schizophrenia patients. *Schizophr Res*. 2001;47(1):77–86.

25. Martin JL, Jeste DV, Ancoli-Israel S. Older schizophrenia patients have more disrupted sleep and circadian rhythms than age-matched comparison subjects. *J Psychiatr Res.* 2005;39(3):251–259.
26. Wulff K, Joyce E, Middleton B, Dijk DJ, Foster RG. The suitability of actigraphy, diary data, and urinary melatonin profiles for quantitative assessment of sleep disturbances in schizophrenia: a case report. *Chronobiol Int.* 2006;23(1–2):485–495.
27. Wulff K, Porcheret K, Cussans E, Foster RG. Sleep and circadian rhythm disturbances: multiple genes and multiple phenotypes. *Curr Opin Genet Dev.* 2009;19(3):237–246.
28. Goldman M, Tandon R, DeQuardo JR, Taylor SF, Goodson J, McGrath M. Biological predictors of 1-year outcome in schizophrenia in males and females. *Schizophr Res.* 1996;21(2):65–73.
29. Hofstetter JR, Lysaker PH, Mayeda AR. Quality of sleep in patients with schizophrenia is associated with quality of life and coping. *BMC Psychiatry.* 2005;5:13.
30. Auslander LA, Jeste DV. Perceptions of problems and needs for service among middle-aged and elderly outpatients with schizophrenia and related psychotic disorders. *Community Ment Health J.* 2002;38(5):391–402.
31. Ehlers CL, Frank E, Kupfer DJ. Social zeitgebers and biological rhythms. A unified approach to understanding the etiology of depression. *Arch Gen Psychiatry.* 1988;45 (10):948–952.
32. Chemerinski E, Ho BC, Flaum M, Arndt S, Fleming F, Andreasen NC. Insomnia as a predictor for symptom worsening following antipsychotic withdrawal in schizophrenia. *Compr Psychiatry.* 2002;43(5):393–396.
33. Gruber J, Miklowitz DJ, Harvey AG, et al. Sleep matters: sleep functioning and course of illness in bipolar disorder. *J Affect Disord.* 2011;134(1–3):416–420.
34. Weinberger DR, Harrison PJ. *Schizophrenia.* 3rd ed. Oxford: Wiley-Blackwell; 2011.
35. Uhlhaas PJ, Singer W. Neural synchrony in brain disorders: relevance for cognitive dysfunctions and pathophysiology. *Neuron.* 2006;52(1):155–168.
36. Richardson G, Wang-Weigand S. Effects of long-term exposure to ramelteon, a melatonin receptor agonist, on endocrine function in adults with chronic insomnia. *Hum Psychopharmacol.* 2009;24(2):103–111.
37. Harrison PJ, Law AJ. Neuregulin 1 and schizophrenia: genetics, gene expression, and neurobiology. *Biol Psychiatry.* 2006;60(2):132–140.
38. Mei L, Xiong WC. Neuregulin 1 in neural development, synaptic plasticity and schizophrenia. *Nat Rev Neurosci.* 2008;9(6):437–452.
39. Stefansson H, Sigurdsson E, Steinthorsdottir V, et al. Neuregulin 1 and susceptibility to schizophrenia. *Am J Hum Genet.* 2002;71(4):877–892.
40. Arguello PA, Gogos JA. A signaling pathway AKTing up in schizophrenia. *J Clin Invest.* 2008;118(6):2018–2021.
41. Brandon NJ, Sawa A. Linking neurodevelopmental and synaptic theories of mental illness through DISC1. *Nat Rev Neurosci.* 2011;12(12):707–722.
42. Johnstone M, Thomson PA, Hall J, McIntosh AM, Lawrie SM, Porteous DJ. DISC1 in schizophrenia: genetic mouse models and human genomic imaging. *Schizophr Bull.* 2011;37(1):14–20.
43. Harrison PJ, Lyon L, Sartorius LJ, Burnet PW, Lane TA. The group II metabotropic glutamate receptor 3 (mGluR3, mGlu3, GRM3): expression, function and involvement in schizophrenia. *J Psychopharmacol.* 2008;22(3):308–322.
44. Chumakov I, Blumenfeld M, Guerassimenko O, et al. Genetic and physiological data implicating the new human gene G72 and the gene for D-amino acid oxidase in schizophrenia. *Proc Natl Acad Sci USA.* 2002;99(21):13675–13680.
45. Verrall L, Burnet PW, Betts JF, Harrison PJ. The neurobiology of D-amino acid oxidase and its involvement in schizophrenia. *Mol Psychiatry.* 2010;15(2):122–137.

46. Tunbridge EM, Weinberger DR, Harrison PJ. A novel protein isoform of catechol O-methyltransferase (COMT): brain expression analysis in schizophrenia and bipolar disorder and effect of Val158Met genotype. *Mol Psychiatry*. 2006;11(2):116–117.
47. Williams NM, O'Donovan MC, Owen MJ. Is the dysbindin gene (DTNBP1) a susceptibility gene for schizophrenia? *Schizophr Bull*. 2005;31(4):800–805.
48. Corradini I, Verderio C, Sala M, Wilson MC, Matteoli M. SNAP-25 in neuropsychiatric disorders. *Ann N Y Acad Sci*. 2009;1152:93–99.
49. Vacic V, McCarthy S, Malhotra D, et al. Duplications of the neuropeptide receptor gene VIPR2 confer significant risk for schizophrenia. *Nature*. 2011;471(7339):499–503.
50. Lipina TV, Wang M, Liu F, Roder JC. Synergistic interactions between PDE4B and GSK-3: DISC1 mutant mice. *Neuropharmacology*. 2012;62(3):1252–1262.
51. Fatemi SH, King DP, Reutiman TJ, et al. PDE4B polymorphisms and decreased PDE4B expression are associated with schizophrenia. *Schizophr Res*. 2008;101(1–3):36–49.
52. Numata S, Ueno S, Iga J, et al. Positive association of the PDE4B (phosphodiesterase 4B) gene with schizophrenia in the Japanese population. *J Psychiatr Res*. 2008;43(1):7–12.
53. Tomppo L, Hennah W, Lahermo P, et al. Association between genes of Disrupted in schizophrenia 1 (DISC1) interactors and schizophrenia supports the role of the DISC1 pathway in the etiology of major mental illnesses. *Biol Psychiatry*. 2009;65(12):1055–1062.
54. Brzozka MM, Radyushkin K, Wichert SP, Ehrenreich H, Rossner MJ. Cognitive and sensorimotor gating impairments in transgenic mice overexpressing the schizophrenia susceptibility gene Tcf4 in the brain. *Biol Psychiatry*. 2010;68(1):33–40.
55. Ripke S, Sanders AR, Kendler KS, et al. Genome-wide association study identifies five new schizophrenia loci. *Nat Genet*. 2011;43(10):969–976.
56. O'Donovan MC, Craddock N, Norton N, et al. Identification of loci associated with schizophrenia by genome-wide association and follow-up. *Nat Genet*. 2008;40(9):1053–1055.
57. Oliver PL, Sobczyk MV, Maywood ES, et al. Disrupted circadian rhythms in a mouse model of schizophrenia. *Curr Biol*. 2012;22(4):314–319.
58. Hughes AT, Piggins HD. Behavioral responses of Vipr2-/- mice to light. *J Biol Rhythms*. 2008;23(3):211–219.
59. Johnson MA, Devay P, Role LW. The cystine-rich domain of the neuregulin-1 gene (crd-NRG-1) is required for survival of a subset of neurons in the suprachiasmatic nucleus (SCN). Washington, DC: Society for Neuroscience; 2002. Online. *2002 Abstract Viewer/Itinerary Planner, 2002*.
60. Maywood ES, Reddy AB, Wong GK, et al. Synchronization and maintenance of timekeeping in suprachiasmatic circadian clock cells by neuropeptidergic signaling. *Curr Biol*. 2006;16(6):599–605.
61. Jeans AF, Oliver PL, Johnson R, et al. A dominant mutation in Snap25 causes impaired vesicle trafficking, sensorimotor gating, and ataxia in the blind-drunk mouse. *Proc Natl Acad Sci USA*. 2007;104(7):2431–2436.
62. Oliver PL, Davies KE. Interaction between environmental and genetic factors modulates schizophrenic endophenotypes in the Snap-25 mouse mutant blind-drunk. *Hum Mol Genet*. 2009;18(23):4576–4589.
63. Lin HF, Liu YL, Liu CM, Hung SI, Hwu HG, Chen WJ. Neuregulin 1 gene and variations in perceptual aberration of schizotypal personality in adolescents. *Psychol Med*. 2005;35(11):1589–1598.
64. Hashimoto R, Straub RE, Weickert CS, Hyde TM, Kleinman JE, Weinberger DR. Expression analysis of neuregulin-1 in the dorsolateral prefrontal cortex in schizophrenia. *Mol Psychiatry*. 2004;9(3):299–307.
65. Snodgrass-Belt P, Gilbert JL, Davis FC. Central administration of transforming growth factor-alpha and neuregulin-1 suppress active behaviors and cause weight loss in hamsters. *Brain Res*. 2005;1038(2):171–182.

66. Bernstein HG, Lendeckel U, Bertram I, et al. Localization of neuregulin-1alpha (heregulin-alpha) and one of its receptors, ErbB-4 tyrosine kinase, in developing and adult human brain. *Brain Res Bull.* 2006;69(5):546–559.
67. Sharif A, Duhem-Tonnelle V, Allet C, et al. Differential erbB signaling in astrocytes from the cerebral cortex and the hypothalamus of the human brain. *Glia.* 2009;57(4):362–379.
68. Lamont EW, Coutu DL, Cermakian N, Boivin DB. Circadian rhythms and clock genes in psychotic disorders. *Isr J Psychiatry Relat Sci.* 2010;47(1):27–35.
69. Yin L, Wang J, Klein PS, Lazar MA. Nuclear receptor Rev-erbalpha is a critical lithium-sensitive component of the circadian clock. *Science.* 2006;311(5763):1002–1005.
70. Roybal K, Theobold D, Graham A, et al. Mania-like behavior induced by disruption of CLOCK. *Proc Natl Acad Sci USA.* 2007;104(15):6406–6411.
71. Mukherjee S, Coque L, Cao JL, et al. Knockdown of Clock in the ventral tegmental area through RNA interference results in a mixed state of mania and depression-like behavior. *Biol Psychiatry.* 2010;68(6):503–511.
72. Kirshenbaum GS, Clapcote SJ, Duffy S, et al. Mania-like behavior induced by genetic dysfunction of the neuron-specific Na+, K+-ATPase alpha3 sodium pump. *Proc Natl Acad Sci USA.* 2011;108(44):18144–18149.
73. Keers R, Pedroso I, Breen G, et al. Reduced anxiety and depression-like behaviours in the circadian period mutant mouse afterhours. *PLoS One.* 2012;7(6):e38263.
74. Mansour HA, Wood J, Logue T, et al. Association study of eight circadian genes with bipolar I disorder, schizoaffective disorder and schizophrenia. *Genes Brain Behav.* 2006;5 (2):150–157.
75. Nievergelt CM, Kripke DF, Barrett TB, et al. Suggestive evidence for association of the circadian genes PERIOD3 and ARNTL with bipolar disorder. *Am J Med Genet B Neuropsychiatr Genet.* 2006;141B(3):234–241.
76. Benedetti F, Dallaspezia S, Colombo C, Pirovano A, Marino E, Smeraldi E. A length polymorphism in the circadian clock gene Per3 influences age at onset of bipolar disorder. *Neurosci Lett.* 2008;445(2):184–187.
77. Tijssen MJ, van Os J, Wittchen HU, et al. Prediction of transition from common adolescent bipolar experiences to bipolar disorder: 10-year study. *Br J Psychiatry.* 2010;196 (2):102–108.
78. Ritter PS, Marx C, Bauer M, Leopold K, Pfennig A. The role of disturbed sleep in the early recognition of bipolar disorder: a systematic review. *Bipolar Disord.* 2011;13(3):227–237.
79. Mattai AA, Tossell J, Greenstein DK, et al. Sleep disturbances in childhood-onset schizophrenia. *Schizophr Res.* 2006;86(1–3):123–129.
80. Miller TJ, Zipursky RB, Perkins D, et al. The PRIME North America randomized double-blind clinical trial of olanzapine versus placebo in patients at risk of being prodromally symptomatic for psychosis. II. Baseline characteristics of the "prodromal" sample. *Schizophr Res.* 2003;61(1):19–30.
81. Chandler R, Wulff K, Foster RG, Goodwin GG. Abstracts from the XXVI CINP Congress, Munich, 13-17 July 2008. *Int J Neuropsychopharmacol.* 2008;11:1–338.
82. Wirz-Justice A. Diurnal variation of depressive symptoms. *Dialogues Clin Neurosci.* 2008;10(3):337–343.
83. Rosenthal NE, Sack DA, Gillin JC, et al. Seasonal affective disorder. A description of the syndrome and preliminary findings with light therapy. *Arch Gen Psychiatry.* 1984;41 (1):72–80.
84. Golden RN, Gaynes BN, Ekstrom RD, et al. The efficacy of light therapy in the treatment of mood disorders: a review and meta-analysis of the evidence. *Am J Psychiatry.* 2005;162(4):656–662.
85. Martiny K, Lunde M, Unden M, Dam H, Bech P. Adjunctive bright light in non-seasonal major depression: results from clinician-rated depression scales. *Acta Psychiatr Scand.* 2005;112(2):117–125.

86. Wu JC, Kelsoe JR, Schachat C, et al. Rapid and sustained antidepressant response with sleep deprivation and chronotherapy in bipolar disorder. *Biol Psychiatry*. 2009;66(3):298–301.
87. Manber R, Bernert RA, Suh S, Nowakowski S, Siebern AT, Ong JC. CBT for insomnia in patients with high and low depressive symptom severity: adherence and clinical outcomes. *J Clin Sleep Med*. 2011;7(6):645–652.
88. Smith MT, Huang MI, Manber R. Cognitive behavior therapy for chronic insomnia occurring within the context of medical and psychiatric disorders. *Clin Psychol Rev*. 2005;25(5):559–592.
89. Myers E, Startup H, Freeman D. Cognitive behavioural treatment of insomnia in individuals with persistent persecutory delusions: a pilot trial. *J Behav Ther Exp Psychiatry*. 2011;42(3):330–336.

INDEX

Note: Page numbers followed by "*f*" indicate figures, and "*t*" indicate tables.

H

T

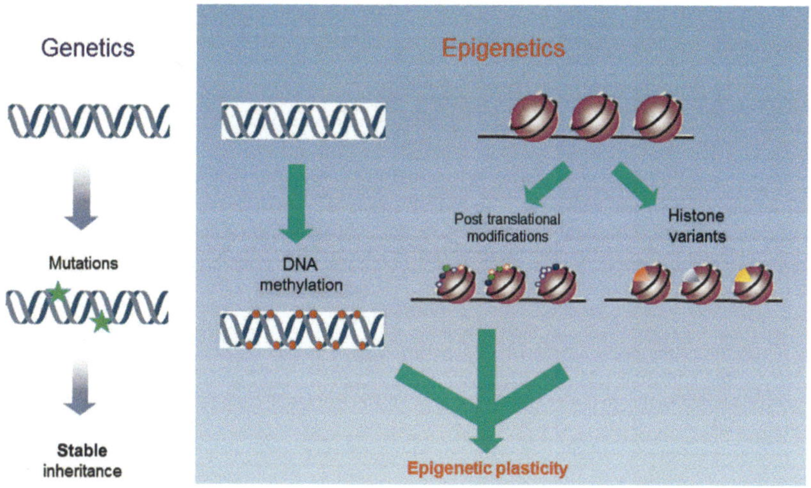

Kristin Eckel-Mahan and Paolo Sassone-Corsi, Figure 2.1

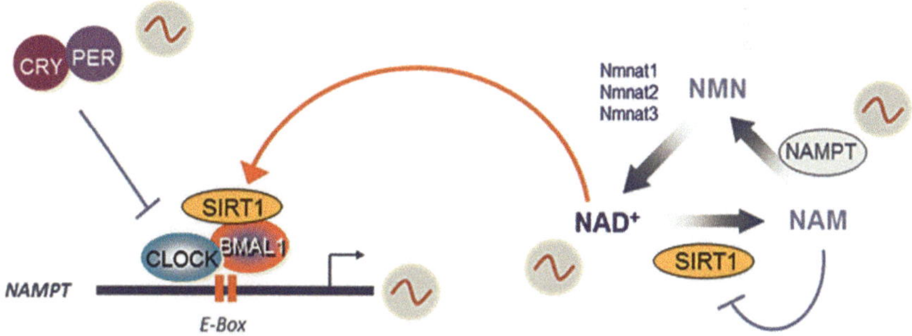

Kristin Eckel-Mahan and Paolo Sassone-Corsi, Figure 2.2

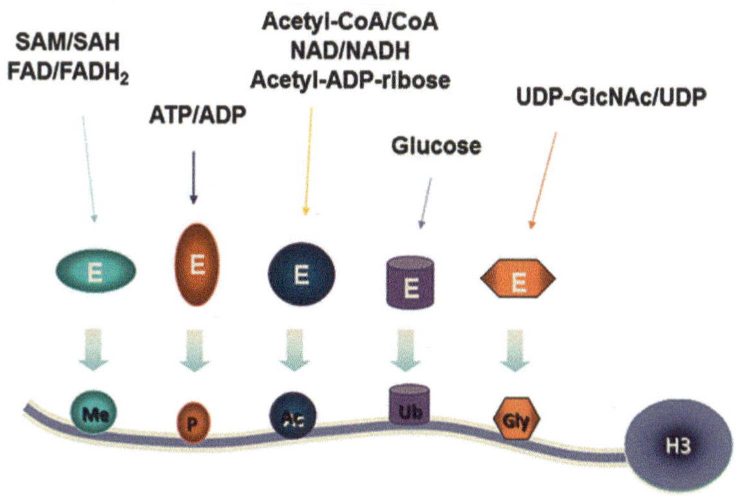

Kristin Eckel-Mahan and Paolo Sassone-Corsi, Figure 2.3

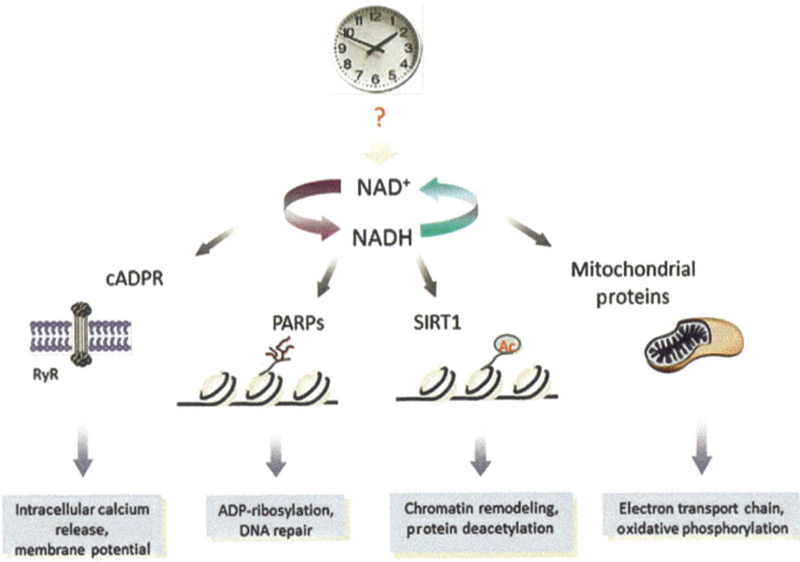

KRISTIN ECKEL-MAHAN AND PAOLO SASSONE-CORSI, FIGURE 2.4

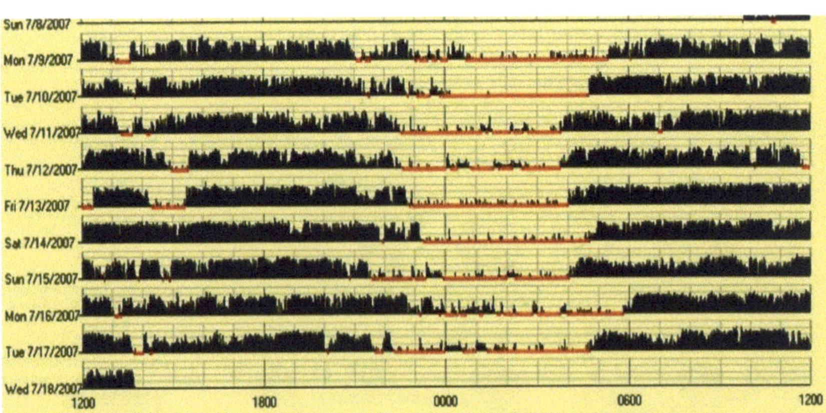

LUOYING ZHANG *ET AL.*, FIGURE 3.2

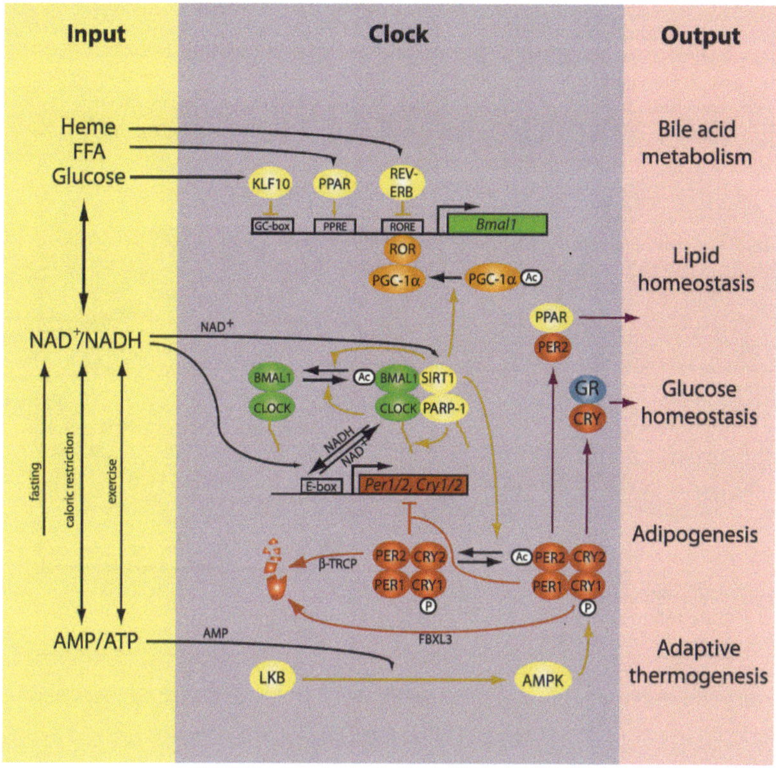

Ruud Buijs *et al.*, Figure 4.1

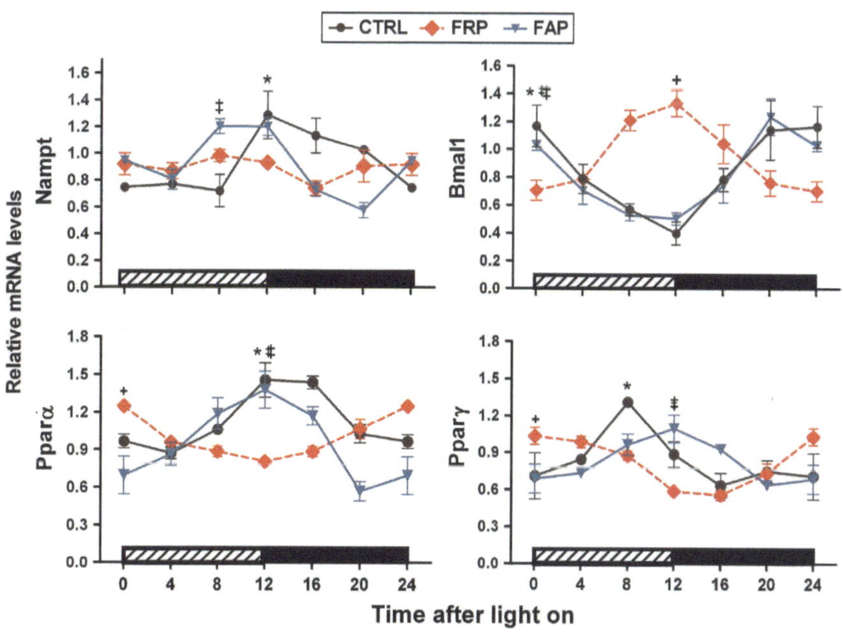

Ruud Buijs *et al.*, Figure 4.2

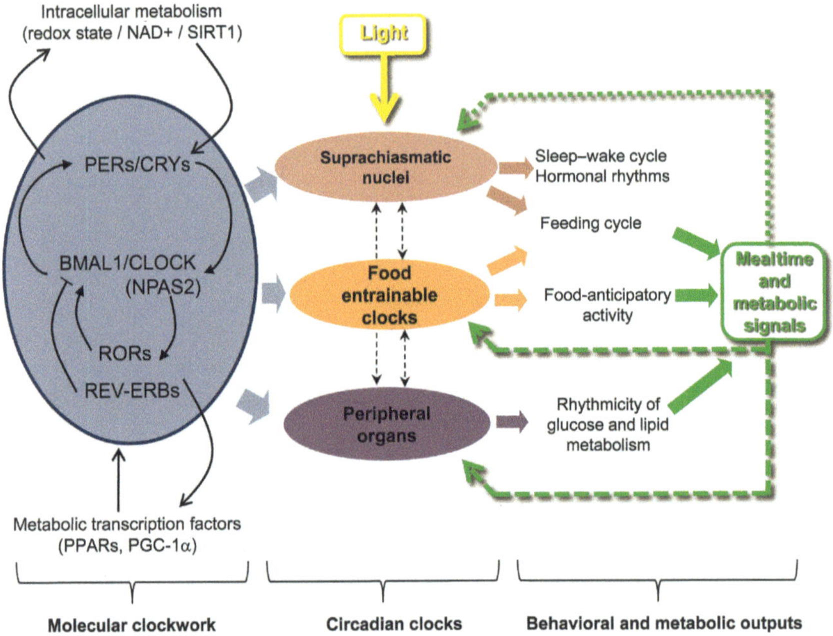

Etietienne Challet, Figure 5.1

Etietienne Challet, Figure 5.2

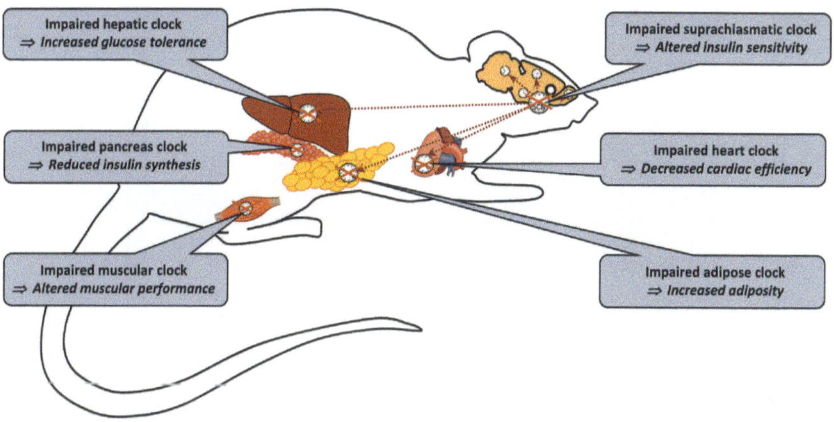

ETIETIENNE CHALLET, FIGURE 5.3

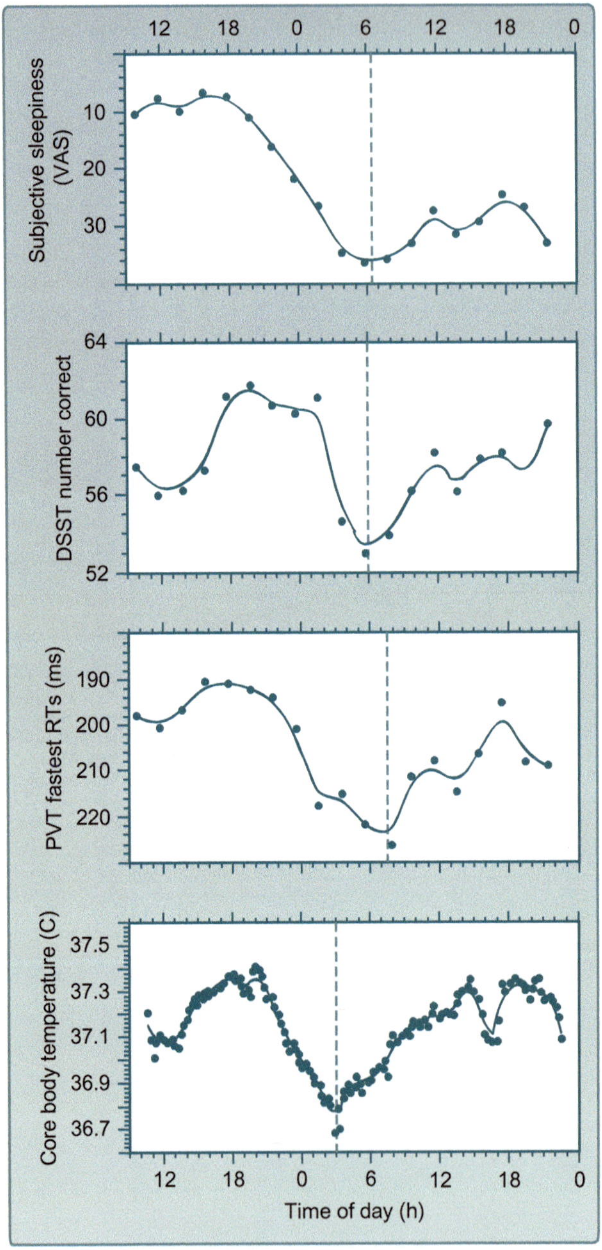

NAMNI GOEL *ET AL.*, FIGURE 7.1

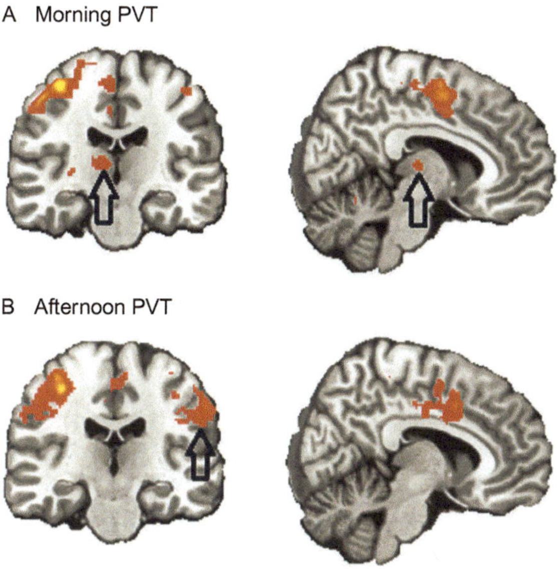

Namni Goel *et al.*, Figure 7.4

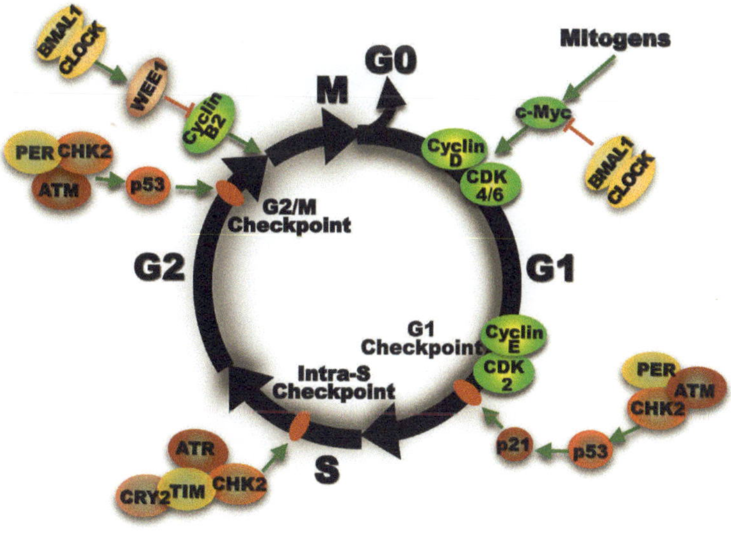

Loning Fu and Nicole M. Kettner, Figure 9.1

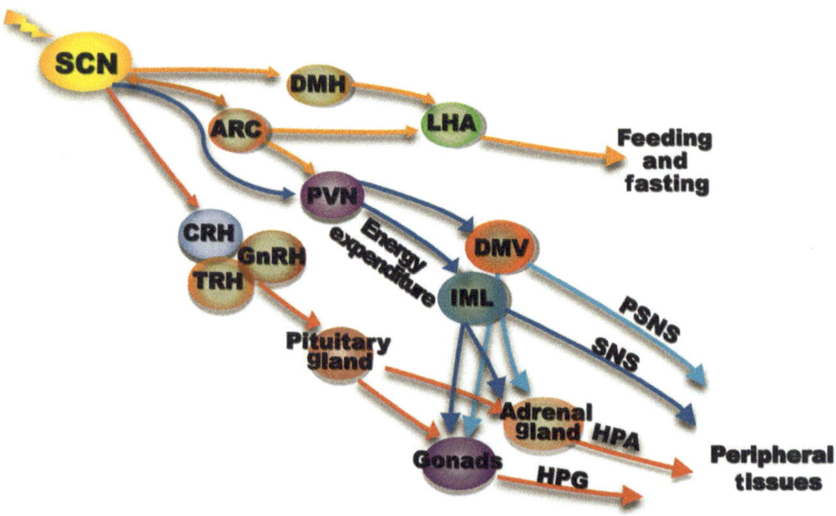

Loning Fu and Nicole M. Kettner, Figure 9.2

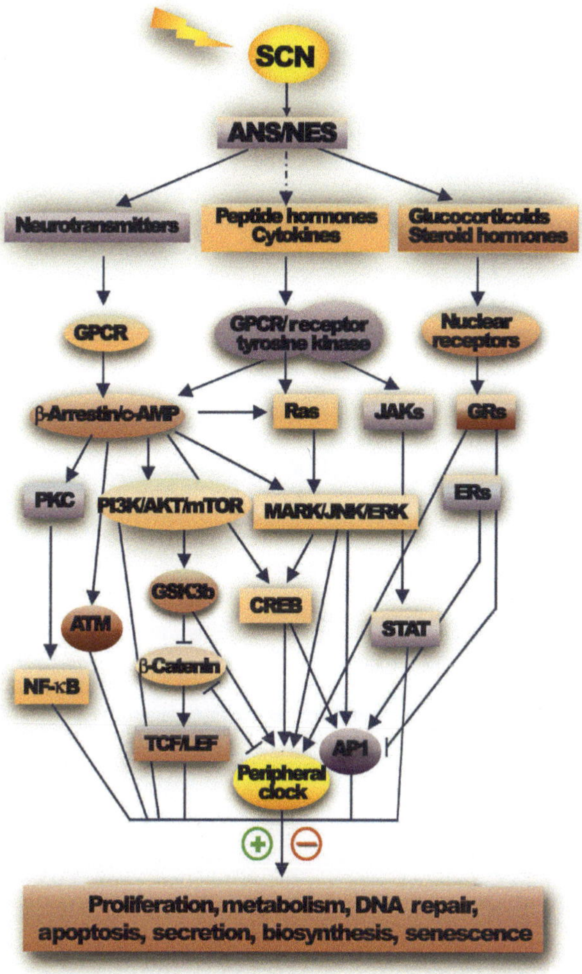

Loning Fu and Nicole M. Kettner, Figure 9.3

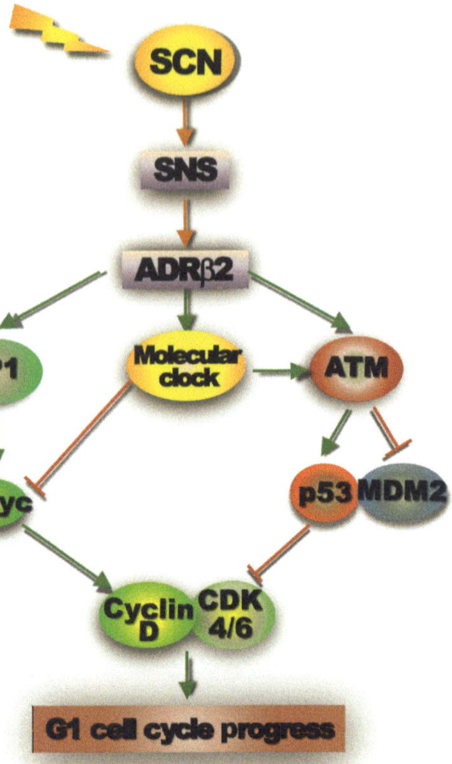

LONING FU AND NICOLE M. KETTNER, FIGURE 9.4

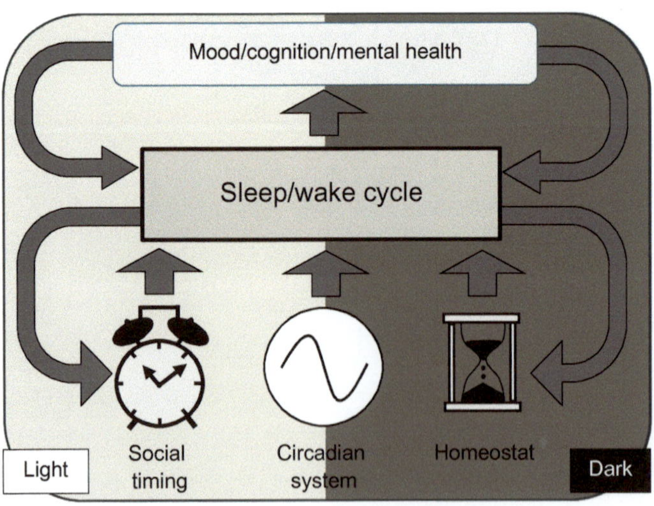

RUSSELL G. FOSTER *ET AL.*, FIGURE 11.2

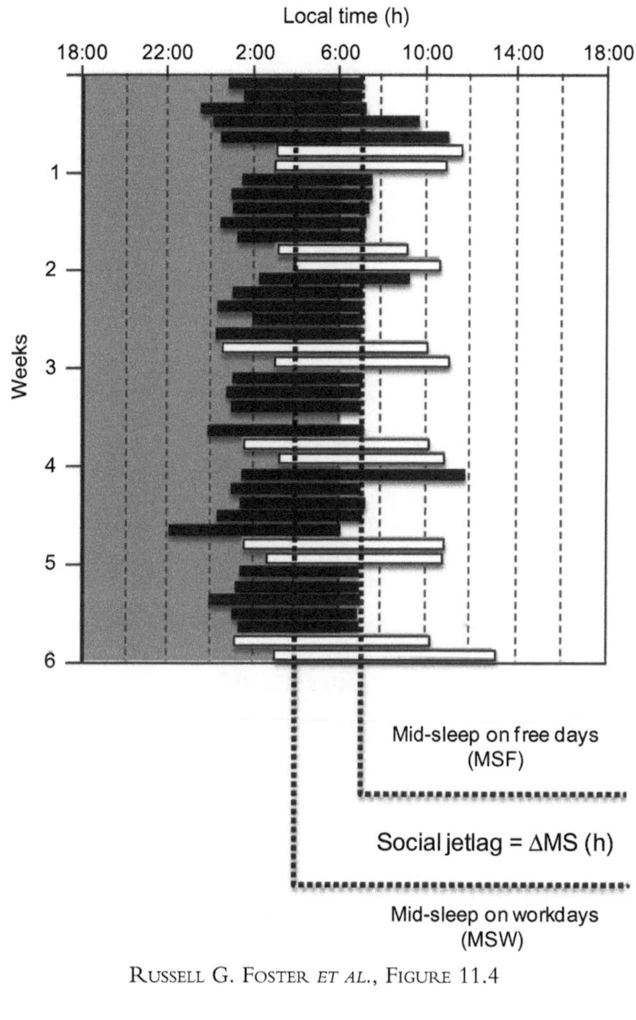

RUSSELL G. FOSTER *ET AL.*, FIGURE 11.4

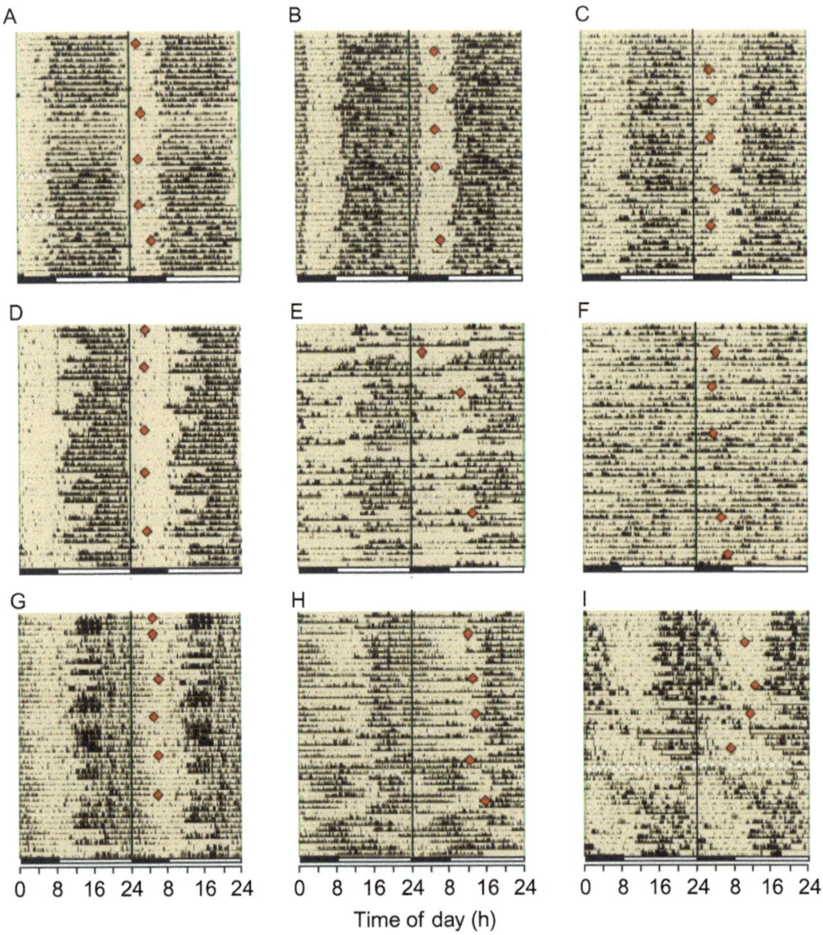

Russell G. Foster *et al.*, Figure 11.6

FSC
www.fsc.org
MIX
Papier | Fördert
gute Waldnutzung
FSC® C083411

Kolibri 360 GmbH
Ferdinand-Jühlke-Straße 7
99095 Erfurt, Deutschland
produktsicherheit@kolibri360.de